Metallurgy of Failure Analysis

Metallurgy of Failure Analysis

A K Das
Formerly Manager (Design) Aircraft
Design Bureau, Hindustan Aeronautics Limited
Bangalore

McGraw-Hill

New York San Francisco Washington, D.C. Auckland Bogotá Caracas Lisbon
London Madrid Mexico City Milan Montreal New Delhi San Juan
Singapore Sydney Tokyo Toronto

McGraw-Hill

A Division of The ***McGraw-Hill*** *Companies*

First published © 1996, Tata McGraw-Hill Publishing Company Limited

1 2 3 4 5 6 7 8 9 0 DOC/DOC 9 0 2 1 0 9 8 7

ISBN 0-07-015804-5

Printed and bound by R. R. Donnelley & Sons Company.

Foreword

An analysis of the in-process and service failures yields valuable information about their causes and the measures to be taken in design, manufacture, quality assurance and maintenance so as to reduce or avoid such failures. Failure analysis has developed over the years as an important scientific tool giving important feedback, and considerable wealth of information is available on it in the form of books, reference handbooks and technical papers. With the advances in materials and technologies, new problems have emerged but more information about how to deal with them is also being continuously generated in the field of failure analysis.

I am very pleased to note that A K Das has compiled a comprehensive book on metallurgy of failure analysis. In doing so, he has put in a wealth of information, experience and knowledge gained over more than two decades of his dedicated work at the HAL, Bangalore, a major part of which related to aircraft accidents and components failure analysis. He has covered a wide spectrum of defects and failures that occur during the processing of materials with a wide range of processes such as casting, forging, welding, machining and heat treatment, in addition to component failures during service such as overload, fatigue, corrosion, fretting and wear. These are supplemented with selected case studies from his own experience. A significant part of the book deals with the sources of failures, viz. defects, their characterisation and detection, and remedial measures. Thus, the book is unique in many ways. I sincerely hope that it will be of interests to the scientists and practicing engineers alike, and of immense use to those engaged in failure, analysis.

Dr C G Krishnadas Nair

Managing Director
Hindustan Aeronautics Ltd
Bangalore Complex

Preface

Eversince the Industrial Revolution, failure has been a constant feature of any engineering industry which combines design, material and manufacture. Over the years, with the advent of flying machines and advanced structural designs, the concept of failure has new dimensions, and failure analysis has gradually evolved into an exact science dealing with the complicated phenomena and intricacies of failure mechanisms. It serves to answer many questions associated with the quality of strategic elements in design features, material characteristics or the manufacturing process of the failed component or system. Thus, failure analysis is important not only from the point of view of a scientific discipline in terms of analysis of failures, but also from that of safety and reliability which are of great concern in the high-technology industries like aviation and nuclear science. Additionally, it is vital in attaining a high degree of product reliability through a systematic study of component failures, and in providing valuable clues to the inherent weaknesses of materials, design and manufacturing processes. Dissemination of such data on failures provides useful information to the designers and engineers.

As a failure analyst in the Central Laboratory and Design Complex of Hindustan Aeronautics Limited, Bangalore, I have conducted a number of failure investigations on components belonging to a variety of aircraft, helicopters and engines and thus gained considerable knowledge and experience in the area.

As failure analysis is a highly specialised field whose practical applications are utilised in the modern engineering industry, I felt definite need to compile my experiences spread over two decades in the form of a comprehensive book, not only for dissemination of knowledge but also for a deeper insight into the metallurgy of failure analysis.

Regarding the contents of the book, I do not believe in incorporating unnecessary material only to add to the volume and hence only relevant chapters concerning the essential features of failure analysis have been included, just enough for achieving success in the world of failures.

Chapter one discusses briefly the current and future trends in failure analysis, so as to give a clear picture of the various concepts and techniques and enable a failure analyst to understand the overall perspective of failure analysis. A defect, whether in the form of a crack, segregation or residual stress, is one of the important causes of failures. Therefore, chapter two has been exclusively the basics of the effect of defects on failure mechanism. Again, to limit the volume to the essentials, failure concept and mechanism have been restricted to a few vital areas namely, fatigue, stress corrosion and hydrogen embrittlement, since other types of failures are statistically insignificant in the aircraft industry. Other important chapters on investigation procedure and types and characteristics of failures have been included so that the beginners do not lose track of the sequence in collecting vital evidence and can equip themselves with the knowledge of basic identification of each failure before proceeding to establish the mechanisms which primarily involve the five *Cs* (completeness, correctness, consciousness, clarity and coherence) as the hallmark of any failure analysis. Chapters seven and eight provide the essence of failure analysis in the form of 37 case-histories, selected to cover every possible aspect of the major failures, involving mostly in-process and in-service failures.

At the end, a concluding chapter on interaction between chemistry composition and metallurgical behaviour in engineering materials has been included with the hope that both the failure analyst and the chemical analyst will go hand in hand in identifying the basic cause of any metallurgical problem, usually associated with the chemistry of the material.

Basically, this book is intended for a wide spectrum of readers including intellectuals, scientists and students at the universities and the research institutions as well as metallurgists and engineers for its wealth of information and practical approach. I hope that my two decades of experience has brought out the best of failure analysis in this book and it will help to solve the day-to-day problems of metallurgical failures in the aircraft industry.

Suggestions for further improvement of the book are welcome.

A K Das

Acknowledgements

Preparing a book on metallurgy of failure analysis with problems and optimum solutions is an extremely difficult task in view of the fast changing global scenario. The advances in this area are the result of dedicated contributions of metallurgists, engineers and designers from every corner of the world. In preparing this book, I have benefited by the useful contribution made by various sources, in addition to my own study of the subject spanning over a period of two decades. The material presented in this compendium is the outcome of investigational studies carried out at the Central Laboratory of Hindustan Aeronautics Limited, Bangalore.

I am extremely grateful to Dr C G Krishnadas Nair (Managing Director, Hindustan Aeronautics Limited, Bangalore Complex) for having kindly agreed to give foreword as well as his constant guidance, and to Prof. E S Dwarakadasa (Indian Institute of Science, Bangalore), and Mr M K Sridhar (Scientist, Aeronautical Development Agency) for their constructive criticism and help in preparation of this book. I am grateful particularly to Dr A C Raghuram and Dr R V Krishnan (Scientists, National Aerospace Laboratories, Bangalore) for their kind permission to use their articles as the basis for the write-up on Introduction and Metallographic Techniques, respectively. At the same time I gratefully acknowledge the authors of the book on Failure Investigations of Aircraft Components (Ref. 14), which primarily served as the basis for this book. I am also grateful to Hindustan Aeronautics Limited, Bangalore, for their permissions to write and publish this book.

Thanks are also due to Mr D Sekaran for his painstaking efforts in typing the manuscript.

A K Das

Contents

Chapter 1

Introduction

1 Introduction

1.1 FAILURE

Recent decades have witnessed a phenomenal growth of engineering knowledge based on the success of high technology applications with their distinct impact on all aspects of human activity. While the growth of scientific knowledge has been steady, technology has advanced at a rapid pace, more often by compulsion than necessity. In the context of current developments of advanced scientific phenomenon and their utilisation in the action of complex engineering machines, an extraordinary demand has been imposed by design engineers on the materials that suit the need of interaction among the elements of safety, reliability, and economics of production, and even those of art and aesthetics. Under these constraints, it is essential that the design engineer takes into account the possibility or probability of failure and suitable measures to avoid them totally or to minimise the damages if they become inevitable. 'Failure analysis' thus comes into real time test which plays a significant role in achieving the target of high system/ structure reliability at a low cost and high safety factor.

When a product goes into service, a long and real test begins. It means that all the theorising, modelling, assumptions and predictions made during the design and development stages are put to test in an actual service environment. It is the in-service behaviour of the structure that bears judgement on all the simulated tests. Failures which occur in service generally leave behind a trail of tell-tale evidence which when analysed brings out the deficiencies in the system—be it the manufacturing stage, or selection of materials, or design criteria, or usage practice, etc. Adequate remedial measures are then taken to prevent or minimise the recurrence of such failures. In the process, failure analysis helps to improve product reliability by suggesting changes in the design, manufacturing process, and materials, and also assists in extending the useful life of the system or structures[1].

By simple definition, failure is the gap between expectation and performance. It represents an adverse situation wherein a component, material or system fails to fulfil its intended function satisfactorily, during its intended service-life. Failures are inevitable in the life span of any system, but the probability of these occurrences may vary to a large extent. In general, the pattern of failure can be compared to that of human life characteristics (Fig. 1.1). The failure curve also known as 'bath tub' curve is distinctly characterised by three failure rate zones. The first is the short initial period during which the failures are called infant or early failure caused generally due to design deficiency, manufacturing defects, assembly and installation lapses. The failures in the second zone are termed service failures or constant rate failures. These failures are random and are fairly evenly distributed. The curve in the third zone is due to the wearing out of parts. The incidence of failures in this zone is high since most of the components would have rendered their service life. The aim of all designers is to minimise the failure in the first zone and extend as much as possible the useful life (second zone) of the components system.

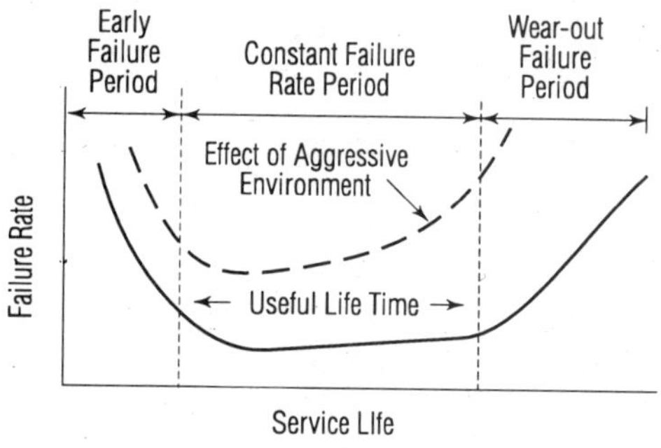

Fig. 1.1 Schematic bath tub curve depicting failure rate as function of service life

With the advent of jet aircrafts, missiles and space vehicles, the highly dynamic element of operational features has added a new dimension to the untimely mechanical failures. More significantly, the environment, the materials used and human errors also contribute to mechanical failures. The occurrence of mechanical failures has significant implications for everyone concerned, from scientists to the policy makers. The cost benefits, design improvements and reliability growth that arise from the analysis of service failures are enormous enough to make the long range investment in it worth the effort.

How best can these failures be prevented, or minimised in the most economical way, is a challenge to all engineers and designers. Essentially no matter what calamities are foreseen or what precautions taken, all objects and devices have a tendency to fail. It should however be the aim of failure analysis to modify and extend the useful service-life with good reliability so as to suit the needs of those it serves.

1.2 FAILURE ANALYSIS

A failure investigation may have various objectives. An investigation may be made to affix blame for the failure but primarily to pinpoint the product reliability so that premature failures are prevented. Occasionally, a failure investigation is conducted purely for academic reasons. In any case, the importance of carrying out failure investigations need not be over-emphasised, since catastropic failures such as aircraft accidents involving loss of human life, enormous cost and property damage are inevitable Fortunately most metal failures can be adequately analysed by the metallurgists. Others are more complex and require experts in other engineering disciplines.

Analysis of service failures is like playing detective with the inanimate objects. Complexity of investigation grows in confusion, being some time unable to establish the missing link on a logical path. However, the following guidelines help the investigator in a systematic manner to reach the root of the cause of the failure.

1. Background information which provides what happened prior to and at the time of failure, the manufacturing history of the part, and discussion with the eye witness.

2. Visual examination includes a visit to the scene of failure, making sketches and taking measurements, notes and photographs. It helps to select parts to be taken to the investigation agency for *prima facie* examination and understanding.
3. Non-destructive examination to obtain defect-information on the surface as well as the core of the failed part.
4. Fractographic examination to identify the type and nature of fracture.
5. Destructive test to ensure soundness of the material in conformity with the specifications in respect of chemistry, properties of microstructure, etc.
6. Examination of all data to arrive at a conclusion of the cause of failure and recommendations for its prevention.

Among all the guidelines mentioned above, *fractography*—a term coined in 1944 to describe the science of studying fracture surfaces—is the key to establish the mechanism and type of failure.

1.3 FAILURE ANALYSIS METHODOLOGY

Failure analysis is a technique by which facts are gathered and analysed to determine the cause of failure. Broadly speaking failures could be attributed to any one or a combination of causes such as improper material selection, design, processing and fabrication, operation, repair and maintenance and exposure to hostile environments not designed for.

Louthan[2] proposed the fault-tree technique in 1978 as a methodology for analysing service failures and failure mechanisms. But this idea is yet to get a strong foothold in the field of failure analysis of mechanical components. A failure analyst faced with a variety of failure problems needs to know the answer to the following questions:

- What facts should be gathered from the failed parts ?
- In what sequence should the facts be gathered so as not to destroy any factual evidence ?
- How should the gathered facts be connected ?
- What information should be sought from the user, which would be of help in the failure analysis ?
- How to prevent the recurrence of such failures ?

Constructing the fault-tree for the failed part provides the failure analyst clues to answer these questions. Though the fault-tree does not pin-point the actual cause of failure, it provides guidelines along which to proceed and conduct the failure analysis in a more organised manner. The general principles involving the construction and evaluation of a fault-tree along with the uses, merits and limitations of the techniques have been described by Shamala and Raghuram[3].

A fault-tree is a graphical technique that provides a systematic description of the combination of possible occurrences in a system, which can result in a fault or undesired event. It is a schematic representation of the interrelationships between the different basic events and the undesired event.

Qualitative and quantitative evaluation of the fault-tree is possible. Qualitative evaluation consists of determining the various minimal cut-sets of the fault-tree. A cut-set is a set of basic events whose occurrence causes the top event to occur. It is minimal when each of the basic events in the cut-set is necessary and whose combination is sufficient to cause the top event. The listing of minimal cut-sets is useful for design purposes in order to determine the weakest links in the system. Since the fault-tree consists of hundreds of gates and basic events, determination of minimal cut-sets by inspection has become impractical. Hence, the use of computer for constructing fault-trees for failures of complicated systems is very essential in order to economise on time, money and effort. Quantitative evaluation of the fault-tree is done to determine the probability of occurrence of the top event. It is also done to determine the importance of different events and cut-sets in causing the top event. However, it has not been possible to fully exploit the potential use of the quantitative evaluation of the fault-tree[4], because of lack of relevant failure data.

The fault-tree analysis has multi-faceted applications. It can be used as a diagnostic tool, a design tool, as a method of system safety analysis, and a management tool. Above all, it can be used as a comprehensive method of documentation and to get a logical understanding of the system.

1.4 TOOLS AND TECHNIQUES OF FAILURE ANALYSIS[5]

Having collected sufficient background information about the failed part, the next step comprises analysing the failure using suitable tools and techniques. In general, the first step consists of observing the fracture surface visually and macroscopically. This helps to view things in their proper perspective. An optical stereomicroscope is a generally used for making a macroscopic examination of the fracture surface. The information gathered from such observation is of great value in analysing the cause of failure and throws light on the following aspects.

- Nature of origin of failure.
- Location of contributing stress concentration.
- Deviation of crack propogation and sequence of failure.
- Failure mode and mechanism.
- Orientation and magnitude of stresses.
- Imperfections contributing to the failure.
- Presence of relevant contaminants on the fracture surface, e.g. scale, paint, corrosion products, etc.
- Dimension and important physical data.

Relevant features are recorded photographically or schematically or both. Apart from the macroscopic examination of the fracture surface, non-destructive techniques such as liquid penetration, ultrasonic techniques, radiography, magnetic particle, eddy current and acoustic emission are used to reveal surface and internal defects.

Microscopic examination of the failed sample takes the investigator a little further in the analysis of problems. The microstructure yields information regarding the heat treatment of the component, crack path, deformation, etc. sustained by the component. A careful choice of specimen is absolutely necessary not only to extract the maximum amount of information but also to reveal the exact reason for the cause of failure. In all aspects of microstructural observation, the surface of study has to be prepared properly in order to assess certain special features, such as flow lines, inclusions porosity, surface finish, etc. under an optical microscope.

However, the optical microscope has certain limitations with regard to magnification, resolution and depth of field. In order to overcome these shortcomings, an electron microscope with a small wavelength of electron beam has been ever increasingly used for fracture examination, because of its excellent resolution power and large depth of field. Two types of electron microscopes are commonly utilised, namely the transmission and the scanning electron microscopes. The main disadvantage of a transmission electron microscope (TEM) is that a special technique (replication) has to be adopted for specimen preparation before the fracture surface can be observed under the microscope. On the other hand in the scanning electron microscope (SEM) the fracture surfaces can be studied directly without any preparation. The large depth of field of the electron microscope also makes it possible record stereoscopic views, which are very helpful in understanding the microscopic topography of a fracture surface.

The novelty of electron microscope lies in that not only can it be used for observing the fracture surface to determine its morphology like the size, shape and orientation of dimples (characteristics of overload failure) but also for observing intricate features such as fatigue striations, their spacing and crack growth rates. It is important to make a proper interpretation of the observed features to define conclusively the character of the fracture mechanism. Based on the spacing and definition of fatigue striations, it is possible to define the type of fatigue under different cyclic loading conditions. When the cyclic loads producing the fatigue crack are of constant amplitude, the striation spacing is uniform. A random loading cyclic condition produces striations that have varied spacing. If the striations spacing is very fine throughout most of the fatigue region, then a low load, high cyclic condition prevails. A wide spacing from beginning to end of the fatigue region indicates a high load, low cycle condition. The clarity or definition of the striation also yields important information. A pure tensile load when responsible for fatigue failure would give rise to well defined striation whereas when compressive loads are encountered, a rubbed appearance on the striation with small cracks would be observed.

Chemical analysis of the material of fracture is also made by SEM using the EDAX attachment. Characteristic X-rays emitted by the surface are analysed either by the energy dispersive mode or the wave length dispersive mode. The energy dispersive mode which is more often used with the SEM is capable of providing an X-ray energy spectrum for identification of elements present in micro or macroscopic areas. With this attachment, inclusions, segregations, contaminant or corrosion products that have contributed to the failure of a part can be easily identified. Computer attachments to both types of X-ray analysis for quantitative analysis, and recent advances in detectors for light elements have greatly extended their usefulness in failure analysis.

Apart from the electron microprobe analysis, there are other surface analysis techniques used to detect contaminants and or corrosion products on fracture surfaces. The need for this sort of surface analysis arises when a component fails due to wear, corrosion, stress corrosion, corrosion fatigue, etc. The basic philosophy in surface analysis is to employ a probing species and examine the energy and intensity of the species emitted as a result of interaction of the probe with the surface. The depth of this interaction is controlled by the energy of the probing species. Spectroscopic techniques vary according to the type of excitation and emission used.

Thus, a wide variety of tools and techniques are at the disposal of the failure analyst. His ingenuity lies in selecting the right one, using it appropriately and above all interpreting the result of the observations correctly. Another fact to be borne in mind is that quite often standard instruments are adequate for analysis, and it may not always be essential to use esoteric instruments and methods such as electron microscopes, surface analysis techniques, and so on. The analysis of results from the facts gathered to determine the actual cause of failure is a very important and interesting aspect of failure investigation and needs to be done with utmost care. It has relevance for those who are concerned with the reliability of a product, as well as the users and future analysts.

1.5 FRACTURE MECHANICS

For quite some time engineers have designed on the assumption that the components they are designing are made ideally from ideal materials. They now know that the former is untrue and metallurgists have warned them of, and demonstrated, the fallacy of the latter. Hence defects will always be present however good be the design, the material and its fabrication into components. When design cannot provide defect-free components or materials, a good defect tolerance should be the aim of any designer or manufacturer. Hence, the concept of fail-safe philosophy in place of safe-life philosophy has revolutionised the design philosophy. However, one can live with defects if there is a rational basis for their evaluation and monitoring. Fracture mechanics plays a vital role in estimating the critical flaw-size at the operating stress levels, the initial flaw-size and the sub-critical flaw-growth characteristics of the materials. But the full exploitation of fracture mechanics in practice depends, to a large extent, on the non-destructive techniques available to detect the defects and provide information on their location and size. Five non-destructive techniques are widely used for defect detection. They are dye penetrant, magnetic particle, eddy current, radiography and ultrasonics. The first three can detect only surface and sub-surface defects, while the other two can also detect remote embedded defects. Ultrasonic is far more sensitive to cracks than radiography, and when used in conjunction with fracture mechanics, it has proved to be a powerful technique for demonstrating component integrity.

Fracture mechanics is fast gaining ground in failure analysis. It is used to determine the defect size which could have caused the failure under given conditions, the criticality of defects present in the component, the residual useful life of a component in service, etc. It is used in failure prevention to assess the defect size which could be tolerated, to organise inspection schedules and to choose the appropriate non-destructive technique for inspection.

Fracture mechanics also plays an important role in predicting, preventing and analysing delayed failures by stress corrosion, cracking, hydrogen embrittlement and other similar causes. As in fatigue crack growth, a threshold value of stress intensity factor exists, below which there is no growth due to stress corrosion in a given environment. This threshold value is generally the lowest when a plain strain condition prevails at the crack tip, and is usually designated KISCC. For any structural component, if the applied or locked-up residual tensile stresses and the environment are known, one can calculate the stress corrosion lifetime as a function of initial crack length, using the crack growth rate versus stress intensity factor. This helps in deciding on the maximum size of the crack or flaw to be detected after manufacture, so that in service there is no risk of failure by stress corrosion.

1.6 FAILURE DATA RETRIEVAL THROUGH FAILURE EXPERIENCE MATRIX

Poor documentation, improper organisation and non-accessibility of failure data are often the causes of under-utilisation of information obtained from the failure experience. In practice, the gap between readily acceptable and rejectable defects can be made smaller only by building up the service experience. Presently the situation is such that detailed computer calculations of crack growth or instability conditions in complex structures can be made. However, uncertainities in design analysis, material characteristics, NDT techniques and service environment continue to complicate the assessment of structural reliability under actual service conditions, and in the long-run experience alone can provide the intuition needed to establish the standards for safety and reliability, and the means by which predictions can be made.

In the past, full use could not be made of the previous failure experiences because of the disorganised fashion in which failure data was stored and the inavailability and inaccessibility of such data. The most important aspect of failure reporting is accuracy and speed. The 5Cs—completeness, correctness, consciousness, clarity and coherency should be the hallmark of any failure analysis report.

Thus there is a need for reporting and storing of failure data in an organised and generalised manner, capable of quick retrieval, so that it can be of use to the entire engineering industry. Failure experience matrix (FEM) provides a method of achieving it.

The FEM is a 3-dimensional cell matrix[5] obtained by plotting elemental mechanical functions along the first axis, failure modes along the second axis and corrective actions along the third axis, as shown in Fig. 1.2. A spatial cell is generated at the intersection of each combination of failure mode, elemental mechanical function and corrective action. Within the cell information on the identification of the parts failed by that particular combination of the three constituents of the matrix, the details of the corrective actions used and their effectiveness are available.

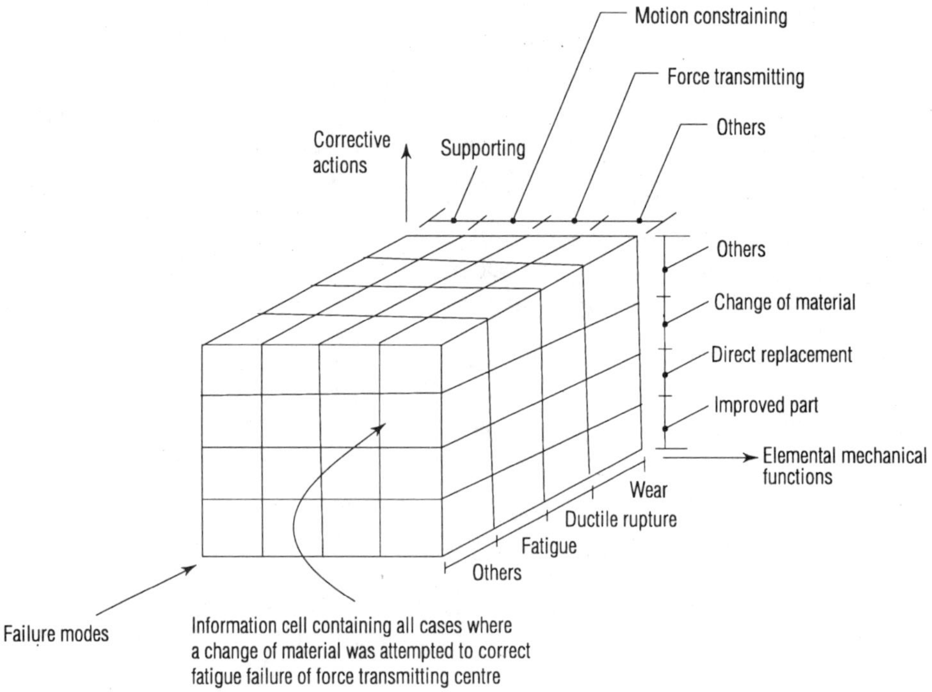

Fig. 1.2 Failure experience matrix[5]

The FEM finds wide applications in the fields of failure analysis, design, reliability and research and development. Service failure problems can easily be solved using FEM. The only task of a field service engineer is to identify the elemental mechanical function of the failed part, enter the matrix and find out the various modes by which the component could have failed, identify the mode by which the component has failed and then finally study the corresponding remedial action that would be most likely to prevent future occurrence of the failure. Any reliability prediction is as good as the data fed into it, FEM serves as a storehouse from which failure data based on real experience can be obtained. It provides the designer with a source of information on design deficiencies in the past and the methods that have been used to counteract them. Using the FEM, the R&D personnel can identify the areas which need a greater understanding and on which resources can be spent more usefully.

1.7 RELIABILITY

However, a designer cannot design a product solely on the information obtained from the past experience. He needs a prediction technique as well, using which the reliability and safety system can be assured. This is more so when a large amount of capital is at stake or there is a risk in human life as in nuclear or aerospace industries. Reliability engineering provides a means of attaining this objective. But economic consideration should always prevail in any failure prevention and reliability estimation programme. There is an optimum degree of reliability at which the total cost achieved is minimal as shown in Fig. 1.3. In practice it is difficult to realise this optimal economic reliability because it involves estimating how production, design, and maintenance costs vary from a given reliability and this can only be done with accuracy if there is a considerable amount of previous experience to draw upon.

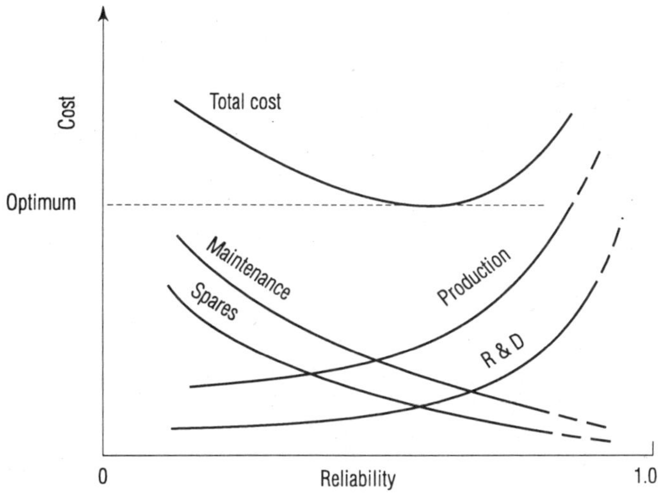

Fig. 1.3 Variation of total cost with reliability[6]

However, one relevant question is often asked by the designers while designing a new project, "reliability at what cost"? To seek an answer to this, Fig. 1.3 shows the cost of quality control which offers a general understanding of reliability cost[6]. Therefore, for the economic production of any component, a compromise has to be made between the cost of quality control and number of defects of failures acceptable. Hence, the crux of any economic failure prevention programme lies in identifying and preventing the failures of the vital few against the trivial many. What needs to be understood with respect to reliability and cost is that one gets what one pays for and by not paying for reliability in the original product, a greater cost may eventually be incurred.

Chapter **2**

Defects: Types and Characteristics

2 Defects: Types and Characteristics

2.1 DEFECTS

Modern technology not only demands superior quality of material but also experienced metallurgists who have an indepth knowledge of parts that are obtained through diffrent processes of manufacture, particularly in the aeronautical field. With the rapid development of materials and process technologies and their increased applications in advanced engineering systems, chances of failures in the complex structural assemblies or parts are largely due to the difficulties in controlling the defects in the finished products.

However, defects which arise due to manufacture are to be strictly controlled within the optimum level, with a view to minimise the chances of abrupt failure that may be brought about during the service period of a highly loaded component or flying machine. As is widely experienced in the field of aeronautical industry, there are too many failures at the primary stage of manufacture as well as during service. The origin of these failure is often attributed to the presence of a variety of defects in the material, either inherent or induced during manufacture. Despite stringent controls exercised during various stages of operation and inspection, some metallurgical defects seem to remain undetected. If the number of such defects exceed the safe level, particularly in the components of critical applications, it would definitely be a potential hazard.

Therefore, it is essential that a failure analyst acquaints himself with defect characterisation processes (including source, identification, cause, detection and remedial measures) involved in the cast and wrought products (both ferrous and non-ferrous) such as rolled bars, extruded bars and profiled sections, forgings, welded components, etc., before and after secondary processing, Further, in order to understand precisely how these defects influence the service behaviour and to assess the acceptability of the defects in a meaningful manner before certifying for use in service, detailed information on effects of the defects is included in this chapter.

Classification of Defects

A non-acceptable defect as distinguished from slight discontinuities or blemishes which are less severe in nature is defined as a gross lack of continuity in a material, in terms of its physical aspects or properties. The defects are also identified in terms of their propagative or non-propagative nature, depending on their linearity and sharpness. This identification is necessary to understand the degree of the effects these two types of defects exert on the ultimate service-life. For example, porosity by its round/elliptical geometry is considered to be of a non-propagative type of defect. It has less deleterious effect as compared to the crack which having both linearity and sharpness can easily propagate at a nominal stress of functioning in assembly. While characterising the defects, it is also necessary to categorise them as surface or sub-surface defects in order to interpret appropriately the severity of their effects. This evaluation is extremely important where hidden defects in the sub-surface cause sudden and sometimes failures in otherwise proven material.

Normally defects in engineering materials fall into the following four broad categories:

1. inherent defects during solidification from liquid to solid state and casting,
2. induced defects during primary processing,
3. induced defects during secondary processing, and
4. defects in service.

Numerous kinds of defects arising out of the above categories of processes adopted during the manufacturing cycle and the service period are shown in Table 2.1.

Table 2.1 Defects during the manufacturing process and the service period[7]

Manufacturing Operation	Process/Product	Possible Defects
Foundry	Cast ingot/billet, sand, die, investment casting	Pipe, blowhole, gas porosity, shrinkage porosity, shrinkage cavity/crack, non-metallic inclusion, oxide/dross inclusion, segregation, internal fissure, ingotism, scabs, ingot crack, hot tear, cold shut, misrun, etc.
Wrought processing	Rolled bars/rods/Plates/sheets	Seam, crack, oxide discontinuity, non-metallic inclusion discontinuity, delamination, orange peel, exfoliation, transverse fine crack, edge crack, etc.
	Forged bars/rods	Forging burst, lap/fold, cold shut, push through defect, crack, rupture, tear cooling crack, thermal crack, assymetrical/undesirable grainflow pattern, orange peel/alligator's skin, surface undulation, surface grain coarsening & discolouration (in Al alloys), hair line crack, decarburisation, overheating, burning, cracking due to segregation, low mechanical strength properties due to improper grainflow/insufficient section reduction, blistering, rough surface, central burst, etc.
	Extruded sections/profiles/tubes	Heat checks, radial and circumferential cracking, central burst, piping, delamination, peripheral coarse grain envelope, longitudinal discontinuity, annular defect, hot and cold shortness, cracking due to inclusion/segregation and gas voids, blistering due to entrapped gas in extruded bar stock/improper lubrication, funnel formation, etc.

	Drawing	Internal burst, transverse cracking, poor surface finish, wrinkling, orange peel, stretcher-strain, edge cracking, season cracking (in brass).
	Swaging	Overlapping dimple, orange peel, surface tearing, micro fissure/crack, wrinkling, fold.
Assembly/ finishing	Welding	Incomplete penetration, incomplete fusion, crack, undercut, slag inclusion oxide film, void, porosity, shrinkage, coarse grained/ widmanstatten structure, nugget cracking, cavitation, intergranular crack (weld decay), distortion, improper weld profile, warpage, excessive concavity, etc.
	Machining	Mechanical defects such as nick, scratch, ridge, etc. grinding crack, burn/discolouration, hot spots, rehardening effect, softening effect, residual stresses, loss of dimension and parallelism, surface irregularities, distortion, undercutting and rough surface (in chemical milling), hydrogen embrittlement/intergranular attack or pitting and corrosion (in electrochemical machining), craters/voids/rough surface, hard chilled white layer, crack, heat affected zone (in electro discharge machining), etc.
	Heat treating	Quenching crack, overheating, coarse grained/ widmanstatten structure, burning, decarburisation, oxidation/scale, hairline crack, retained austenite, soft spots, untempered martensite, distortion, warpage, low strength properties (in steels), crack, high temperature oxidation, discolouration, surface blisters, diffusion of cladding, distortion, overheating, low strength properties, orange peel effect (in Al alloys).
	Plating	Imperfect adhesion, blisters, flaking of coating, discolouration, crack, embrittlement, rough deposits (dark or granular), etc.
	Pickling	Embrittlement cracks, pits, etc.
	Forming	Crack, wrinkling, distortion, etc.
Service	All products	Fatigue crack, corrosion pits, intergranular corrosion cracking, corrosion fatigue, fretting corrosion, stress corrosion cracking, hydrogen embrittlement, exfoliation, galvanic corrosion, crevice corrosion thermal fatigue, hot corrosion fatigue, creep, brittle fracture, etc.

2.2 DEFECTS IN CASTING

Casting defects are usually not accidental (Fig. 2.1). They occur because some steps in the manufacture process are not properly controlled. A defect may arise due to single clearly defined cause, or more frequently as a result of a combination of factors. Among the numerous defects normally encountered during conventional casting processes, viz. sand casting, die casting and investment casting, only the common ones are selected here for their characterisation.

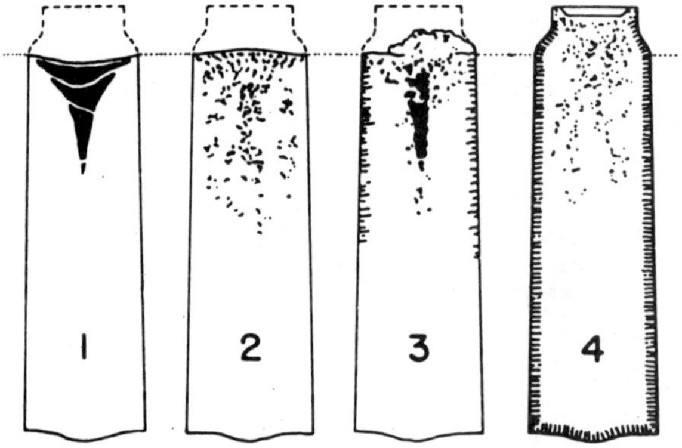

Fig. 2.1 Schematic illustration of shrinkage cavity in cast steel ingots

Solidification Defects

Solidification defects[8], such as shrinkage, hard spots, macro/micro segregation are the major category of defects in the casting process. Shrinkage defects may be in the form of shrinkage cavity or pipe (Fig 2.1), centreline shrinkage, discrete porosity and surface sink. These defects arise from failure to compensate for the contraction of metal during solidification. The most common examples are those of primary shrinkage cavity or pipe resulting from inadequate feeding. Relatively large jagged cavities occurring in heavy and isolated sections are called hot spots. Secondary shrinkage defects may occur in the form of a filamentary network (centreline shrinkage) in the central zone. Scattered shrinkage porosity defects, which are usually in the form of intercrystalline cavities, are generally revealed on machining or sectioning. These are due to long freezing characteristics of an alloy, particularly as experienced in aluminium-copper alloy system.

Shrinkage Defects

The cause of shrinkage of defects are many and include:

(i) lack of sufficient metal feed;
(ii) isolated heavy sections and abrupt changes in section;
(iii) failure to obtain directional solidification towards desired heat centres, such as raisers or in-gates;

(iv) mould dilation due to soft or uneven ramming;
(v) unsuitable metal composition, such as a high phosphorus content or low carbon equivalent; and
(vi) high pouring temperature.

Gas Defects

Gas defects are commonly observed in the form of blow holes or surface or sub-surface pin holes or intergranular porosity caused by the generation and/or accumulation of gas or entrapped air in the casting. Gas holes generally occur in the riser of runner top. Gas defects are generally caused due to:

(i) presence of high moisture content, fume, carbonaceous and gas producing material in the sand mixture, and its poor permeability and grain size distribution;
(ii) insufficiently vented, undried, hard rammed and non-uniformly rammed moulds;
(iii) low permeable, hard, unbaked and inadequately vented cores and excessive or wet core paste and wash (cores absorb moisture during storage);
(iv) cold and damp ladles, sluggish and gassy metal, interrupted and slow pouring, pronounced stirring, overheating, etc.; and
(v) use of rusty, improperly coated, oxidised and damp chills and wires.

Pin Holes

Pin holes are caused mainly due to the evolution of hydrogen during solidification. However, moulds having a high moisture content and/or gas generating ingredients and metal-mould reaction also produce pin holes. Gas porosity is also caused due to improper degassing.

Misrun

Misrun or short-run occurs due to lack of completeness of the process, as a result of the failure of the metal to fill the mould cavity. Cold shut is caused by faliure two streams of metal liquid flowing from different directions to unite, leaving crack-like indication but with smooth rounded edges. The causes of these defects are:

(i) insufficient fluidity of the metal.
(ii) low pouring temperature,
(iii) interrupted pouring, improper sizing of the gate, runner and sprue; and
(iv) thin sections.

Hot Tears and Cold Cracks

Hot tears, cold cracks, warpage, etc. are defects that form during the cooling of the casting from solidus temperatures to room temperature. *Hot tears* are crack-like defects that occur while cooling in the mould or just immediately after solidification. These cracks are very irregular with heavily oxidised surfaces. *Cold cracks* are continuous in contrast to hot tears, which are ragged or irregular. Hot tears occur due to:

(i) internal stresses arising from hindered contraction due to mould or core resistance;
(ii) internal stresses arising from differential cooling due to improper casting design; and
(iii) high pouring temperature.

All the above mentioned casting defects are usually detected by radiographic examination. A hot tear can also be detected visually (Fig. 2.2).

Fig. 2.2 A typical hot tear in aluminium alloy casting

2.3 PRIMARY PROCESSING DEFECTS

Rolling

Rolled bars of ferrous and non-ferrous alloys are initially cast as ingots or billets and then hot-rolled to various sizes and shapes. The top and/or bottom, and the portions which are usually prone to defects like pipes and blow holes, are cropped off. However there is still a possibility of retention of some of the defects due to inadequate cropping.

These defects are likely to collapse and elongate in the direction of rolling. These discontinuities mostly in the form of films of oxides invariably run in the direction of rolling up to any depth from the outer surface. The most efficient method of detection is a micro-examination of a transverse disc cut at different locations of the rolled length. The ultrasonic flaw detector is used for the detection of this defect throughout the length of the bar.

Adequate cropping of the top and bottom ends of the rolled length would ensure freedom from this defect.

Seam

Seam is another common defect, generally observed as one of the rolling defects which is a discontinuity running along the direction of rolling at the surface. It is caused by oxidised surfaces being folded and rolled (Fig. 2.3). Cracks are likely to occur in the rolled bar surface due to the thermal gradient during rolling. Although it is a practice to cut-off the ends which are prone to cracking in rolling, residual cracks may persist in the cut ends.

In steel bars, seams and cracks on the surface or sub-surface can be easily and precisely detected by the magnaflux check. In light alloy bars, visual and/or dye penetrant check followed by light etching reveals these defects.

Non-metallic Inclusion

If the melt is not clean, the ingots on rolling result in discontinuous stringers of inclusions of silicates, sulphides and chains of oxide particles.

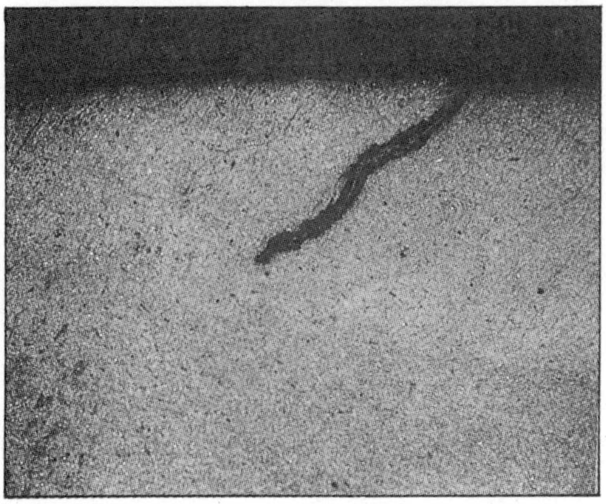

Fig. 2.3 A section through the seam (× 15)

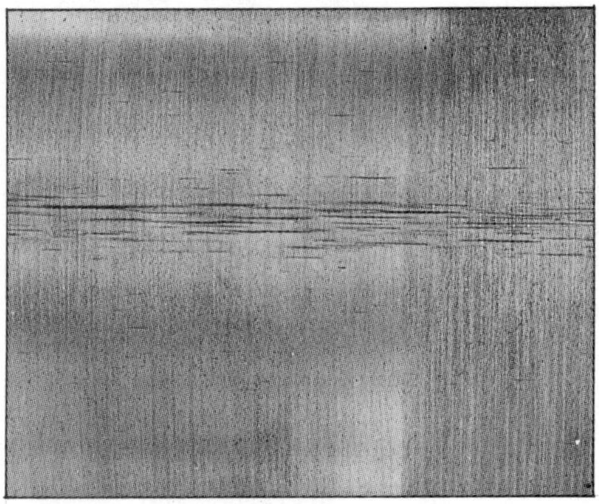

Fig. 2.4 Discontinuous straight line indications on magnafluxing a steel bar after turning off a 2 mm skin layer (× 3/4)

In steel bars, magnetic particle inspection reveals these defects in the form of discontinuous straight lines (Fig. 2.4). A series of step grinding and magnafluxing processes indicates the severity and distribution of these inclusions. Straight line indications may not be continuous, but globular oxide particles, present in the form of chains, show up as form of continuous flow. These longitudinal inclusions cannot always be termed as harmful. It depends on the end use of the product.

Delamination

It is another major defect encountered in a sheet and plate. Large blow holes in the ingot/billet get flattened on rolling and do not get welded together due to the presence of oxide layers in the blow hole surfaces (Fig. 2.5).

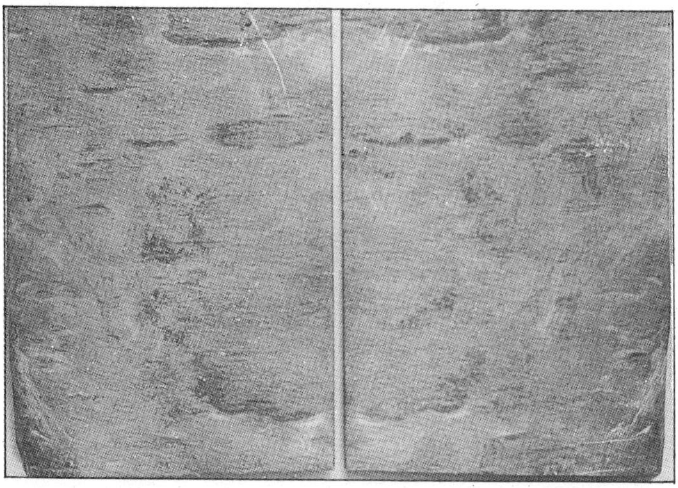

Fig. 2.5 A stainless steel sheet showing lamination opened out. Note the oxidised surface layers caused due to large blow holes in the original ingot (× 1/8)

If required, the defect can be detected by macro-examination of the edge after polishing and light etching. Ultrasonic check is more applicable for thicker sheets and plates, and is not practicable in the case of thin sheets. Before discarding the remedial measures adopted to ensure freedom from the defect, are to adequately crop the ingot rolled billet at the top and bottom ends and to examine it ultrasonically and by macroetching.

Exfoliation

Although not a frequent defect, it still occurs commonly in rolled/extruded products of certain aluminium alloys under certain temper conditions. The defect shows up as the voluminous powdery products[9] (Fig. 2.6) of corrosion, with the directional attack along the grain boundary (running parallel to the direction of working). This results in a leafing action (Fig. 2.7). This defect is susceptible to growth when the grain deformation attains a critical level with respect to grain size versus width, especially if the materials processed are in the T4 or T6 condition. To detect the onset of exfoliation cracking, an ultrasonic inspection check can conveniently be used. Visually or under a magnification of ×10, this defect can be very easily detected from the presence of its distinct characteristics. However, micro-examination will confirm the presence of exfoliation. To overcome this, the Al-Cu/Al-Zn-Mg alloys are heat treated in the T73 condition, i.e. the double ageing process.

Fig. 2.6 A typical example of exfoliation corrosion in Al-Cu alloy rolled bar with powdery products. (× 1)

Forging

It is a widely used manufacturing process, for making complicated parts, as well as meeting the demands of modern superior service characteristics. Although great care is exercised in producing a sound forging, various types of defects, such as folds laps, push-through defect, thermal crack, burst, undesirable grain flow, etc., are often encountered. Ultimately these defects result in failure.

Folds/Laps

These are the major defects[10] that occur in the blanks, due to the oxidised surface/ears folding over and getting pressed in subsequent forging without being welded. A fold is distinctly characterised by the presence of a 'V' shaped angular discontinuity filled with oxides in the microsection (Fig. 2.8) of a fold (Fig. 2.9)

Undesirable Grain Flow

Abnormal or undesirable grain flow pattern due to improper forging constitutes a serious hidden defect. This is because a part of the major deviation from the flowlines which are otherwise made to orient in such a way that they are at right angles to the major sustained load in service. An abrupt change in the smooth flow direction of grain

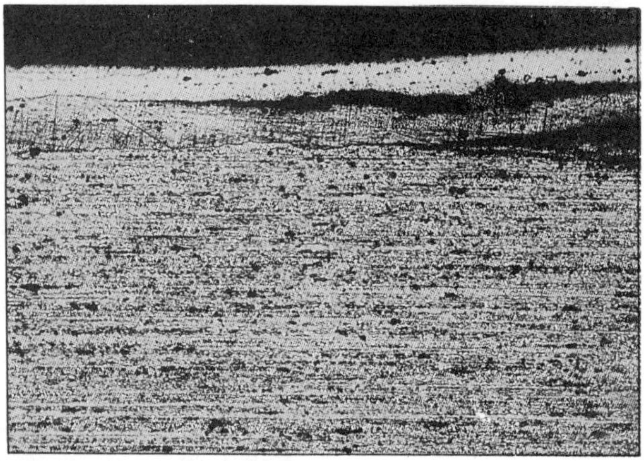

Fig. 2.7 A section through the exfoliated area in Fig. 2.6 showing the leafing action (×50)

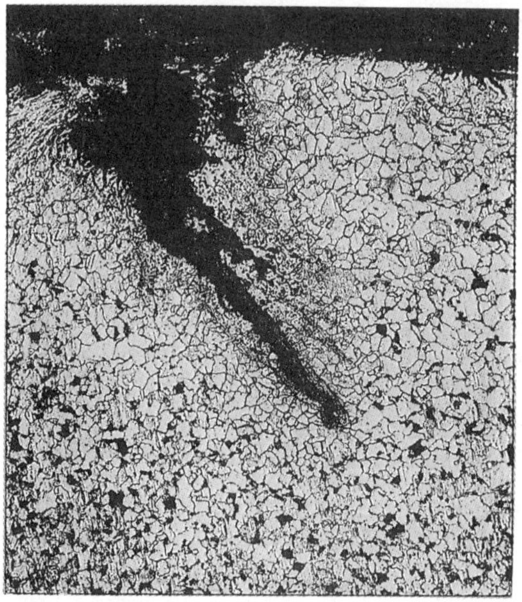

Fig. 2.8 A section through the fold (2.9) showing grain distortion surrounding the defect (× 100)

fibres tends to form a kink/re-entrant (also known as push-through defect) which serves as a potential source for premature fatigue fracture due to pronounced anisotropy in strength properties.

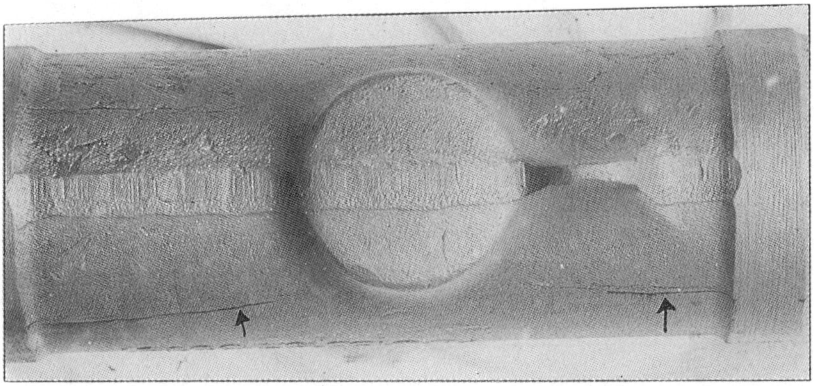

Fig. 2.9 Typical fold like discontinuity on a forged steel part after magnafluxing (× 1)

These defects are caused due to (i) improper size and volume of forging stock, (ii) improper preforging techniques, and (iii) improper positioning of forging stock. Only the macro etch test of the section of the forged blanks reveals the defect.

Thermal Cracks

Cracks are also liable to develop in the forged blanks, mainly along the flash line areas and are due to improper or uneven post cooling, particularly as is evident in the case of the large triple alloy (Ni-Cr-Mo) steel blanks. This is due to its extreme sensitivity to thermal shock (Fig. 2.10).

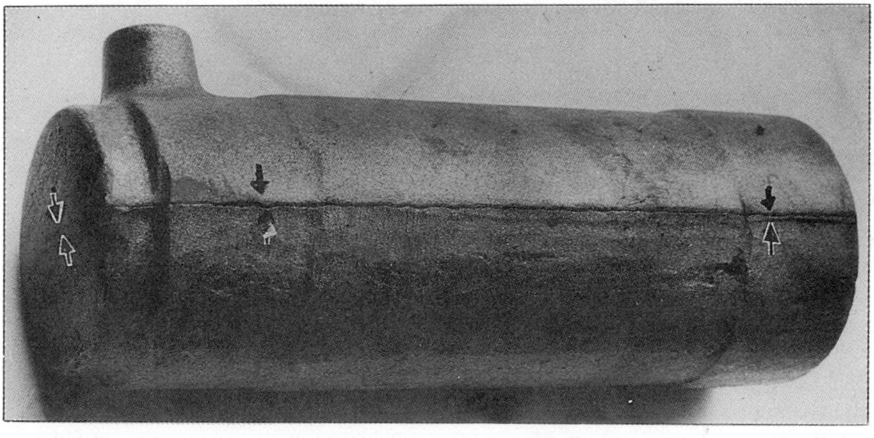

Fig. 2.10 Crack revealed on magnaflux check of a solid cylindrical forged blank out of a triple alloy along the flash line due to improper post cooling process. (× 1/2)

Remedial measures This type of cracking, due to thermal sensitivity may be overcome by directly cooling the forged blanks down to 550 °C and maintaining this temperature for a sufficient time before bringing it to room temperature. Magnaflux check is an effective means of detecting such defects in the case of steel forging, preferably after sand blasting. Cracks also arise due to insufficient or excessive forging temperature.

Forging burst occurs in the mid zone area of large blanks. Bursts are often visible to the naked eye and can be seen at the end portion. Otherwise, light etching and macro-examination reveals them clearly.

Orange Peel

The orange peel/alligator skin is commonly observed on the surface of hot worked metals as a rough finish (sometimes in a pebble-like form in steel forgings) (Fig. 2.11). The reason for the roughness is due to the inherent variation in the degree of grain deformation due to severe and non-uniform hot working effects. This is particularly noticeable in over-heated metals because of the large grains.

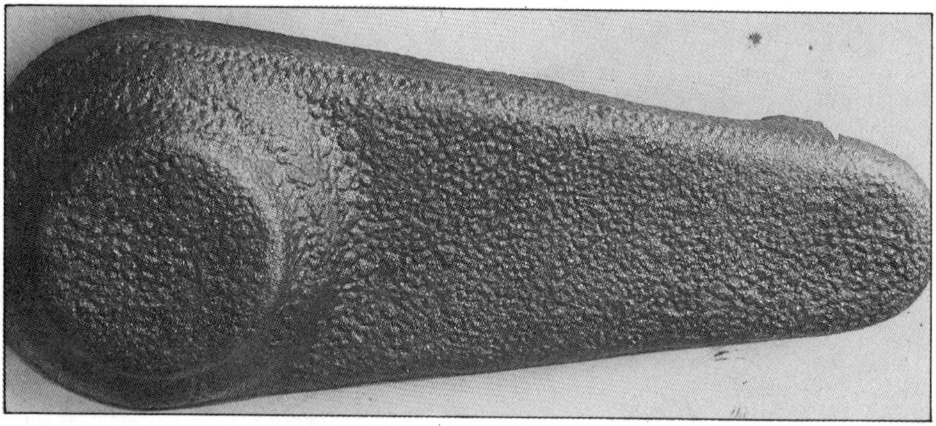

Fig. 2.11 Alligator skin in a forged steel blank (× 1)

Extrusions

Major defects that are normally observed in wrought aluminium alloy products that go through the extrusion processes are blisters, heat checks, delamination, longitudinal discontinuities, peripheral coarse grained envelope, blisters, burst, crack, overheating, etc. Occasionally defects such as segregation, inclusion, gas voids, shrinkage cavity/pipe are found carried over from the casting, increasing the chances of failure of the worked component during the metal deformation stage itself.

Heat Checks

In extrusion and profiled section heat checks defect is frequently observed during improper extrusion and identified as parallel, discontinuous surface cracks in the transverse direction (Fig. 2.12). They are caused by frictional overheating and incipient fusion with the die under the combined effect of undue extrusion pressure and speed. Severe surface cracks are visible to the naked eye. On the other hand, if they are fine they can be detected by a magnifying glass after light etching.

To overcome this problem a careful control of the extrusion parameters, such as speed, pressure and temperature of the billet, is required, so as to minimise the rise in temperature due to severe friction.

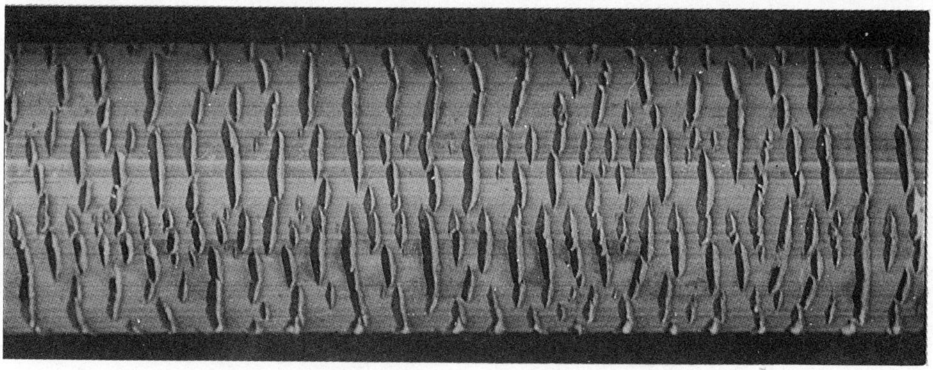

Fig. 2.12 Heat check in the Al-Cu alloy extruded bar (× !)

Longitudinal Discontinuities

These can occur deep inside the extruded product due to blow holes or non-metallic inclusions originally entrapped in the ingot/billet. These defects can be detected by ultrasonic checks or macro-examination of transverse discs cut at different locations.

Blistering

This is another major defect commonly observed in the aluminium alloy extruded products and constitutes a serious problem, particularly in aeronautical applications. This appears as tiny bubbled zones on the surface (Fig. 2.13) and is caused due to the presence of gas holes/voids or due to eutectic melting/grain boundary cracking.

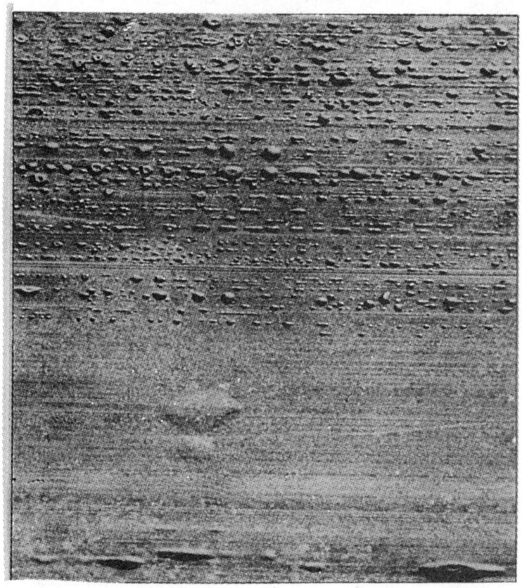

Fig. 2.13 Surface blisters of an Al-Cu alloy extrusion (× 1)

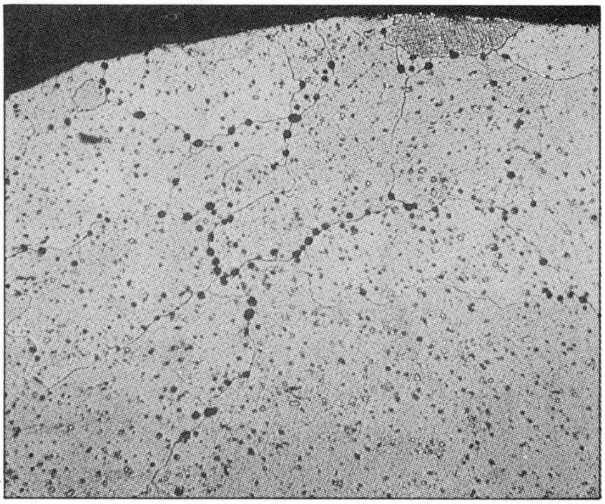

Fig. 2.14 Hign temperature oxidation (× 100)

Three important reasons[11] for blistering are: (i) gas entrapment during the original melt (ii) high temperature oxidation, and (iii) over-heating during heat treatment (Figs 2.14 and 2.15). Besides these, improper lubrication during extrusion also causes blisters (Fig. 2.16). Based on their characteristic appearance there are three different types of blisters normally encountered in aluminium alloy products. These defects show up usually on high temperature solution treatment. When blisters are opened up with a pointed tool, the inner surface exhibits three distinct colours: (i) grey. (ii) dark brown or black, and (iii) white. These colour differences generally help to identify the source of blister formation, respectively, from: (i) incipient melting, (ii) improper lubrication, and (iii) dissolved gases.

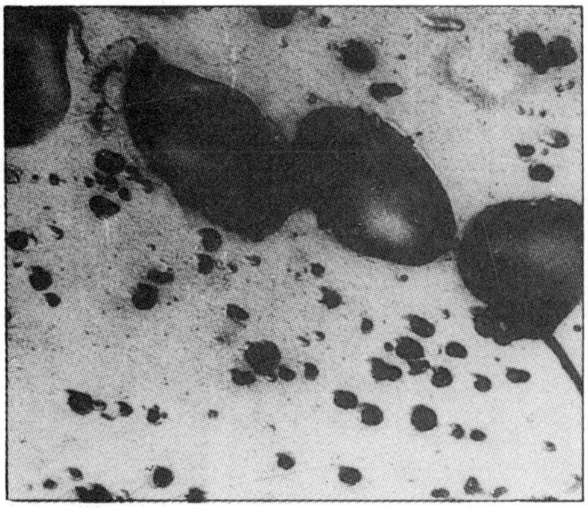

Fig. 2.15 An over-heated structure (× 100)

Fig. 2.16 A section through the surface blistered zone of an Al-Cu alloy extrusion due to the lubricant (× 200)

The first type of blisters is generally associated with the phenomenon involving too high a solution-heat-treating temperature. The second type occurs from improper lubrication causing the surface metal to seize and shear away from the sub-surface metal. Lubricants trapped below the surface tend to volatilise during heat treatment causing blisters. These blisters are invariably found in the edges or the portions of extrusion characterised by the rapid flow. The mechanism of formation of the third type of blisters is rather complex but is believed to be from the concentration of accumulated atomic hydrogen (during melting fabrication and heat treatment) in the lattice imperfections, followed by a change over to a less soluble, molecular form. Development of this type of blister is favoured by the pressure-expansion of the trapped gases during the solution annealing cycle.

The most common source of gas entrapment, which usually occurs during the melting/casting process and which is carried over into the extruded products, can be eliminated or at least minimised by carrying out the proper degassing treatment of the melt prior to pouring.

Peeling/Delamination

This is another defect commonly noticed in the extrusion of tube products and is caused due to contamination of the surface with oxides prior to extrusion. When the surface of the tube material is heavily oxidised and the billets used are comparatively long, oxides or films of oxides can flow into the subsurface towards the end of extrusion, resulting in the peeling effect [(Fig. 2.17 (a) and (b)].

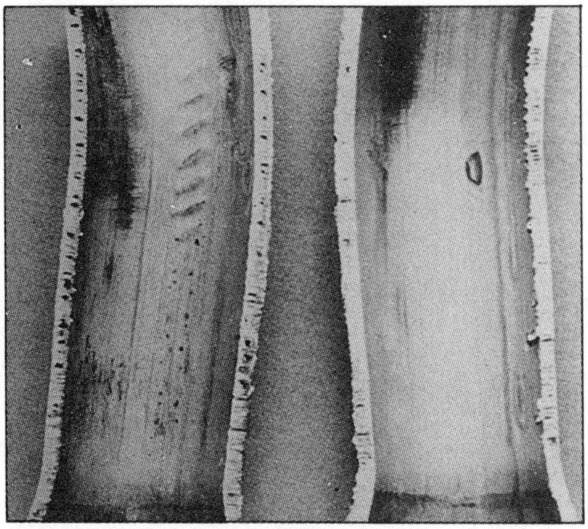

Fig. 2.17(a) Cracked flared ends of an Al alloy extrusion tube, on splitting (× 2)

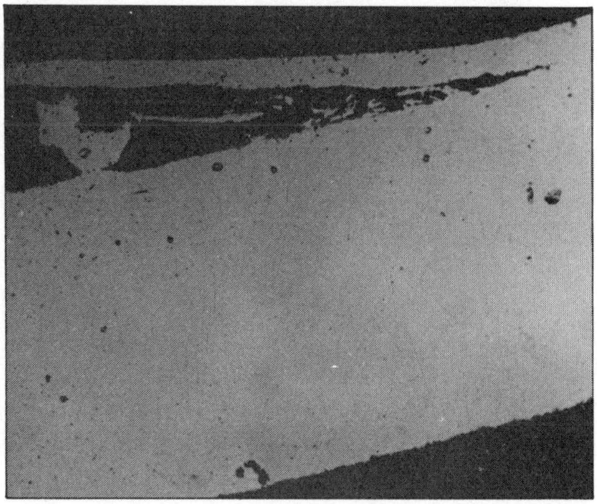

Fig. 2.17(b) Peeling off of layer in the base surface on micro-examination of a section across the peeled off area in Fig. 2.17(a) (× 50)

This defect can be visually recognised and avoided by controlling the process para meters, such as use of lowest possible billet temperature, shortest heating time, minimum billet length and maximum possible extrusion ratio. Scalping or blasting the oxidised surface with water during the period the billet passes from the furnace to the container helps to produce a defect free surface.

Peripheral Coarse Grain Envelope

This is another defect commonly observed as a peripheral layer occurring during maximum concentration of the shear deformation process. The coarse grain layer forms either spontaneously during extrusion or in the emerging product on subsequent heating for further hot working process. The heavily deformed zone of the extrusion favours formation of a peripheral coarse grain envelope and is distinguished clearly from the less severely deformed central portion by an assymetrical crescent-shaped layer. Macro test of a cross-section of the extrusion reveals the defect (Fig. 2.18).

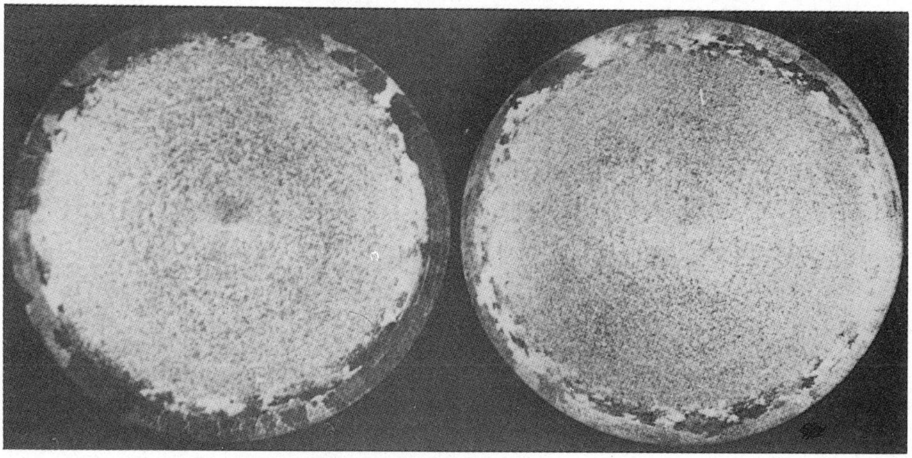

Fig. 2.18 Peripheral coarse grain envelope (× 1)

Coarse grain formation during the extrusion of aluminium alloys is prevented either by raising the recrystallisation temperature to such an extent that both core and peripheral zone recrystallisation is reduced, or lowering the recrystallisation temperature where in the whole cross-section recrystallises with a fine grain structure.

Annular Defect

It commonly occurs in extrusion towards the back end, due to a special profile of the flow of the metal in the extrusion die, and originates from the inclusion or foreign material. These materials include, aluminium oxide and lubricant or dirt from the surface of the unmachined billet which flows into form a part of the extrusion. Macro- and micro-examination of a cross-section of the extrusion reveals the presence of the defect (Fig. 2.19).

The defect is minimised by machining the billet atleast 1-2 mm into the radius so that the surface oxides and other contaminants are removed, and by avoiding the use of lubricants during extrusion.

2.4 SECONDARY PROCESSING DEFECTS

Welding

A solidified weld metal is nothing but a miniature casting. The defects produced in welding are thus similar to the ones observed in casting.

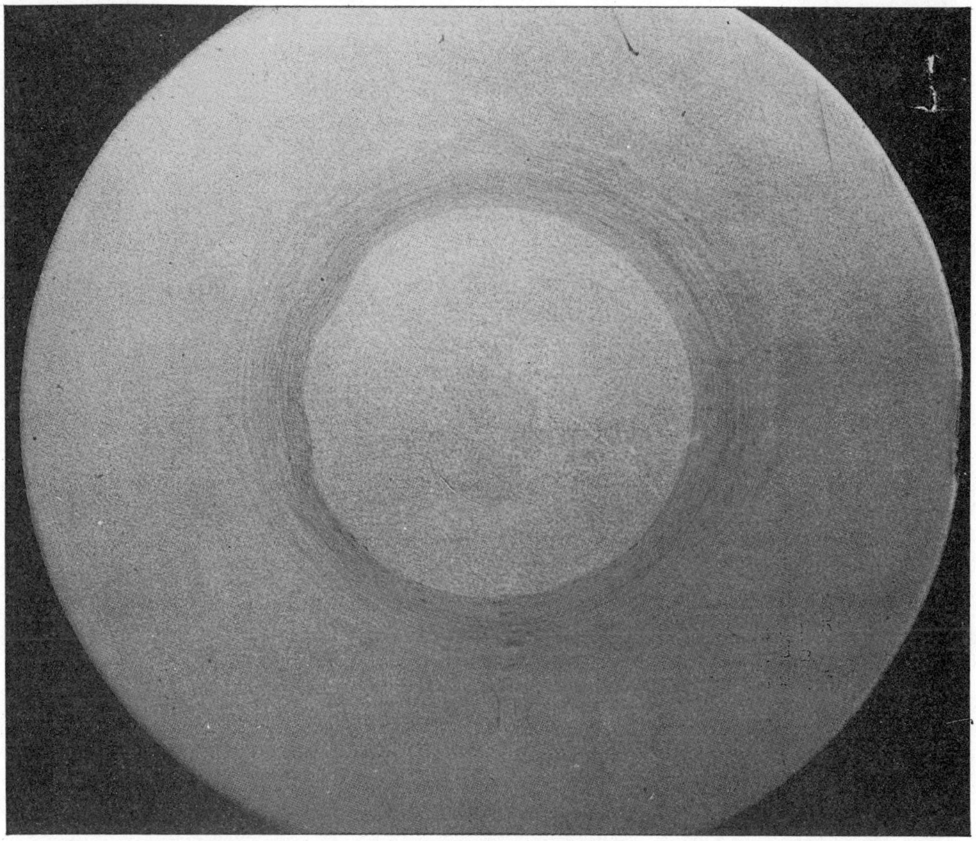

Fig. 2.19 Typical annular defect in the midzone of a cross-section of the back extruded rod (×2)

Cracking

It is the most common defect in the fusionwelded joints. Cracks may occur along or across the bead, along the weld bead-parent metal interface, and in the parent metal near the fusion zone. These are caused either by thermal gradients or the shrinkage phenomenon involved in the solidification process of the weld pool. This defect is often due to lapses in the welding techniques, i.e. too high a welding temperature, improper fixture or jigs (imposing undue constraints), inadequate pre-heating and post-heating, and poor skill of the welder.

Radiography of the welded joints is a good NDT method of detecting and identifying the weld cracks.

Other Defects of Welding

Porosity, blow holes, flux *inclusion.inadequate fusion and penetration* (Figs 2.20 – 2.23) are some of the major defects that are commonly encountered in weldments. In butt welds, lack of fusion and penetration constitutes a serious defect, adversely affecting the joint strength as well as giving rise to the notch effect. Under cut is another serious defect

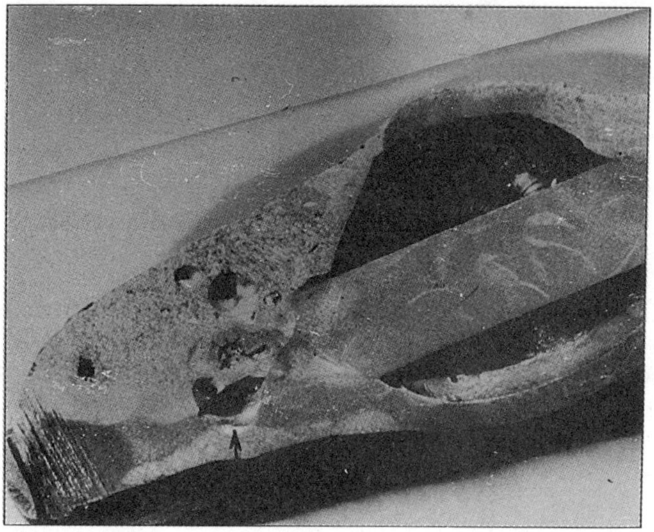

Fig. 2.20 Blow hole and cavity in a gas welded component (× 1)

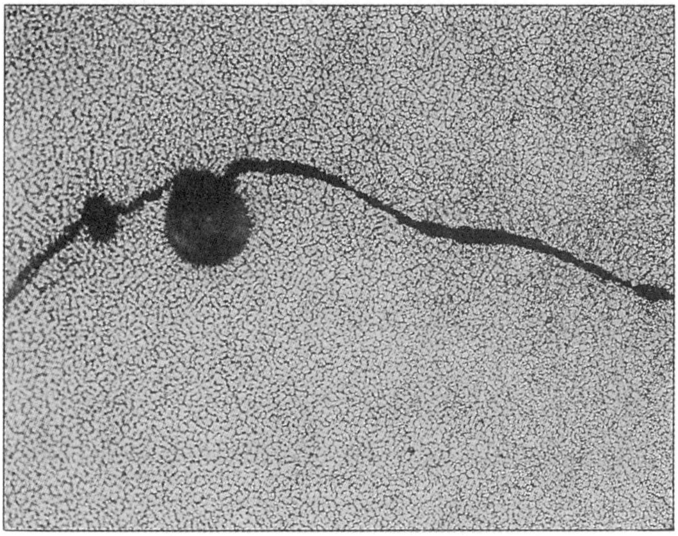

Fig. 2.21 Oxide film in the weld zone of a magnesium alloy (× 100)

caused by the melting of the parent metal and acts as a severe localised notch. Improper weld joint preparation and joint design can be detected by radiography.

In multiple-weld joints, *embedding of oxide layers* often occurs due to improper cleaning of the surfaces before each pass. It is rather difficult to detect this defect in thin weld

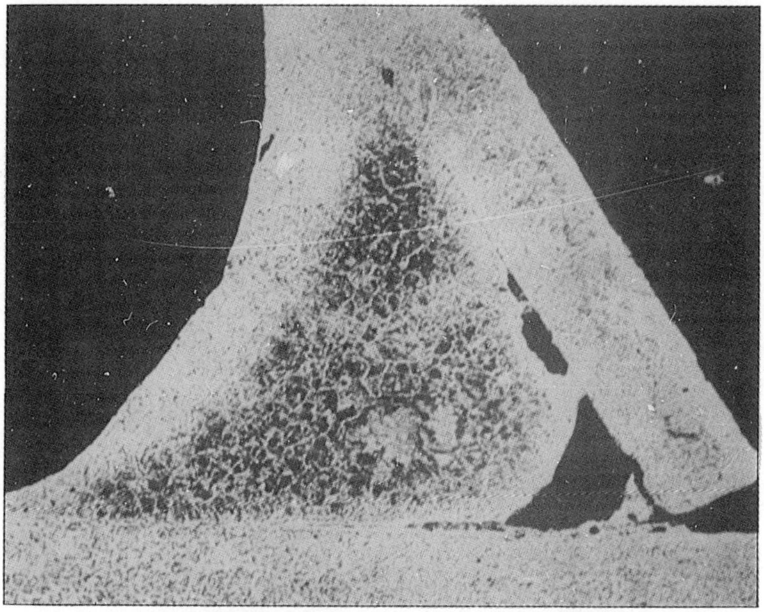

Fig. 2.22 Lack of penetration in a welded component (× 3.5)

Fig. 2.23 Undercut in a welded component (× 3.5)

beads by any conventional NDT method, but macro examination of a section of the joint reveals it clearly.

The most common defect in resistance welding is *cracking*. It consists of a transverse crack running through the nugget and extending to the outer surface of the outer most member or terminating in the nugget (Fig. 2.24).

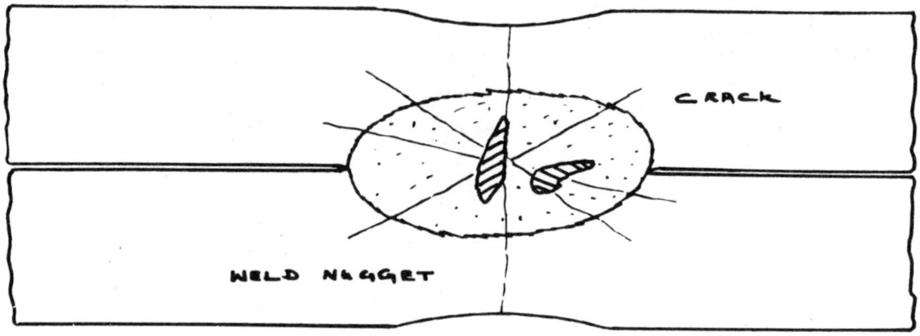

Fig. 2.24 Transverse cracks and cavities in resistance welding

In some cases, radial cracks extend beyond the nugget into the parent metal, *Cavitation* is another defect in resistance welding and occurs within the nugget. It is due to explusion of the metal as a result of undue pressure exerted while mating surfaces under the electrode tips when they are in a molten state. Visual examination at low magnification after light etching and X-ray examination are the methods commonly adopted for detection of these defects.

Machining

Various machining processes, both conventional as well as non-conventional, are extensively used for shaping parts to their intended geometry. Invariably, each of these machining processes[12] produce certain characteristic machining faults, rendering some of them into metallurgical defects. In addition, machining introduces different types of stresses such as residual tensile stresses, compressive stresses and thermal stresses. Metallurgically these defects would exert harmful effects on the structural behaviour of the material, mostly in terms of fatigue strength properties during service. Particularly for aeronautical and aerospace applications the presence of these defects constitutes a potential hazard in view of the unforeseen magnitude of stresses involved during service.

Grinding Cracks

In machined components, *grinding cracks* are the most common defects and they occur in hard or brittle steel parts. The intense heat generated by grinding produces severe thermal stresses in the material and the materials sensitive to grinding heat develop shallow cracks during or after grinding, on drastic quenching. Undue heat may also be generated due to improper grinding parameters such as the grinding wheel, feed, speed, depth of cut or coolant. Grinding cracks may also be due to material fault because of the presence of hard constituents like cementite or alloy carbide in a thick network from. They may also occur due to the low thermal conductivity characteristics of the materials such as high carbon steel and stainless steel.

Grinding cracks have a distinguishable network pattern running parallel and at right angles to the direction of grinding (Fig. 2.25).

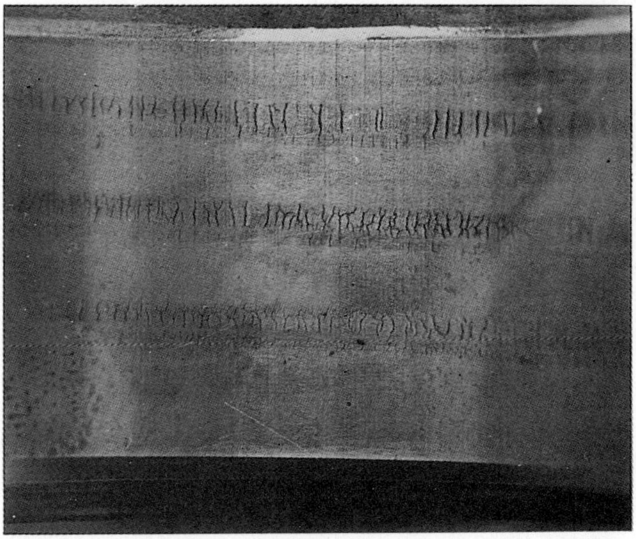

Fig. 2.25 Typical grinding cracks on the ground bore surface of a case hardened steel part, on magnafluxing (× 1)

These shallow cracks, if invisible to the naked eye, can easily be detected by the magnaflux method. In the case of non-magnetic materials such as stainless steel, dye penetrant and fluorescent penetrant tests are used to detect them.

To avoid occurrence of grinding cracks, it is essential to use proper coolant in an appropriate manner, maintain grinding parameters appropriately and use materials with proper microstructural constituents. Grinding burns are identified as surface discolouration in various shades of straw, brown or blue caused by the formation of a thin oxide film on the ground surface of the metal. The thickness of the film depends on the extent of rise in temperature. The grinding defect can be detected visually or macroscopically using a low powered stereomicroscope and also by the chemical etching method.

Hot Spots

Hot spots are localised soft zones due to decarburisation which occurs due to high temperature. They appear as bright zones against a black background (Fig. 2.26). and are detected by the visual or chemical etching method and also by the micro-hardness check.

Fig. 2.26 Bright hot spots and discolouration effect in a steel component on etching (× 1)

Residual Stresses

Residual stress or distortion is caused mainly due to thermal effects produced in the metal by the machining operation. The residual stress systems after removal of the metal are mostly reflected in the form of distortions. The distortion is related to the distribution of the residual stresses, their magnitude and location of the metal removed. Development of surface residual stresses, either compressive or tensile, depends on the type of tool used, pressure applied and the method of cooling employed.

Problems due to distortion during machining are of a severe nature in the case of parts containing unrelieved high quenching stresses. The drastic warpage that can result from unsymmetrical removal of metal in a thin heat-treated section of an aluminium alloy during milling is shown in Fig. 2.27.

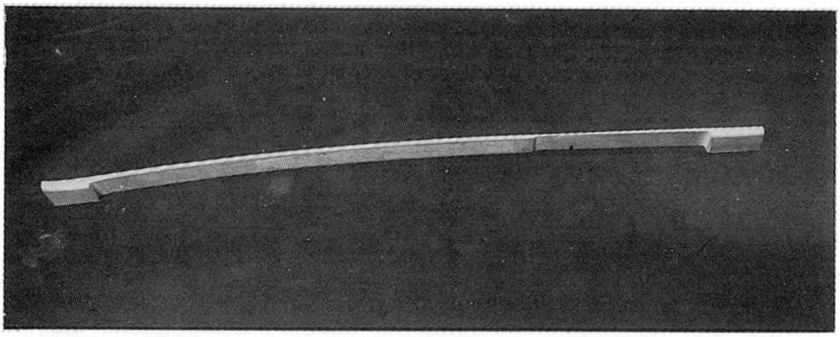

Fig. 2.27 Warpage in the Al Alloy part during milling (× 1/4)

Residual stresses can be detected and evaluated by X-ray diffraction techniques as well as by mechanical methods such as sectioning or progressive machining.

Any of the following stress relieving methods, either thermal or mechanical, may be employed to remove or replace residual tensile stresses *viz.* part or full annealing, plastically deforming, ultrasonic vibration, shot peening, cold rolling, stretching, etc.

Undercutting Defects

Undercutting defects are normally encountered in the chemical milling operation and are usually due to the seepage of chemical solution (etchant) below the maskant, where the thickness of the material at the etching edge gets depleted. The defect is easily identified schematically as shown in Fig. 2.28.

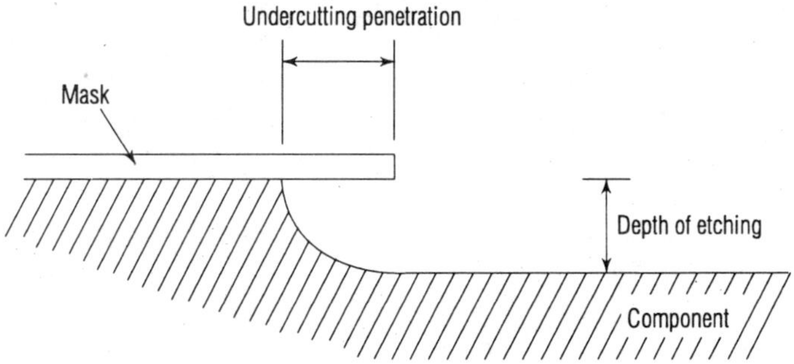

Fig. 2.28 Under cutting phenomenon

In order to prevent such an abnormality extra allowance is made while marking out for undercutting penetration which occurs at the edge of the mask during etching.

Rough Surface

A rough surface is another defect commonly found in chemically milled material with a coarse-grained structure. Unfavourable etching conditions may also add to the roughness. The surface is affected by the etching of the grain boundaries due to action of the etchant leading to the dissolution of grain boundaries and subsequent bodily removal of each grain. Also due to severe local deformation, the grain structure may change the etch ratio causing an irregular depth of cut and an uneven surface. A typical example of a very rough surface due to presence of coarse grains is shown in Fig. 2.29.

Hydrogen Embrittlement

It is a serious defect more common in titanium alloys and is caused due to absorption of hydrogen into the material during chemical milling. Hydrogen absorption depends on the formulation of the etchant as well as on the types of structures of the material. Those alloy compositions which produce beta structures are prone to hydrogen absorption. These are dissolved in the structure interstitially causing the embrittlement phenomenon to occur.

Fig. 2.29 Original extrusion die marks, fine smooth grit blasted and rough chemically milled surface (from left to right) (× 1)

Hydrogen embrittlement may be detected by conducting sustained load tests on the standard notched test specimen and also by measuring the hydrogen content by the gas analysis method.

Intergranular Attack

In electrochemical machining, microscopic surface defects like intergranular attack and pitting are observed. These defects are caused due to preferential or selective attacks on certain constituents (e.g. intermetallic compounds), precipitates at the grain boundaries, particularly in the heat sensitising alloys. Among the variables that can produce selective etching pits or intergranular attack at the grain boundaries are incorrect electrolytes, different work pieces, material heat-treatment or low current densities.

Intergranular attack is caused by potential differences between the grain boundary regions and the grain bodies. Intergranular attack can easily be identified on micro examination as a corrosion path oriented randomly along the grain boundary. This can be detected non-destructively on the surface by the magnaflux method in steel parts, and by the dye penetrant test in the aluminium alloys. Internal corrosion attack is checked ultrasonically.

Craters

In electrodischarge machining process the most common defects are formation of craters, voids, rough surface, burns, chilled layer, macro and micro cracks. Craters, voids, rough surfaces and irregularities, burns (Fig. 2.30) are formed due to the application of higher current densities. Crater size as well as surface/roughness increase with the rate of metal removal. Voids are created when any molten metal after stoppage of sparking or discharge tends to resolidify in the form of a sphere.

Craters are randomly arranged, similar to an etched surface, and lack the usual marks associated with machining.

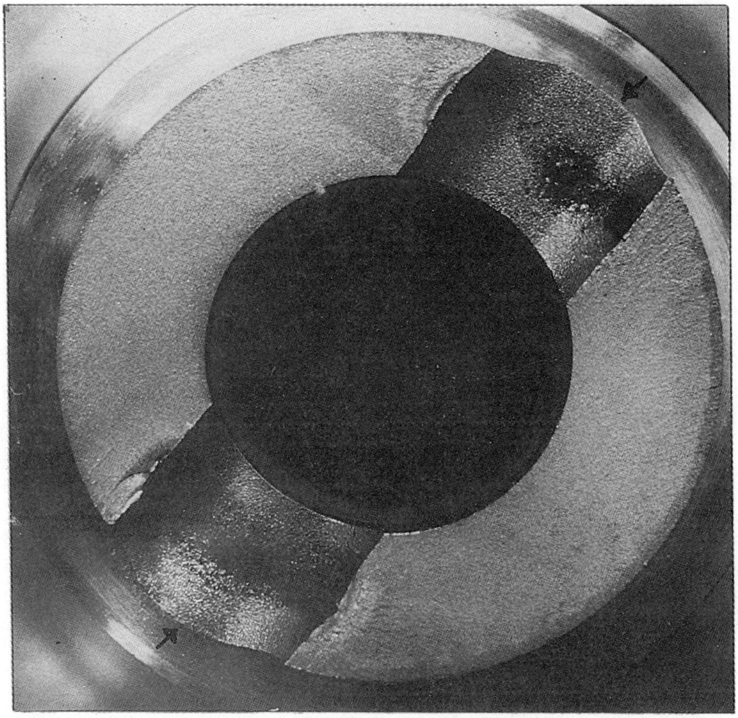

Fig. 2.30 Typical defects such as burn, craters, rough surface during improper EDM process (× 3)

Heat Treatment

Wrought products of both ferrous and non-ferrous materials are heat-treated to achieve higher strength properties, including fatigue, etc. Depending on the heat-treatment technique adopted various defects are produced. These defects weaken the soundness of the material and severely impair the service life of the part. Apart from the normal recognisable defects like crack, distortion, and burning, degradation of mechanical strength properties due to the introduction of various factors from the heat-treatment lapses also constitute metallurgical defects within the chemistry of the metal, as they result in the lowering of specific properties such as fatigue. In this chapter, heat-treatment defects are dealt with separately mainly for aluminium alloys and steels.

Defects in Aluminium Alloys

In heat-treated products, cracking and structural deterioration are the two most common defects. They occur in aluminium alloys which are noted for their sensitivity to excess temperature during solution treatment. When the temperature rises beyond the critical point, even for a short period, overheating or heat deterioration takes place in the form of rossettes. In more severe cases, surface blistering and eutectic melting along the grain boundaries occur, when such overheated parts are quenched in water, cracking takes place on account of grain boundary embrittlement. Examples of cracking and the heat-deterioration effects are shown in Figs 2.31 and 2.32.

Fig. 2.31 Flash line cracking during heat-treatment of an Al-Cu alloy forged component, due to overheating (× 1/2)

Overheating and heat deterioration can be due to (i) the heating elements being placed in the vicinity of the work, (ii) overshooting of the furnace temperature as a result of equipment failure, (iii) non-uniform distribution of temperature within the furnace due to improper air circulation, or (iv) failure to check accuracy of the prevailing furnace temperature.

Cracking can be detected by liquid dye penetrant or chromic acid anodising methods. Sometimes light etching helps to detect invisible cracks. Structural deterioration can be assessed by destructive tests such as micro examination. Direct checking by means of the potentiometer type of temperature measuring instrument is required to ensure correctness of the temperature employed.

Surface blisters (Fig. 2.13) occur either during improper heat treatment as stated earlier or during processing. These surface blisters are a result of (i) overheating, (ii) presence of gas voids in the product arising from dissolved gases in the original melt, or (iii) ingress of foreign particles during the fabrication process.

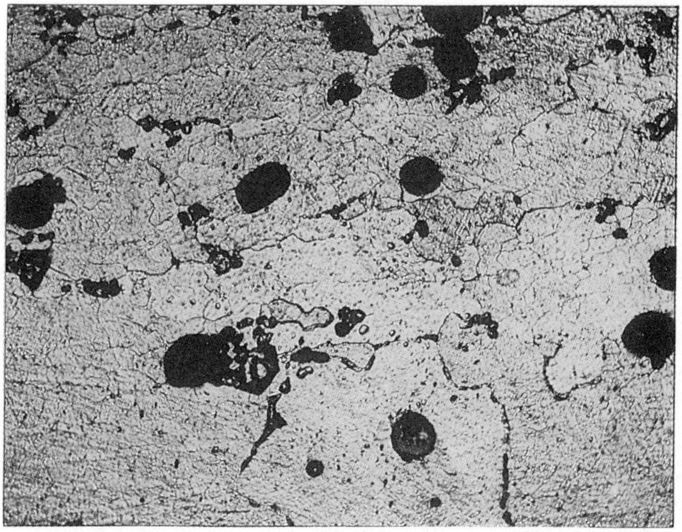

Fig. 2.32 Micro structure of an overheated Al-Cu alloy part (× 100)

Surface blisters can either be detected visually, or by using a magnifying glass or a stereomicroscope.

High temperature oxidation or grain boundary void formation (Fig. 2.14) is another type of defect commonly observed in parts. It is due to the furnace atmosphere contaminated with water vapour, sulphur or ammonia and prolonged exposure to high temperature. Voids are caused due to the formation of atomic hydrogen by reaction of moisture with the aluminium surfaces, diffusion of the atomic hydrogen through the aluminium lattice and recombination to form molecular hydrogen at the locations of lattice discontinuity, grain fragment and constituent boundaries. This defect may be controlled by maintaining a moisture free atmosphere or by using volatile fluoride-containing salts.

Discolouration or surface blemish occurs due to the presence of contaminants either in the salt bath, quenching media or furnace. They also occur because the parts are not cleaned adequately before loading or because of the decomposition of volatile fluoride salts.

Lubricants such as oil and grease decompose at the heat-treatment temperature and stain the surface. The use of an atmosphere free from moisture and contaminants is the best remedial measure.

Distortion is a common problem encountered in the heat-treatment of aluminium alloy. This can result during quenching from the high temperature solutionising due to stresses set up from complex geometry, severity of quenching, faulty quenching method, improper loading or packing of parts in the furnace. Proper care at all stages can reduce the extent of distortion.

Failure to achieve satisfactory strength properties on heat treatment is a common problem in the aluminium heat-treatment shop. In the case of precipitation hardening heat-treatment, the lower tensile strength properties together with unsatisfactory elongation and reduction of area values can either be due to heat deterioration during solid solution treatment or overageing. A micro-examination of a section of the component confirms the heat deterioration effect, if any in which case the product is to be rejected. In the case of overageing, a reheat-treatment will restore the required properties.

Low mechanical properties with a higher percentage of elongation and a reduction in area can be due to (a) underageing, (b) lower percentage of hardening elements, (c) inadequate solution treatment and (d) delayed quenching ('delay quench' effect is more pronounced in quench-sensitive alloys like the aluminium-copper alloy group). In the case of (a) an extended period of ageing will improve the properties. With regard to (b), nothing can be done. And in the case of (c), it can be either due to low solutionising temperature or lower soaking period. A reheat-treatment will enable attaining the required properties in the case of (c) and (d).

Low yield strength with higher percentage elongation which is rather a common observation in the heat-treatment of high strength aluminium-copper alloy extrusion, is attributed to lower extrusion temperature employed during extrusion process. In such a case, the extrusion temperature should be maintained above 430 °C and allowed to cool rather slowly.

Defects in Steel

Quenching cracks (Fig. 2.33) are most frequently observed in steel parts and are caused due to sudden volume changes during hardening, internal stress—thermal as well as residual, high temperature effect, localised carbon enrichment due to segregation. overheating, failure to temper the parts immediately after hardening specially in the case of high alloy steels, improper quenching media and the presence of a sharp geometry. A quenching crack normally follows an intergranular path and is associated with the oxide layers.

Quenching cracks can be detected either visually or by magnetic particle inspection. Fine quenching cracks can be observed under a stereoscopic microscope.

Using a proper quenching media, avoiding sharp stress concentration factors with liberal section change, radius, etc. and by maintaining a uniform rate of heating, quenching crack defects can be minimised.

Decarburisation (Fig. 2.34) is another major defect. It occurs due to the loss of carbon from the steel surface when it reacts with the oxidising atmosphere of the furnace, resulting in a lower surface hardness. The defect is easily identified as a white zone on the etched surface of the cross-section when examined under an optical microscope. Alternatively, a hardness traverse will also reveal the presence of decarburisation more effectively in the case of high carbon bearing steels where structural distinction between white and tempered (coloured) zones becomes difficult.

Avoiding undue soaking at high temperature, using a controlled atmosphere and nickel plating before heat-treatment help to minimise the phenomenon of decarburisation in steel during heat-treatment.

Overheating of steel gives rise to a very coarse grained structure. In severe cases, it is known as widmanstatten structure (Fig. 2.35) which is brittle.

Fig. 2.33 Quenching cracks in a steel part on heat treatment (× 1)

Fig. 2.34 White decarburised layer in a heat treated steel part (× 100)

The characteristic coarse grained fracture is a visual evidence of overheating. Micro-examination confirms overheating through the presence of coarse grain and the widmanstatten pattern structure. An overheated structure can be salvaged by reheat-treatment through normalising or a combination of the annealing and the normalising

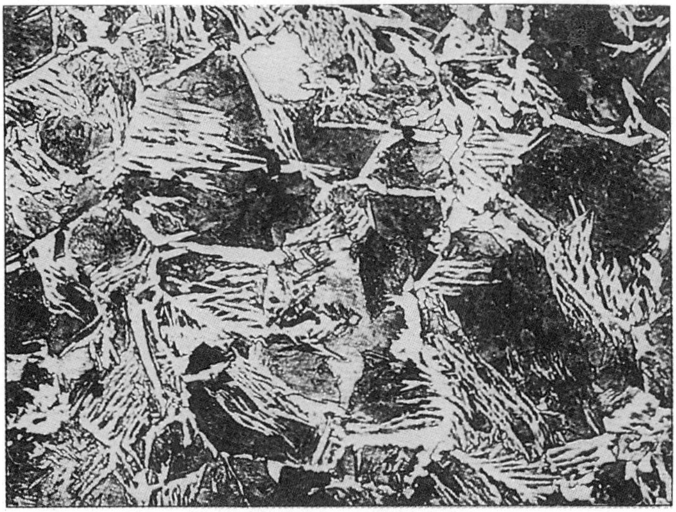

Fig. 2.35 Widmanstatten structure in a medium carbon steel after high temperature hardening (× 100)

refining process. Coarse grained steel produces a characteristic rough appearance known as 'orange peel'.

Burning is a dangerous defect resulting from melting and is caused by heating the metal to a high temperature, close to its melting point, so as to cause permanent damage to the metal by way of intercrystalline penetration of oxidising gases or incipient melting.

The burning defect can be detected visually by the presence of black surface colouration or solidified oxide lumps. Micro-examination will reveal burning as intergranular cracks filled with oxide film products and voids at the grain boundary region.

Distortion is also a common defect in steel. It is caused by improper stacking of the parts in the furnace, abrupt change in section geometry, residual stresses, drastic quenching and improper quenching media.

Low Mechanical Properties

Non-attainment of strength properties desired on completion of the prescribed heat-treatment process, is regarded as a defect and is a major problem in the heat-treatment shop. When the causes for the defects are identified, the heat-treatment parameters can be modified in a majority of the cases to achieve the desired results. Thus the finished components can be saved not only from a costly rejection but also from inadvertant failures in service, due to the use of material with low strength properties.

For a better appreciation of the causes for non-attainment of strength, the factors[13] that can lead to different strength properties than desired are classified as:

- Chemical composition of the material
- History of the product

- Cleanliness and microstructure
- Shape, size and batch quantity of parts
- Type, size, temperature and atmosphere of the furnace
- Rate of heating and cooling
- Soaking temperature and time
- Quenching media and quench delay
- Time lapse between quenching and tempering
- Tempering techniques and parameters
- Sequence of different heat-treatments

Apart from the above, preparation of test specimens is also an important aspect involving a proper selection of specimen location as per the drawing or test sheet, type of specimen to be prepared, dimensions, and the finish quality. Any deviation from these or a defective specimen can render the test results incorrect or unreliable.

Defect *Low yield and tensile strength with satisfactory elongation and toughness.*

Probable causes	Remedial measures
(a) Tempering temperature being higher than the specified.	Renormalise, harden and temper at a lower temperature depending on the strength obtained.
(b) Hardening temperature lower than the specified.	Renormalise, harden at a temperature of 25 °C higher than the one adopted and temper suitably.
(c) Slower cooling rate and delayed quenching.	Renormalise, harden and temper. While quenching after hardening a faster quenching technique and a faster agitation of the bath may be used.
(d) Excessive retained austenite	Sub-zero treatment or multi-tempering treatment.

Defect *High yield and tensile strength with low elongation and toughness.*

Probable causes	Remedial measures
(a) Inadequate tempering temperature/ soaking period.	Retemper at a temperature exceeding the one employed at least by 25 °C depending on strength properties, or reharden and temper at a suitable higher temperature.

(b) High percentage of carbon/alloying elements.	Suitably modify heat-treatment parameters to obtain satisfactory elongation and toughness values.

Defect *Satisfactory yield and tensile strength with low elongation and/or toughness.*

Probable causes	Remedial measures
(a) Excessive inclusion orientation along the grain fibres with respect to the axis of the test specimen.	Higher tempering temperature may be adopted to obtain better elongation/toughness properties.
(b) Austenitic grain size being too coarse.	Homogenising for a longer period followed by normalising, rehardening and tempering.

Defect *Satisfactory yield and tensile strength and elongation but low toughness properties.*

Probable causes	Remedial measures
(a) Austenitic grain size being coarse.	Renormalise, harden and temper.
(b) Susceptibility of the material to temper-brittleness.	Renormalise, harden and temper followed by rapid quenching to avoid a temper brittleness effect that usually occurs during slow air cooling. Adopt rapid heating if the section size is not abnormal and also not of a complex shape.
(c) Excessive inclusion orientation along the grain fibres with respect to the axis of the test specimen or improper grain flow pattern.	Higher tempering temperature may be adopted to obtain better impact properties without sacrificing strength properties. No salvage action for improper grainflow pattern.
(d) Excessive untempered martensite caused by the presence of retained austenite in the quenched state.	Adopt double tempering or triple tempering if necessary.
(e) Banded structure.	Homogenise for a longer period depending on the severity of the banding.

Defect *Low hardness in the case-carburised parts.*

Probable causes	Remedial measures
(a) Decarburisation caused by improper atmosphere.	No salvage action.

(b) Excessive retained austenite in the case.	Controlled sub-zero treatment of the case-carburised parts.
(c) Banded structure.	Prolonged (3-4 h) stabilising treatment in neutral bath, followed by quenching and tempering.

Defect *Excessive carbides in the network form in case-carburised parts.*

Probable causes	Remedial measures
(a) Higher carbon potential in the carburising atmosphere.	Stabilising treatment at 670 °C for 1-4 h.

Defect *Excessive white layer in nitrided parts* (Fig. 2.36)

Probable causes	Remedial measures
(a) Improper ratio of percentage of ammonia dissociation.	(a) No salvage action.

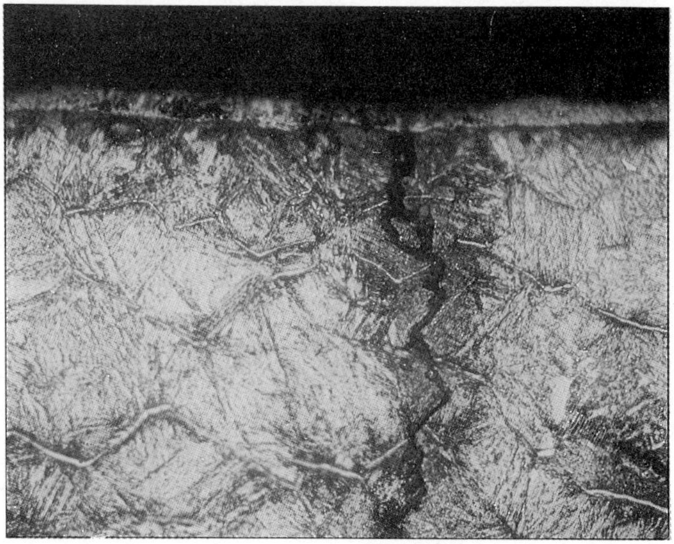

Fig. 2.36 White layer in a nitrided steel component, (cracks seen originating from the excessive white layer) (× 200)

(d) Lower carbon percentage than minimum specified.	If confirmed on chemical analysis, no salvage action.

(e) Mass effect	Rough machine/bore, re-normalise and harden and temper.
(f) Insufficient soaking	Renormalise and harden with increased soaking time at the rate of 2½ m/mm of thickness.
(g) Excessive amount of retained (untransformed) austenite / austenite-rich banding effect (Fig. 2.37).	Sub-zero treatment or multitempering.

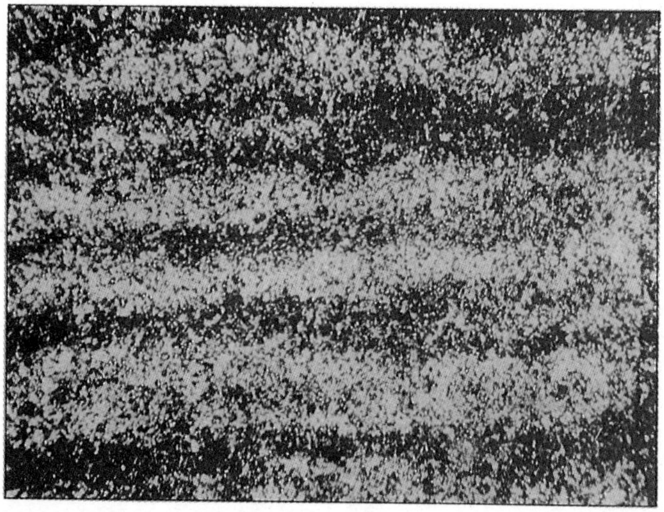

Fig. 2.37 Austenite-rich bands in Ni-Cr-Mo steel (× 50)

Defect *Low yield or tensile strength with satisfactory elongation and toughness.*

Probable causes	Remedial measures
(a) Quench delay.	Faster quenching technique and cooling media.
(b) Abnormal microstructure like widmanstatten type overheated structure (Fig. 2.35).	Re-normalise to refine the structure followed by hardening and tempering.

2.5 EFFECT OF DEFECTS ON SERVICE PROPERTIES

Sources of Service Failure

A failure occurs when a component is unable to carry out its intended function satisfactorily due to fracture, deformation or deterioration. The failure mechanism is generally dictated by the entire environment and history.

The basic sources of failure can be traced back to design, material selection, imperfection, fabrication and processing, assembly, quality control, service condition, unanticipated exposure to overload and mechanical or chemical damage in service. Often, ignorance of operator, assembly man or analyst at the time of fabrication and inspection is a major source. The occurrence of a failure is often due to more than one source. Broadly, sources of service failure, are grouped into five distinct classes:

1. design deficiencies,
2. material imperfections arising from manufacture or fabrication,
3. processing deficiencies,
4. assembly errors, and
5. improper conditions due to environmental and service conditions.

Defects and Their Effect on Service Properties

Design Deficiencies

Deficiencies in design often result in catastrophic service failure. This emphasises a need for a well coordinated engineering effort to eliminate the failure-prone conditions from the design features. Design deficiency characteristics[14] along with their respective effect on service failures are given below.

Defects	Effect
Wrong material/heat treatment/joining technique.	Lowers mechanical properties including fatigue resistance, toughness and wear, increases corrosion susceptibility.
Inadequate stress analysis of component or providing inadequate fastener/stiffner/shimming.	Increases the chances of failure to withstand the service loads as well as susceptibility to fatigue.
Mechanical notches, *viz.* sharp corners, sharp/abrupt section change or abnormal stress raisers.	Increases the stress concentration factor and reduces fatigue strength drastically.
Improper location of oil holes, vents, etc.	Introduces weakness for increased accumulation of stress intensity.
Improper assembly design/dissimilar material.	Increases load factors due to mismatch and imparts galvanic corrosion.
Unanticipated service condition.	Induces an element of overloading.

Material Imperfection

Failures originate from the sites of imperfection which occur in the finished products during the stages of manufacture/fabrication[14]. These imperfections include segregation, inclusion, lamination, seam, fold, and crack. These internal as well as surface imperfections generally tend to reduce the overall strength of the material, develop preferential paths for propagation of cracks and act as notches. These imperfections also provide an opportunity for the pitting type attack or the propagation of intergranular corrosion.

Defects	Effects
Improper casting—porosity, voids, shrinkage cavities, pipes, hot tear, cold shut, cracks, etc.	Causes unsoundness of the material, lowers the strength capacity to withstand the service load and can serve as a notch for a propagative type of crack.
Inclusion	Acts as a source for fatigue crack propagation and enhances corrosion susceptibilities.
Banding / segregation	Renders the material discontinuous and susceptible to stress sensitivity by decreasing the fatigue resistance / toughness. Presence of micro and macro segregation causes the inherent problem of cracking either during heat-treatment or forging / rolling.
Improper forging–folds/laps cold shut, thermal crack	Is highly dangerous as a state of precracking which easily becomes a stress concentration factor for fatigue failure and susceptible to sudden overload failure by virtue of the reduced section thickness.
'Push through' defect	Acts as a discontinuity or crack in the normal grain flow pattern and is critical, particularly in the fillet radius and section thickness change area for easy development of fatigue crack.
Unfavourable grainflow / (kink / re-entrant) defect	Serves as a potential source of premature fatigue failure in service primarily due to pronounced anisotropy in strength properties and the intended strength not being realised in the location particularly end grains in the fillet areas can lead to initiation of fatigue cracks under fluctuating loading conditions in service.
Grain coarsening/grain boundary oxidation (or burning)	Renders material susceptible to brittleness in service. Parts bearing burning defect are totally wasted. Reduces the fatigue resistance property.
Decarburisation / scaling	Reduces surface fatigue properties drastically.
Improper rolling—seam, longitudinal discontinuity, crack, delamination, etc.	Acts as a mechanical notch and reduces fatigue strength property.

Improper extrusion—heat checks, blisters, etc.	Serves as a mechanical notch for fatigue crack development, induces brittleness and acts as a potential notch creating a localised area of weakness.
Delamination	Results in the peeling effect.
Longitudinal discontinuity	Acts as an oxide film discontinuity and reduces fatigue resistance property.
Coarse grain envelope	Causes cracking during forging of the extruded bar stock.

Processing Deficiencies

Failures also occur as a result of defects arising from improper/unsuitable processing procedure, ambiguous specification or modification of specification without proper evaluation. Failure may also result from either any lapse on the part of the operator in carrying out the process schedule or accidental occurrences. Various areas showing deficiencies[14] in processing with their corresponding effects are given below.

Defects	Effects
Improper machining:	
Grinding crack	Acts as a harmful defect/discontinuity, whether shallow or deep, by way of stress raiser for development of premature fatigue failure.
Hot spot	Constitutes localised surface weakness due to either the formation of brittle (overheated) zones from the effect of high temperature or reduction in hardness due to decarburisation. Such surface anomalies exert adverse effects on the fatigue as well as wear resistance property, particularly in case-hardened parts.
Improper heat treatment:	
Overheating (coarse grained/widmanstatten structure in steel, blistering, eutectic melting/grain boundary cracking) (in aluminium alloys)	Renders structural deterioration, induces brittleness and reduces fatigue strength as well as corrosion resistance property.
Burning	Renders product unacceptable for use.
Oxidation/scaling	Reduces effective thickness and induces decarburisation.
Quenching crack	Reduces section thickness and fatigue strength significantly.
Undertempering	Increases susceptibility to brittle fracture.
Overtempering/overageing	Softens the strength capacity and reduces fatigue strength.
High temperature oxidation (in aluminium alloys)	Causes structural weakness.

Decarburisation	lowers the surface hardness and reduces significantly the endurance limit.
Excessive case thickness (case carburising)	Develops micro crack in the transition zone between the case layer and the core.
Carbide network	Causes embrittlement and crack.
White layer (case nitriding)	Causes embrittlement and crack.
Retained austenite	Softens the strength property as a localised weakness but develops brittleness/crack when converted into untempered martensite during tempering.
Improper cleaning, plating and coating:	
Pitting, hydrogen embrittlement galvanic corrosion, crack.	Deteriorates fatigue strength of the parts which are susceptible to hydrogen absorption during the plating cycle or associated acid/alkaline cleaning cycles. Induced hydrogen in the parts causes the hydrogen embrittlement phenomenon. Also, high residual tensile stresses produced during plating reduce fatigue strength of the base metal and the corrosion resistance of both the plated surface as well as the base metal. Pits formed due to improper cleaning also act as stress raisers for fatigue crack propagation.
Residual stress	Is potentially dangerous particularly in the critical service applications. Causes geometrical distortion, fatigue and stress corrosion cracking.
Surface irregularities (geometric variation, local waviness, surface texture, tool marks, rough surface, etc.)	Causes stress concentration effect and adversely affects the fatigue strength.
Under cutting	Causes loss of dimension and a severe notch effect.
Etch pits/corrosion	Acts as notch for initiation of either corrosion fatigue or stress corrosion cracking.
Craters/rough surface	Are highly stress forming zones and adversely affect the fatigue strength.
Hard-chilled layer	Causes surface crazing, crack, etc. by virtue of extreme brittleness/internal stresses and accelerate low cycle fatigue/brittle failure.
Crack	Reduces section thickness as well as strength bearing capacity affecting the fatigue resistance property drastically as a severe notch.

Heat affected zone/softening	Lowers the hardness value in the subsurface zone with loss of yield/tensile strength. Fatigue cracks usually originate in the transition zone between the heat affected zone and the hard white layer due to the existance of high tensile stresses.
Residual tensile stresses	Helps to diminish the fatigue resistance property.
Improper welding:	
Incomplete/inadequate fusion	Reduces the weld joint strength and fatigue strength.
Underbead cracks, hot or cold cracks	Affects fatigue strength and acts as a point of stress concentration.
Undercut	Reduces the weld thickness and strength properties. Also acts as a severe notch for development of fatigue crack.
Non-homogeneous microstructure (coarse and widmanstatten structure, oxide film, decarburised layer, carburised skin, alloy segregation, martensite in the heat affected zone, inclusion, gas porosity, etc.	Weakens the weld metal strength properties, develops cracks (in the case of formation of martensite in the heat affected zone) and reduces fatigue strength property.
Weld decay	Develops intergranular/stress corrosion cracking particularly in austenitic stainless steel.

Assembly Errors

Failures may also result from trivial assembly errors. These are the errors in the assembly stage which escape detection during assembly inspection and initially do not interfere with the normal operation of the assembled parts. Defeciencies of this type[14] are given below.

Defects	Effects
Operator's negligence	Results in costly damage interrupting seriously the progress of production.
Inadequate/inaccurate or ambiguous specification of procedure for assembly, e.g. non-uniform and improper tightening.	Introduces a relative movement between the parts and initiate fatigue cracking at the point of stress concentration due to fretting
Misalignment/mismatch	Imposes additional load causing overstressing and overheating. Also initiates fatigue cracks. Stress-corrosion crack may result from undue straining due to mismatch in high strength aluminium alloys.

Improper Service Conditions

Service failures are not uncommon due to abnormally severe conditions of speed, loading, temperature and chemical environment. Absence of regularly scheduled maintenance, inspection and monitoring are the causes of such failures. There are different types of deterioration[14] associated with the failure system and these are given below.

Defects	Effects
Wear-adhesive and abrasive type	Causes dimensional deviation, introduces stress-concentration and susceptibility to fatigue.
Corrosion-oxidation, pitting and intergranular corrosion	Causes dimensional deviation and stress concentration and reduces strength and fatigue resistance.
Thermal-high temperature scaling, microstructural changes, grain growth, incipient melting.	Reduces strength and thermal fatigue resistance and affects creep property.
Inadequate/bad maintenance and improper repair.	Results in premature failure or undetected defects during overhaul.

Chapter 3

Procedural Steps for Investigation

3 *Procedural Steps for Investigation*

3.1 FEEDBACK SYSTEM

Establishing the correct cause and sequence of a failure requires a systematic and scientific approach. The amount of useful information gathered from the examination of a failure depends as much on the experience, knowledge, analysing skill and basic approach of the investigator as on the availability of details and the necessary infrastructure. Experience shows that a seasoned metallurgical engineer can handle any type of failure investigation in metal products using the same instruments and facilities as are ordinarily available in the metallurgical laboratories, provided he has same ingenuity and the knack to tackle complex problems. However, sophisticated equipment like scanning electron microscope and microprobe analyser certainly provide an edge by giving the investigator the fracture details that are normally beyond the conventional microscopes.

Before we go into the steps for investigation, it is logical to emphasise the importance of a feedback system of gains resulting from the failure investigation in any failure analysis oriented organisation, (Fig. 3.1). Depending on the particular feedback system of a failure investigation, procedural steps may be at times changed or modified to suit the needs of the investigator.

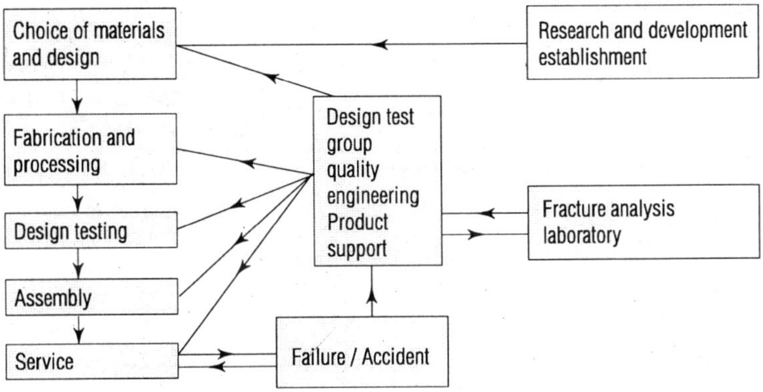

Fig. 3.1 Feedback system of failure investigation

3.2 INVESTIGATION PROCEDURE

Central Laboratory of Hindustan Aeronautics Limited, Bangalore, has conducted failure investigations of both aeronautical as well as general engineering components for the last three decades. Over the years, a considerable amount of knowledge and experience has been gained on the subject. The analysis of the available data and systematic presentation of the same may be of special interest to those who are concerned with failure analysis. With this in mind, the steps/sequences[14] adopted in the Central Laboratory, HAL, Bangalore, for conducting failure investigations are given in Fig. 3.2, with a need to identify the areas where long-term research and development may arise out of failure sequences. However, these steps vary depending on the needs of a specific case.

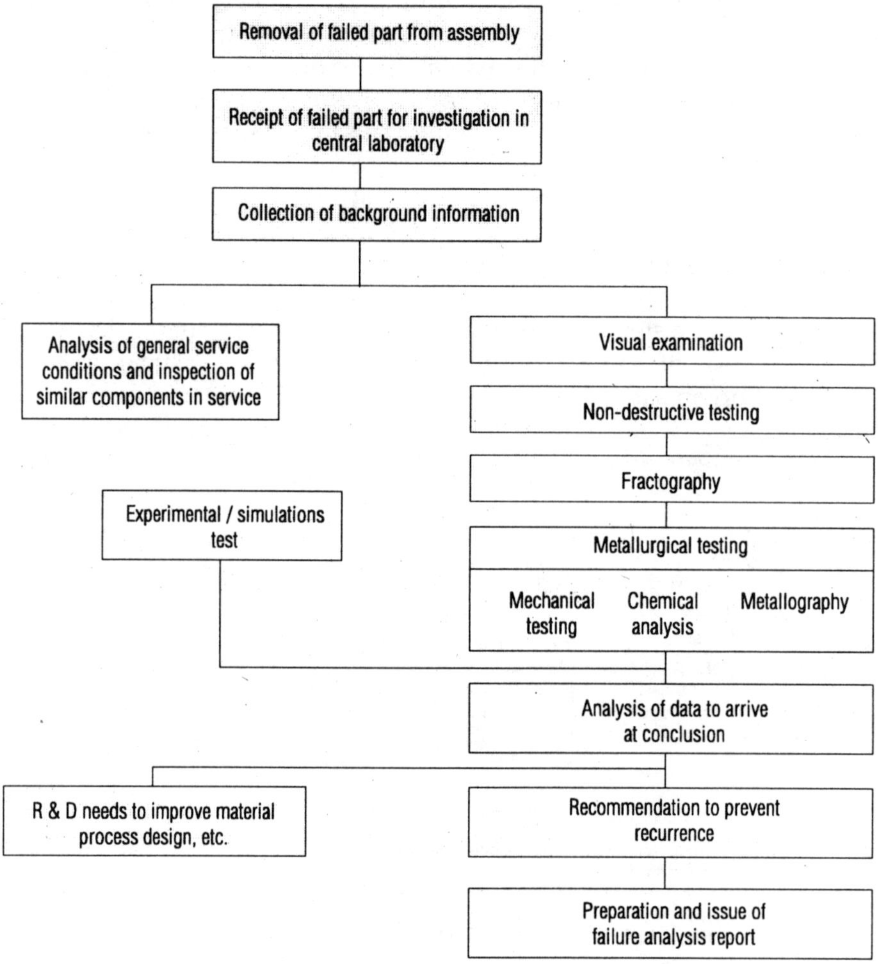

Fig. 3.2 Sequence of failure investigation

3.3 BACKGROUND INFORMATION

Collection of background information constitutes one of the first important steps towards understanding the basic factors that may be associated with the ultimate failure.

Therefore, wherever possible, details as regards part name and number, type of equipment (mobile or static and in the case of an aircraft whether trainer, fighter or domestic, depending on the service function), location of failure with respect to the whole unit of the equipment, relevant drawing, specification, service/maintenance history, flight log data (in case of flying machine), previous accident report and statement and evidence offered by eye witnesses are carefully collected.

However, if the background provided is considered inadequate, the investigator should personally visit the assembly unit to familiarise himself with the service parameters of the failed component. Firstly, the functional aspects should be studied in relation to design, operating parameters *viz.* static loads, service stresses and assembly conditions. Secondly, the environmental conditions such as temperature variations, operation in a corrosive environment, etc. should be analysed and recorded. Finally, discussions should be held with the concerned field personnel with reference to the failure details as observed by the operator and the witness.

In case of accident investigations,, photographic records of the actual location of the accident are of vital importance.

Advantages

A fairly complete background information has several advantages.[14,15]

(a) It puts the failure analyst in the correct perspective to concentrate on the main problem. Before any failure investigation is taken up, the analyst has to answer several questions to identify the areas where he should mainly concentrate:

- What is the material?
- What are its physical, chemical and mechanical properties?
- How was it made?
- How long was it used?
- What was it designed to do?
- What environment was it exposed to?

(b) Failure of specific manufactured components and assemblies such as shafts, bearings, springs, dies, gear, etc. have a tendency to categorise themselves in some characteristic manner. For example, in the case of gear failures, it has been noticed that failure is normally due to either wear, surface fatigue (spalling), plastic flow or breakage due to overload, fatigue, defective materials, etc.

(c) In the case of accidents, statements of eye witnesses are of extreme importance and could lead to the actual cause. Based on the statements and photographic records, the accident scene is reconstructed.

(d) Another major advantage of background information is for the record purposes. Properly classified and carefully documented case histories of past failures contain a wealth of information and act as a guide to the analyst to solve many investigations.

Importance

The importance of background information will be evident from the following case where incorrect background information, caused serious concern to the user.

A number of imported aluminium alloy upper housing parts used in the landing gear assembly of a bomber aircraft, failed repeatedly after a few hours of service. Results of investigation emphasised the need to look into the background information of the failure and the failed part.

On macroscopic examination the fracture surface showed clear evidence of progressive beach marks, typical of fatigue failure. The conclusion of this investigation became extremely difficult, as to how a newly manufactured component could show such progressive beach marks after only a few hours of service. This led the analyst to examine all the newly procured upper housings by ultrasonic examination. Surprisingly many of the newly procured parts showed indications of cracks at the location where failures had taken place. On fractographic examination, these cracks were also identified as fatigue. Based on this, it was evident that the parts supplied as new were actually serviced parts reconditioned and fraudulently supplied as new.

Handling of Parts

Failed components are carefully dismantled from the assembly and transferred to the investigating department without incurring any type of obliteration of the fractures or introducing new defects such as bending, denting, etc. This is particularly important where the sources of debris or corrosion products may provide clues to the failure or where examination of the fracture at high magnification, with the help of the scanning electron microscope, is required. Thus, it is preferable that the investigator himself should collect the failed components from the assembly. However, in actual practice, this is not possible. Failed parts however should be sent for investigation within a reasonably short period of time so that no corrosion sets in prior to investigation. At the initial stages, cutting of specimens is neither necessary nor desirable. Particularly in legal case the prohibition against cutting of any kind often turns out to be a blessing in disguise.

3.4 VISUAL EXAMINATION

Visual examination is carried out in two stages:

(a) Failed parts, including all its fragments, are subjected to a thorough unaided examination in the 'as received' condition, i.e. before any cleaning is undertaken. Often soil and debris found on the part provide useful evidence in establishing the cause of failure or determining the sequence of events that lead to the failure.

(b) The failed part is then thoroughly cleaned (if considered necessary) with non-corrosive solvents and subjected to further examination with the aid of a low power (magnification ×10-30) stereomicroscope.

The following points[14,15] are to be borne in mind when analysing the defective areas:

(a) Reported main defect, its nature and freshness.
(b) Location of the main defect with respect to the part.
(c) Parameters of the defect, its severity, frequency and area covered (i.e. extent of damage).
(d) Related defects in the vicinity of the main defect which have initiated or contributed, directly or indirectly to the failure.

If the investigator finds it necessary, he should take a general photograph of the component with special emphasis on the main defect. During visual examination, the investigator concentrates on the following defects in the area close to the main failure:

(a) Fabrication defects such as sharp fillets, deep machining marks, fold, or any abnormal stress-raisers.
(b) Inherent manufacturing defects such as cracks, surface porosity, shrinkage in castings, undercut, sitting gap, etc. in welds.
(c) Assembly defects such as defective assembly, e.g. improper flushing, misalignment, mismatch of mating parts, rubbing, etc.
(d) Defects due to environmental or service conditions such as corrosion pits, wear marks, dents as well as rubbing marks, erosion, abrasion, overheating, heat tinting, etc. during service or post failure.
(e) Bad handling, operation or maintenance.
(f) Deviations from drawing specifications, general assessment of basic design and workmanship including dimension.

However, in many cases, it may not be possible to determine the true length of the main crack and also the existence of secondary cracks just by visual examination. Hence, non-destructive testing is vital to supplement these informations for correct assessment of defects quantitatively.

3.5 NON-DESTRUCTIVE TESTING

Non-destructive testing is necessary to correlate internal, sub-surface or surface defects of the failed component to that of the main defect. Several NDT methods are extremely useful in the failure investigation, notably, magnetic particle inspection for surface cracks, inclusions, discontinuities in ferrous materials; dye penetrant examination and etch inspection for surface cracks in non-ferrous and ferrous materials; eddy current testing for surface or sub-surface defects, ultrasonic testing for internal cracks or discontinuities, radiography for internal flaws in castings and weldments as well as corrosion inside a bore assembly, misalignment, eccentricity, etc., and experimental stress analysis for determining residual stresses in components.

3.6 FRACTOGRAPHIC EXAMINATION

Fractography is one of the most important steps in failure analysis. The fracture surface of a broken part displays markings which constitute a characteristic topographical map. This map enables the investigator to identify the type of failure and reveals the history of events preceding the failure. In most instances, the knowledge of fracture surface topography is invaluable and expedites the analysis of failure. Information concerning the various features of a fractured surface helps us in three ways: firstly the crack origin can be located and the data indicating the mode and type of fracture can be determined. Secondly, the fracture surface constitutes a visible record of freshness of the fracture, direction of crack propagation and stages and extent of crack growth. And finally, fracture zones which are present as well as their respective sizes, qualitatively indicate the toughness of the material related to its application.

Failed parts may either be simply cracked or parted into pieces. Fractured parts are cleaned only when considered absolutely necessary. Cracked parts are carefully opened up without damaging the fractured surfaces.

Examination of a fracture begins with visual scrutiny, followed by scrutinising through a low-powered magnifying glasses, stereomicroscope and then a detailed study using the high-powered optical microscope (up to ×2000). For cases requiring a still more detailed study at very high magnifications, transmission electron microscope (TEM) or scanning electron microscope (SEM) with magnifications up to 50,000 and resolution of 100Å can be better used.

Visual scrutiny with or without the use of magnifying glasses establishes the gross aspects of the fracture such as evidence of mechanical abuse or deformations, obvious secondary fracture, direction of crack propagation and relation to the axis of stress, identification of the origin of the crack. These findings lead to deductions concerning the service conditions that existed at and prior to the time of failure.

The above findings can then be extended by an examination of the fracture surface at low magnification (×10 to ×30) with a stereomicroscope. A survey at low magnification is important for identification of those portions which are not distinguishable by visual examination and to identify areas that require further examination at higher magnification.

Examination with the aid of a high-powered optical microscope and scanning electron microscope is useful in examining specific small areas. As far as the electron microscopy is concerned the scanning electron microscopy is extremely popular for the study of fracture surfaces, as it is capable of magnifications of about 5 to 50,000 with high resolution (100Å), increased depth of field, and permits direct examination of the fracture surface without any preparation of the films or surface replicas as required by transmission electron microscopy. It also enables instant chemical analysis of microconstituents on the fracture surface and enables the investigator to assess the role of segregation, second phase particles, inclusions, corrosion, etc. in the initiation/propagation of the fracture.

The principal categories of fracture features revealed by SEM are: 1. cleavage features (tongues, microtwins and location of cleavage-crack origins; 2. quasi cleavage features;

3. dimples from microvoid coalescence; 4. tear ridges; 5. fatigue striation; 6. separated grain facets; 7. mixed fracture features; and 8. features of fracture resulting from chemical and environmental conditions.

3.7 METALLURGICAL TESTS

Metallurgical tests are conducted first to verify whether the failed component was made out of the material specified in the drawing. This is done by spot tests or detailed chemical analysis. If the material of the component conforms to the specifications, then it is to be examined whether the failure was due to any inherent defect in the material or due to processing faults. This is examined by metallography and mechanical testing.

Chemical analysis of the surface deposits, scales, corrosion products or the medium with which the affected materials has been in contact, is of vital importance in establishing the primary cause of failure in the case of stress corrosion, corrosion fatigue, general corrosion, etc.

Metallographic examination of selected sections close to the failure in polished and etched conditions, using optical microscopy gives a considerable amount of information and is carried out as a routine procedure. Fabrication history of failed material *viz.* cast, wrought, sintered, welded, etc. is determined with particular attention to their quality aspect as regards inherent defects like inclusions, microdefects, etc. Material imperfections caused during processing and in service operation conditions and environments, which might have contributed to the failure, are assessed. Seam, overheating, decarburisation, carbon pick up, improper heat treatment, untempered white martensite, intergranular corrosion, high temperature oxidation/burning are among the many undesirable defects that can be detected. Additional information regarding case depth, thickness of plated coatings, grain size, microhardness are measured and compared with the specification requirement wherever applicable. Nature of crack propagation following intergranular or transgranular path along or across the grain flow are observed and recorded. In certain cases, origin of crack, location at corrosion pits, seams, fillets, etc. may be identified.

Generally, a hardness survey is carried out as a routine test on the component close to the failed part to have a knowledge of the strength and heat treatment condition of the material used and to compare it with the specified limits. Other mechanical properties such as tensile, impact, shear strength are assessed at the discretion of the investigator, depending on his previous findings and additional information required by him.

3.8 ANALYSIS OF SERVICE PARAMETERS AND SIMULATED TESTS

If it is considered necessary, after gaining a fairly good idea of the type of failure, the investigator should review the service conditions under which the failure of the component has occurred and compare the same with the general service conditions to which similar components are subjected by studying these in function.

Simulation tests are conducted during the concluding stages of investigation, if it is considered necessary, to reproduce the conditions under which the failure is believed to have occurred. In many cases, it is not practical due to the elaborate equipment required and inadequate knowledge of the actual service conditions.

3.9 CONCLUSIONS AND RECOMMENDATIONS

Based on the results of the detailed investigations discussed above, a sequence of the total failure is established i.e., the primary failure and correlation of the same with the secondary effects. The mechanism of failure, its origin, propagation and causes (material, design, process, assembly, service conditions) are identified. Where the cause of failure is particularly elusive, negative evidences may be helpful for eliminating some causes of failure from consideration. Previous reports of similar failures help to arrive at in certain cases, unnecessary tests, can be avoided by referring to earlier reports of similar failures.

Once the primary causes of failure and the contributory factors are established and assessed, recommendations for prevention of similar failures or for correction of similar components in service are made. Wherever there is a doubt regarding the actual causes of failure and the failure is of a repetitive type, further experiments are carried out on various contributory variables to establish the cause and suggest suitable remedial measures.

Research and Development

Based on failure analysis, appropriate conclusions and recommendations may provide an effective insight into the serious inadequacies either in current material/process technologies or design philosophy. Therefore, an R&D effort is absolutely necessary to improve upon the existing technologies for controlling the recurring factors responsible for failures. This imperative is increasingly felt, specially in dynamic machines like aircraft where safety is a more important consideration than the cost.

Chapter 4

Failure: Types and Characteristics

4 *Failure*: *Types and Characteristics*

4.1 TYPES OF FAILURE

Before undertaking the analysis of any failure, it is important for the investigator not only to be well versed in the identification processes of fracture characteristics but also sufficiently knowledgeable in interpreting the results in a correct way. Any misinterpretation or wrong identification of the characteristic features of fracture would invariably lead to an incorrect conclusion and thus the very significance of failure analysis will be defeated. It is thus imperative that we learn in full detail the fracture types, and their individual characteristics with relation to each other. A brief mechanism of each mode of failure is also incorporated in order to assess the fracture in its total perspective, with an in-depth description given in the respective areas.

Various types of failure that commonly occur in service are given below with their possible modes.

1. **Overload failures**
   - Tensional
   - Bending
   - Impact
   - Torsional
   - Shear
   - Tear
2. **Distortion failures**
   - Warping
   - Bending
3. **Fatigue failures**
   - Push pull
   - Flexural (reverse bending)
   - Torsional
   - Brinelling
   - Bending (unidirectional)
   - Rotary-bending
   - Spalling (contact)
   - Thermal
4. **Corrosion failures**
   - General corrosion/pitting
   - Stress corrosion
   - Fretting fatigue
   - Exfoliation
   - Intergranular corrosion
   - Corrosion fatigue
   - Hydrogen embrittlement
5. **Creep failures**
6. **Wear failures**
   - Adhesive wear
   - Erosive wear
   - Surface fatigue
   - Abrasive wear
   - Corrosive wear

4.2 OVERLOAD FAILURES

Every structure has a load limit beyond which it is considered unsafe. An applied load that exceeds this limit is known as overload[15]. When a component fails because due to a single application of a load greater than the strength of the component, it is termed as overload failure. The nature of fracture arising due to overload failure could either be ductile or brittle or a combination of the two. This depends on the condition of the material, geometry of the member, types and rate of loading and temperature conditions. In general, ductility of the part is decreased by (a) increasing the strength of the metal by heat-treatment or cold working, (b) sources of stress-concentration such as notches, fillets, holes, scratches, inclusions, porosity, etc., (c) increasing the rate of loading, and (d) decreasing the temperature.

In general, ductile fractures are associated with metal flow at failure zone due to plastic deformation and fibrous-surface appearance. In brittle fractures plastic deformation is almost absent and the surface shows irregular bright facets of a cleavage type. The general features that are associated with overload failures are summarised below. They have been considered both from the macroscopic as well as the microscopic analysis point of view for the purpose of easy identification of the mode of overload failures.

Tension Overload Failure

Ductile	Brittle
(a) Geometry of fracture[16] is characterised as shown in Fig. 4.1. Thick sections > 6 mm ϕ of thickness	

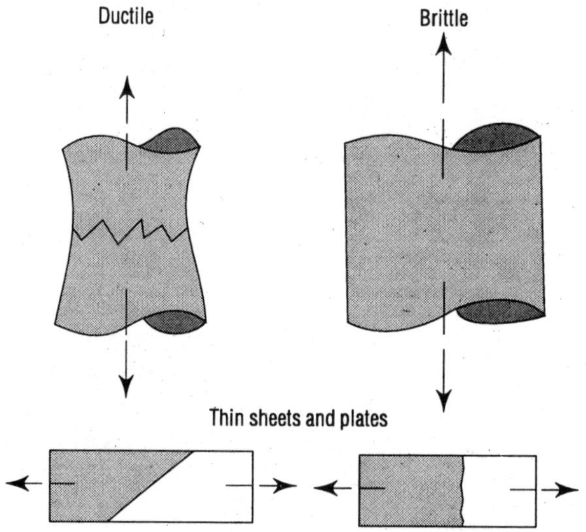

Fig. 4.1 Geometry of fracture of tension overload failure

Ductile	Brittle
(b) Excessive plastic deformation and necking adjacent to fracture (Fig. 4.1).	Little or no plastic deformation adjacent to fracture (Fig. 4.1).
(c) Fracture surface predominantly characterised by fibrous features with a pronounced shear lip zone at the periphery of the final fracture (Fig. 4.2). Fibrous feature is indicative of slow crack propagation. The origin of fracture is centrally located in the fibrous zone.	Fracture surface predominantly characterised by radial zone with a limiting shear lip. Radial feature is indicative of fast crack propagation.

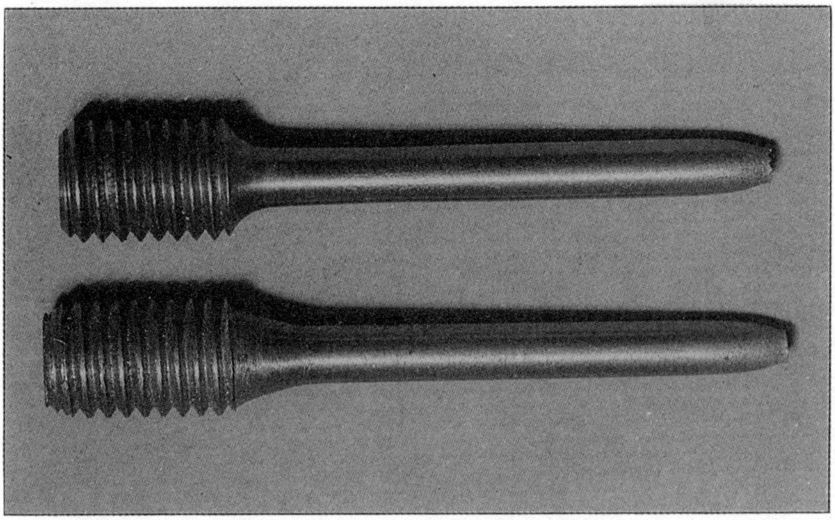

Fig. 4.2 Typical cup and cone fracture showing necking (× 1)

(d) Grey fibrous appearance or a silky texture is common.	Irregular fracture surface composed of a bright crystalline facet.
(e) Presence of microvoid coalescence indicates separation by means of plastic deformation. Equiaxed dimple is the positive signature of overload fracture in tension when examined under high magnification by scanning electron microscope (SEM) (Fig. 4.3 a)	Presence of cleavage step including river pattern, feather marks revealed by SEM is indicative of brittle fracture (Fig. 4.3 b).
(f) Transgranular or intergranular mode of cracking.	Transgranular mode of cracking.

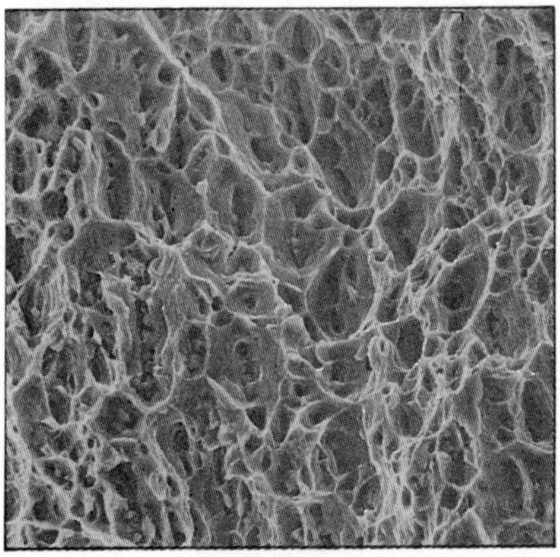

Fig. 4.3a Equiaxed dimple structure of an overload fracture (× 200)

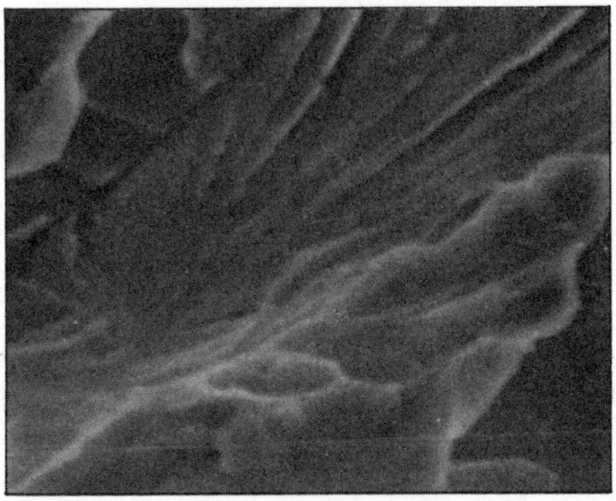

Fig. 4.3b Brittle fracture showing river pattern and feathery marks (× 450)

Torsion Overload Failure

Ductile	Brittle
(a) Geometry of fracture[16] is shown in Fig. 4.4.	

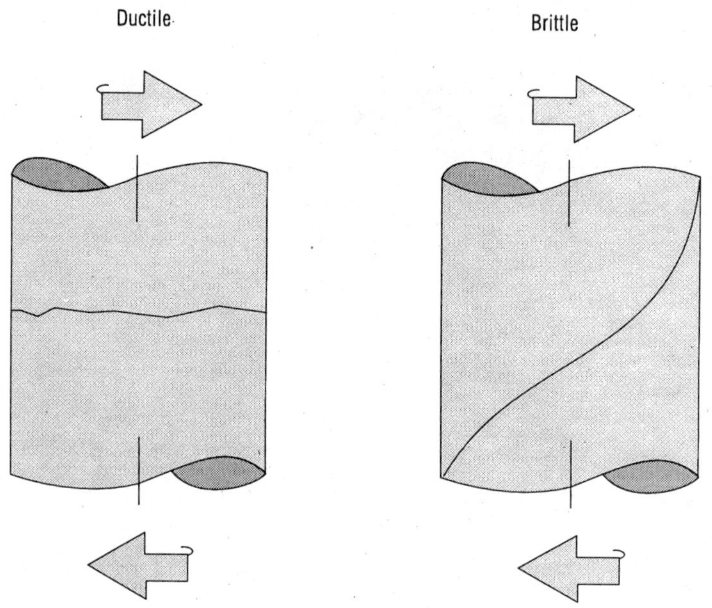

Fig. 4.4 Geometry of fracture of torsion overload failure

(b) Fracture is usually perpendicular to the direction of maximum tensile stresses, i.e. transverse shear plane.	Fracture follows the 45° direction of maximum tensile stress.
(c) Outer surface/geometry of the part shows deformation in the direction of rotation, which would not occur if the fracture were caused by fatigue. Fracture surface shows clearly rotary deformation marks usually with metal smearing effects at macroscopic level—a definite characteristic of a torsion overload fracture of ductile material (Fig. 4.5).	A spiral type fracture surface of a brittle material without any deformation on the fractured surface.

Bending Overload Failure

Ductile	Brittle
(a) Deformation adjacent to fracture—top surface similar to tension and the bottom surface that of compression.	No deformation adjacent to fracture.

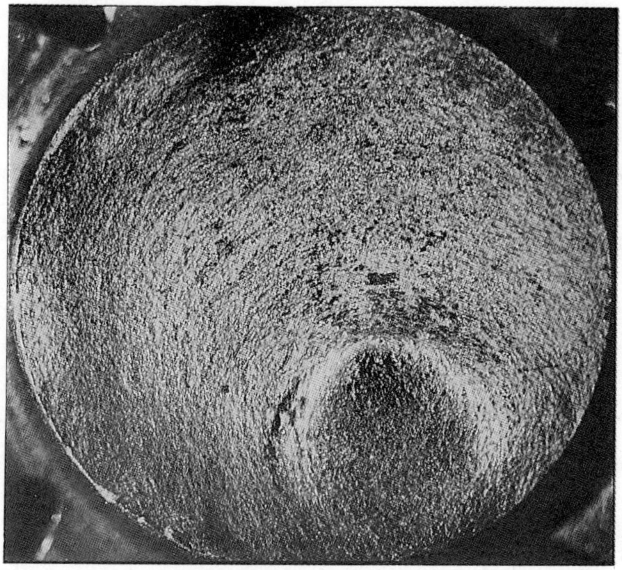

Fig. 4.5 Fracture surface of the failed splined shaft by a single torsional overload (× 4)

(b) Other characteristics similar to tension failure.	Other characteristics similar to tension failure.
(c) Reveals flat-face regions (perpendicular to the direction of principal normal stress) and shear face regions (at 45° to the direction of principal normal stress).	
(d) Microstructure in the longitudinal axis shows distinct bent fibres.	

Shear Overload Failure

Ductile	Brittle
(a) Deformation adjacent to fracture with shift in central axis (Fig. 4.6).	No deformation or shifting of central axis (Fig. 4.6)
(b) Presence of elongated or parabolically shaped dimples under scanning electron microscope.	
(c) Bright smooth crescent shaped zones in the fracture initiated surface.	

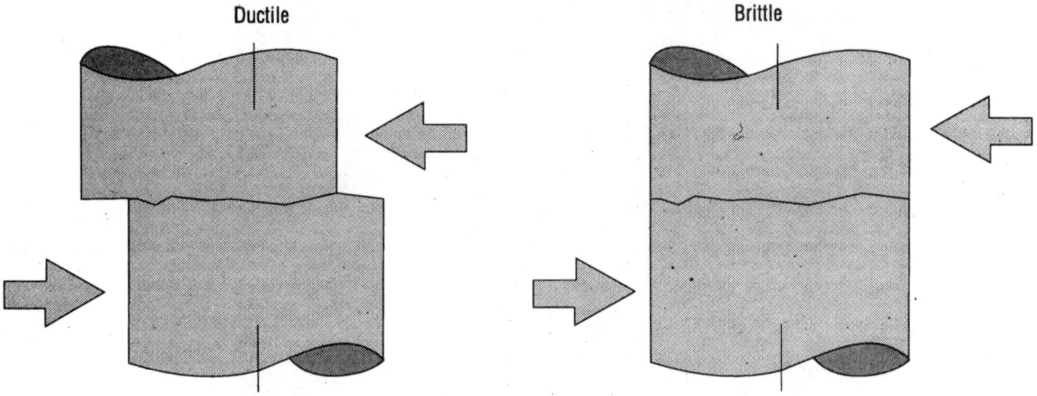

Fig. 4.6 Geometry of fracture of shear overload failure

Impact Overload Failure

Ductile	Brittle
	(a) The radial zone of a rectangular specimen, that has width much greater than the thickness, exhibits a chevron or herring-bone pattern (Fig. 4.7) and is associated with unstable and relatively rapid crack propagation.

Tear Overload Failure

Ductile	Brittle
(a) The shape of dimples produced by tensile tearing is short, almost equiaxed dimples with a high ratio of plastic strain at the tip of the crack to plastic strain in the material ahead of the crack and long dimples in case of low ratio yielding.	

Fig. 4.7 Surface of high velocity fracture showing chevron pattern (× 1)

4.3 FATIGUE FAILURES

Fatigue fracture results from the simultaneous action of repeated or fluctuating cyclic stress, tensile stress and plastic strain. No fatigue crack starts or grows in the absence of any of these three active components. Cyclic stress initiates the crack and tensile stress produces the crack growth. There are many variables in service that influence the fatigue behaviour or characteristic pattern. These include the magnitude and frequency of application of the fluctuating stress, the presence of a mean stress, temperature, environment, part size and shape, state of stress and residual stresses, surface finish, surface damages and microstructure.

The occurrence of fatigue may be considered as a three-stage process (Fig. 4.8a).

1. Initiation of the surface or sub-surface fatigue crack under a fluctuating load.
2. Crack propagation under tensile stress until the section thickness critically bears the imposed load.
3. Sudden fracture under overload.

Characteristics of Fatigue

On macro-examination the most characteristic feature of a fatigue-fracture is the presence of beach/conch shell or ratchet marks which are centred around a crack origin (Fig. 4.8b).

Normally the presence of these marks indicates a fatigue phenomenon but the reverse need not be true, i.e. all the beach marks need not be indicative of fatigue. This is true in the case of high speed steel brittle failure which exhibits a characteristic pattern of beach marks. There is no evident macroscopic plastic deformation near the fracture/crack. Fatigue-fracture surfaces are smooth textured near their origins carrying 'beach marks' which become coarser in texture as the crack grows and finally fails by ductile, shear or cleavage. Most fatigue cracks are transcrystalline without much branching, although intercrystalline fatigue is not uncommon. At high magnification (SEM or TEM), the most prominent fracture feature observed is patches of fully spaced parallel marks, called fatigue striation (Fig. 4.9).

Striations orient themselves perpendicular to the direction of crack propagation and with uniform loading increase in spacing as they progress. Each striation is the result of a single cycle of stress. Striation is confirmation of a fatigue. However, striations are not evident in high strength steels.

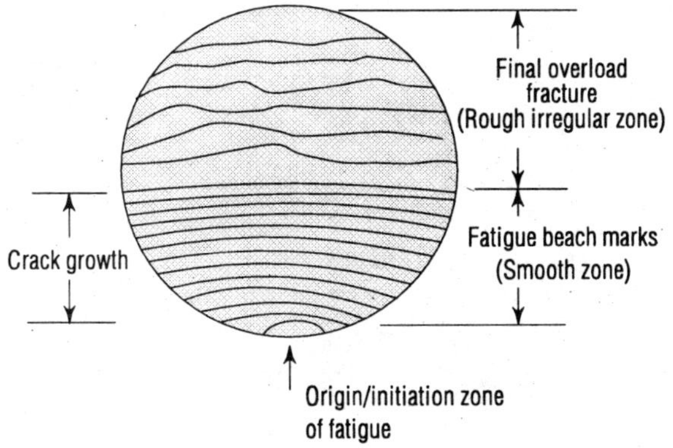

Fig. 4.8a Schematic representation of the fatigue crack (three-stage) phenomenon

Fig. 4.8b Distinct beach marks due to fatigue, originating from the root radius zone of gear tooth (× 4)

Low cycle fatigue and high cycle fatigue are almost with the following exceptions.

High cycle (above 10^6 cycle)	Low cycle (below 10^5 cycle)
Smoother surface.	Rather coarse surface.
Beach mark spacings are fine, uniform and narrow.	Beach mark spacings are coarse, non-uniform and wide and far apart.
Arrest line are clear and distinguishable under suitable magnification.	Feeble arrest lines—sometimes difficult to distinguish between low fatigue and brittle fractures.

In the presence of a severe stress concentration multiple crack origins are formed which eventually unite to form a single crack front.

When the stress concentration on the surface is absent cracks propagate more readily near the centre of a section than at the surface. While there is a stress concentrating notch, W-shaped crack fronts are observed. For a given material, zone of the final fast fracture increases with increasing applied load. On rotating or bending the final fracture region is often rotated or offset towards the origin in the direction opposite to the direction of rotation. Also, with increasing nominal stress, the final fracture moves towards the centre.

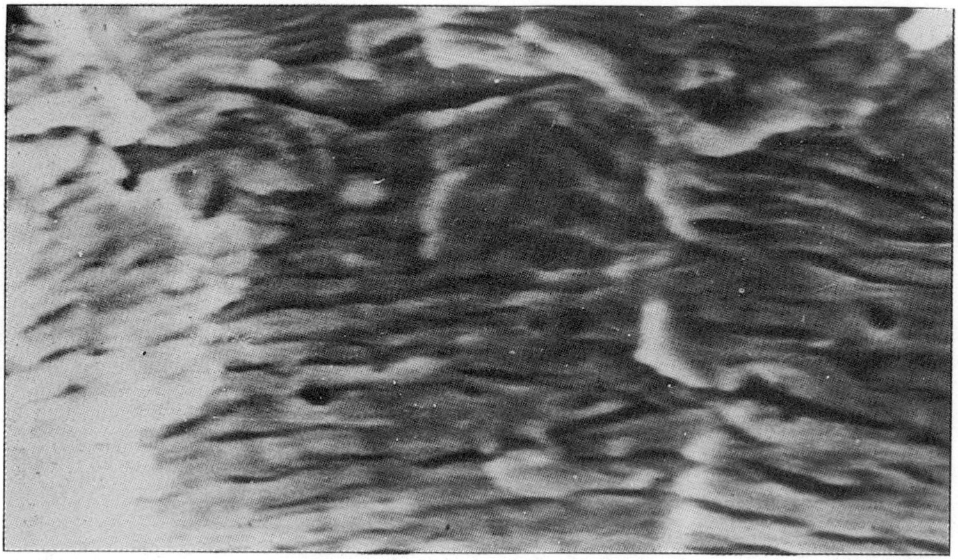

Fig. 4.9 Typical fatigue striations in a mild steel failure at high magnification

Bending Fatigue

Fatigue can result from three types of bending loads, *viz.* unidirectional (one way), reversal (two way) and rotating. When a shaft is subjected to a bending stress, the convex surface experiences the tensile component of the stress while the concave surface remains compressed.

Unidirectional Bending Fatigue

Beach marks, indicative of a fatigue crack, show a single origin (at the point indicated by the arrow in Fig. 4.10 (a). The crack front with its beach marks shows symmetry with respect to the origin and retains a concave form throughout. Both the single origin and the smallness of the final fracture zone in Fig. 4.10 (a) are suggestive of a low nominal stress. The larger final fracture zone in Fig. 4.10 (b), indicates a fatigue failure under a higher (moderate) nominal stress.

However, fatigue crack under high nominal stress seems to originate as several individual cracks that ultimately merge to form a single crack front (Fig. 4.10c). Usually characteristic 'ratchet' marks are present between crack origins. Figures 4.10 (d) (e) and (f) indicate typical beach marks, under moderate stress concentration, provided by a change in section in a uniformly loaded shaft. With a nominal stress, the crack front changes from concave to convex prior to rupture (Fig. 4.10(d)) At higher nominal stresses, the crack front flattens as Fig. 4.10(e) and (f). With severe stress concentration produced by a change in section in a uniformly loaded shaft, beach marks show a

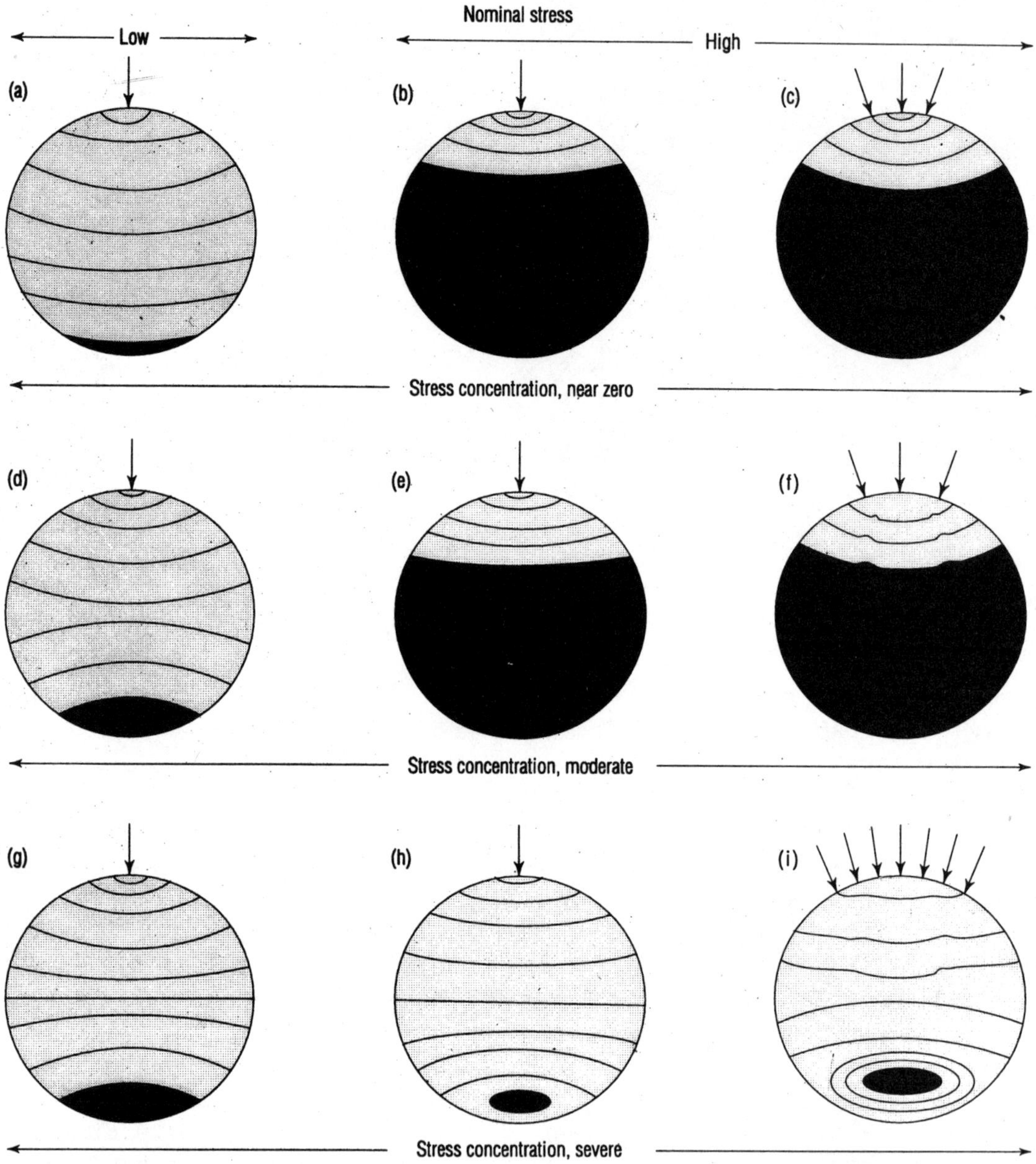

Fig. 4.10 Schematic illustration of typical fatigue induced marks on fracture surfaces of a uniformly loaded shaft, produced from a single origin when subjected to unidirectional bending at low and high nominal stresses and those features obtained from multiple origins at high nomial stress. Origins are indicated by arrows and final fractures as hatched areas, (for additional reading ref. *ASM Metals Hand Book*, Failure Analysis and Prevention, Vol. 10, page 376)

typical pattern as in Fig. 4.10(g), (h) and (i). Multiple crack origins, high nominal stress and unidirectional bending producing the fatigue beach marks are shown in Fig. 4.10(i).

Reverse Bending (Flexural) Fatigue Failure

If the applied bending moment is reversing and maintains the same magnitude in either direction, two cracks of almost equal length are identically developed from the origins located diametrically opposite to each other and usually in the same transverse plane. However, two cracks differ in length in case the bending moment is found greater in one direction than in the other. Figure 4.11 shows typical fatigue marks produced on the fracture surface of a uniformly loaded non-rotating shaft subjected to reversal bending stresses. The pattern shown in Fig. 4.11 (a) is typically that of a single diameter shaft with no stress concentration, indicating equal bending moment in both directions. Fig. 4.11 (b) shows the typical pattern of beach marks due to a moderate stress concentration imposed by a large radius fillet at a change in the shaft diameter. Figure 4.11 (c) gives the typical pattern due to a severe stress concentration by a small radius fillet at a change in the shaft diameter.

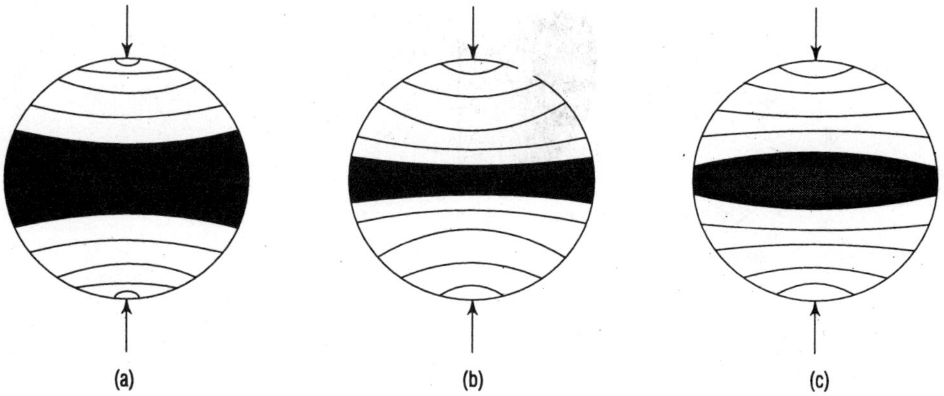

Fig. 4.11 Schematic illustration of typical fatigue induced marks on fracture surface of a uniformly loaded non-rotating shaft failed by reverse bending stresses under three conditions of (a) no concentration, (b) moderate stress concentration and (c) severe stress concentration, origins are indicated by arrows and final fracture as hatched areas (for additional reading refer *ASM Metals Handbook* 'Failure Analysis and Prevention,' Vol. 10, pp 377).

Rotary Bending Fatigue

In contrast to a stationary shaft, where tensile stress is only confined to a portion of the periphery, the rotating shaft sustains a tensile stress at every point on the periphery, with the asymmetrical development of the crack front from a single origin. In such a case, a crack may be initiated at any point on the periphery [Fig. 4.12(b) and (d)]. The crack front which tends to extend in a direction opposite to that of rotation, usually swings around by 15° or more [Fig. 4.12(a) and (c)].

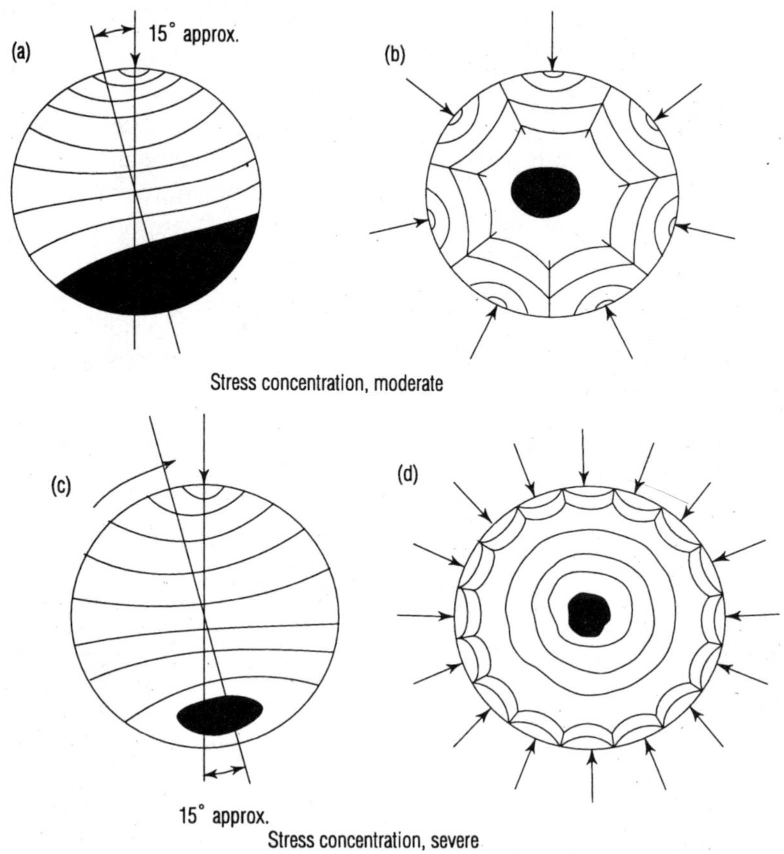

Fig. 4.12 Schematic illustration of fatigue induced marks on fracture surfaces of a uniformly loaded rotating shaft, originating from single and multiple origins under moderate and severe stress concentrations. The shaft rotation is in the clockwise direction. The origins are indicated by arrows and final fractures as hatched areas, (for additional reading ref. *ASM Metals Handbook*, Failure Analysis and Prevention, Vol. 10, page 377)

Torsion Fatigue Fracture

Fatigue cracks resulting from torsional stresses show the same beach marks and ridges as those produced by bending stresses (Fig. 4.13). Under torsional loading of a shaft, the maximum local tensile stresses are at 45° to the axis of the shaft. Under a fluctuating torsional load, fatigue crack may develop normal to the tensile stresses. Under alternating (reversed) torsional loading, two sets of fatigue cracks perpendicular to each other, may develop.

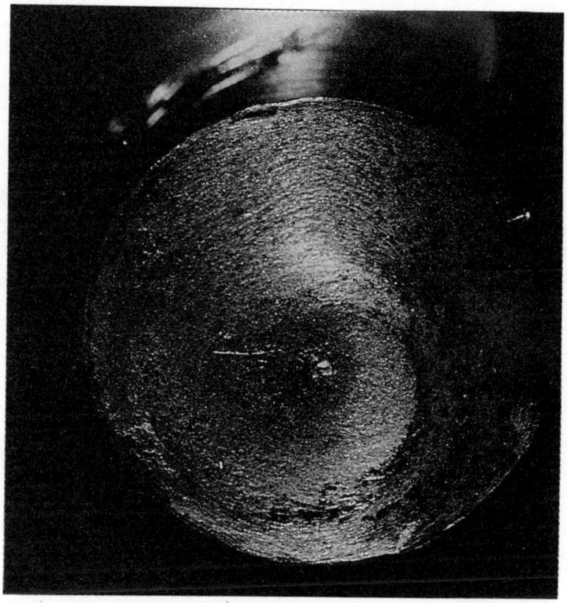

Fig. 4.13 Fracture surface of a shaft that failed by torsional fatigue (× 3)

Contact Fatigue Fracture

This type of failure is commonly observed in parts (such as bearing) involving rolling, or rolling and sliding operations against each other under high contact pressure and cyclic loading. The progressive development of cyclic stresses, at or near the contact surface during operation results in a gradual deterioration of the material by damage mechanisms leading to initiation and propagation of fatigue cracks. In some failures, small pits are initially observed on the polished contact surfaces before the spalling starts.

Spalling is characterised by surface deteriorations like pits, crater-like cavities in the highly stressed surfaces of raceways or rolling elements (Fig. 4.14). The irregular cavities are sharp edges with steep walls and flat bottoms. Spalling is also recognised as a severe form of flaking.

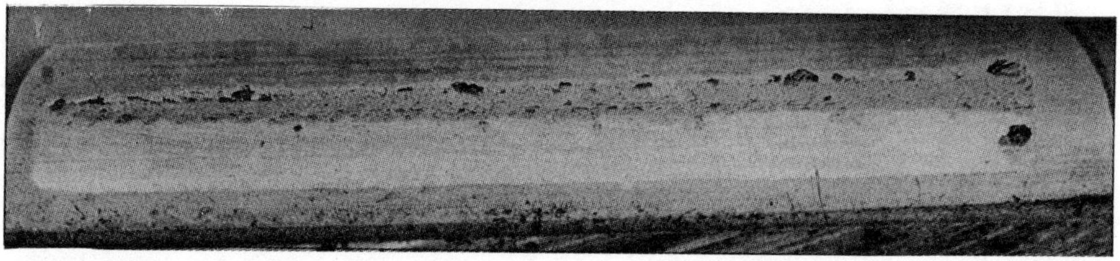

Fig. 4.14 Spalled tooth surface of a failed gear in design testing

Generally, spalling (Fig. 4.15) in a roll is attributed to initiation of a subsurface fatigue crack of undetermined origin or is a result of surface induced microscopic cracking. Spalling may originate below the surface as a result of localised stress concentration.

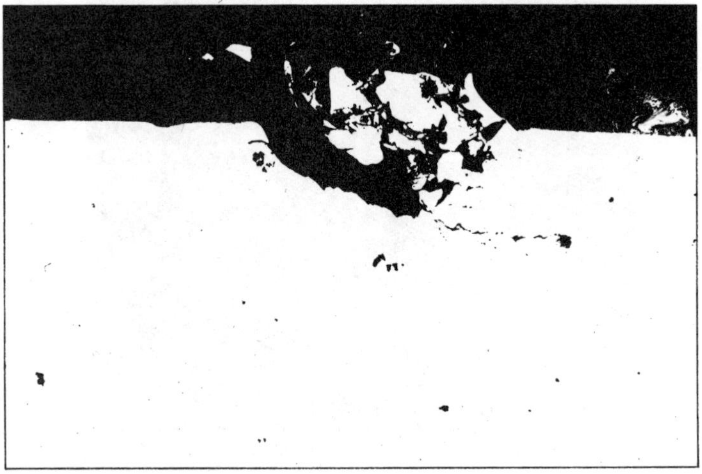

Fig. 4.15 A typical spalling characteristic phenomenon induced by a sub-surface fatigue crack in the section of a case-carburised steel (× 400)

Brinelling

Fatigue failures due to brinelling effect are commonly noticed in a stationary bearing or engine component under excessive pressure or impact loading. Permanent deformations, like hollows or dents, are caused by plastic flow of metal and identified as indentations corresponding to ball spacing in the ball bearing component.

These characteristic features differentiate brinelling from false brinelling (produced by vibrations or oscillation) and dents made by debris (in which case dents are spaced randomly).

Thermal Fatigue Failure

Thermal fatigue failure is the result of repeated thermal cycling which generates repeated stress gradients within a free component, or stress cycling within a component constrained to fixed dimensions.

Thermal fatigue fractures are characterised by surface crazing formed by multiple initiation sites that join randomly by edge sliding to form the main crack as shown in Fig. 4.16. Other features to identify thermal fatigue are:

- Fractures are planner and transverse with no visible plastic deformation.
- Fracture are mostly transgranular.
- Oxidised fracture surfaces and oxide wedge filled cracks further characterise thermal fatigue failures.

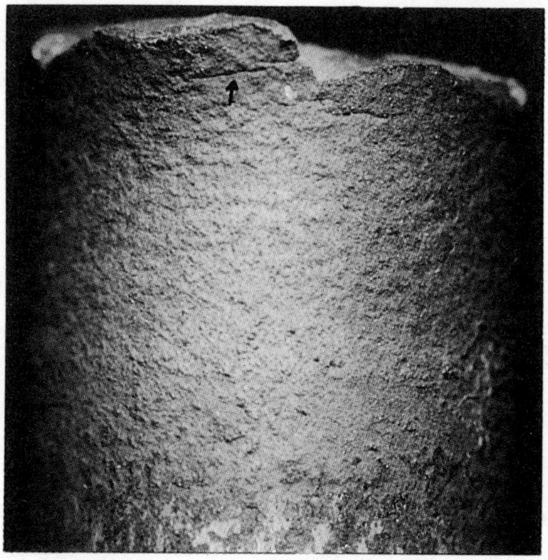

Fig. 4.16 Shows multiple initiation zones of thermal fatigue on the surface of an engine inlet valve (×6)

4.4 CORROSION FAILURES

'Corrosion', by definition, is a from of degradation or deterioration of some material due to chemical or electrochemical activity. There are several basic forms of corrosion. Each form is classified by external appearance or altered physical properties. In general, most types of corrosion occur by an electrochemical mechanism.

The corrosion process which converts the metal into a metallic compound such as oxide, hydroxide or sulphate involves two chemical changes: anodic and cathodic. However, neither characteristic products nor weight loss of the parent metal due to corrosion may necessarily be present in corrosion-aided failures.

Uniform (Surface) Corrosion

It is surface phenomenon which occurs through uniform attack of the metal (of homogeneous chemistry and structure) resulting from contact with certain strongly acidic or alkaline electrolytes as well as conditions of high humidity or moisture laden atmosphere. This results in discolouration as the oxide film thickens slightly. On a polished surface this type of corrosion produces a powdery deposit on the metal surface. After the surface is polished, it will regain its normal lusture.

Uniform corrosion attack usually proceeds along specific crystallographic planes. Uniform corrosion is measured by weight loss or decrease in thickness. The rate of attack usually is expressed in mils per year (mpy).

Characteristic Features

- Corrosion attack vary from superficial etching and staining to rapid dissolution of metal (Fig. 4.17).
- Discolouration of the metal surface appearing as iridescent, white, brown or even black depending on the alloy composition and the degree of oxidation.

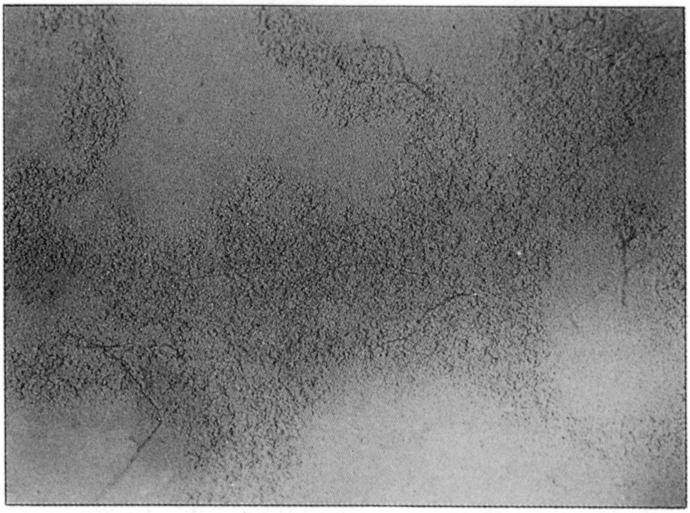

Fig. 4.17 Typical patches of general corrosion in an aluminium alloy shut (asreceived) (× 1)

Pitting

It is the most common form of corrosive attack and is extremely localised in nature (Fig. 4.18). It results from chemical or galvanic action which usually takes the form of cavities on the surface. It may occur in any metal system, but is mostly observed in the aluminium and magnesium alloys. Pitting first appears as a localised white powdery deposit on aluminium, but on buffing the surface tiny pits/holes will be revealed.

Pitting occurs when one area of a metal surface becomes anodic with respect to the rest of the surface, or when highly localised changes takes place in contact with the metal[9]. It usually starts at localised discontinuity sites such as, constituent particle, inclusion, grain boundary or any other discontinuities of the surface. Causes of pitting are mainly due to local inhomogenity on the metal surface, local loss of passivity, mechanical or chemical rupture of protective oxide film, galvanic corrosion. An edge attack involves directional form of pitting associated with a concentration gradient in the metal or a grain structure deformed by a working metal. This edge attack is observed as an undermining form of pitting (Fig. 4.19).

Fig. 4.18 Localised pitting corrosion (× 1)

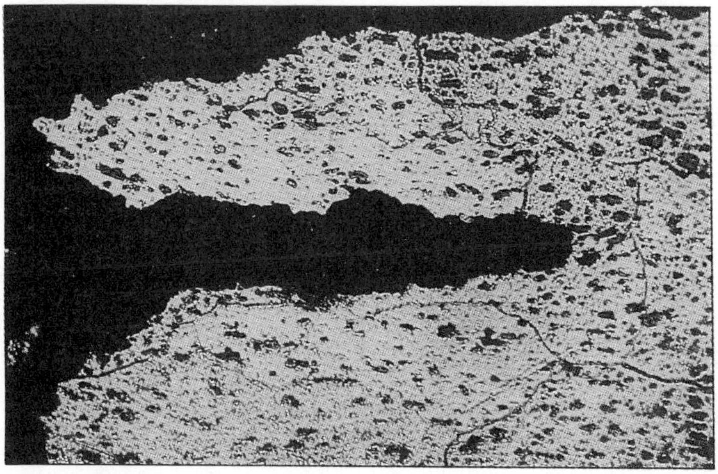

Fig. 4.19 Typical undermining form of pitting in the microstructure of an aluminium alloy wrought product (in-service) (× 100)

Microscopically, the presence of a cubic characteristic of the attack indicates a preferential attack on certain crystallographic planes, resulting from an acid-chloride medium. Besides these, there are three other common forms of specialised pitting corrosion. One of them is fretting corrosion, caused by slight relative motion between two contacting surfaces, which results in small pits. The other is cavitation erosion resulting from the forming and collapsing of vapour bubbles at a dynamic metal liquid interface. It has small, deep fissures and resembling a honeycomb. The last one is a special type of pitting corrosion (corrosion erosion/impingement attack) which occurs in those metals that are subjected to liquids flowing at high velocities.

Characteristic Features

- Pits are identified as sharply defined holes.
- Pits are generally small and often remain undetected. Size may vary with wide scattering depending on the alloy composition, the quality of the oxide film and the character of the corrodent.
- When pits are accompanied by slight or moderate general corrosion, the corrosion products often mask them.

Parting / Dezincification

It is a special type of corrosion which occurs when one or more reactive components of the alloy corrodes preferentially, leaving behind a porous residue. Dezincification is a part of the parting corrosion process that attacks alloys containing zinc in which zinc corrodes preferentially.

The characteristic feature of this type of corrosion, apart from its porous nature, is that the surface tarnishes without any apparent damage. However the tensile strength and the ductility of the affected alloy will be drastically reduced.

Filiform Corrosion

It is commonly found electrolytically initiated in a high humidity environment, in the presence of moisture, oxygen and corrosive ions. Its usual place of occurrence is around fasteners, at skin butt joints or at breaks in coating films.

This type of corrosion is seen as fine trenches under the paint that spread out from the fastener, skin butt joint or coating break in a filiform or thread like manner. However, filiform corrosion involves only mild surface corrosion with no structural damage.

Intergranular Corrosion

It is preferential dissolution of the zones adjacent to the grain boundary, with slight or negligible attack on the main body of the grain. The preferential attack is caused by potential differences between the grain boundary region and the abutting grain bodies. It is enhanced by the segregation of specific elements or compounds, or enrichment of one of the alloying elements in the grain boundary areas.[9] When the crack is severe failure occurs, under residual or applied load, due to resultant loss of strength and ductility.

Characteristics

- Intergranular corrosion is more or less randomly oriented and does not have any highly localised propagation (Fig. 4.20), unlike in the case of stress corrosion in which cracking progresses in a direction normal to applied or residual stress.

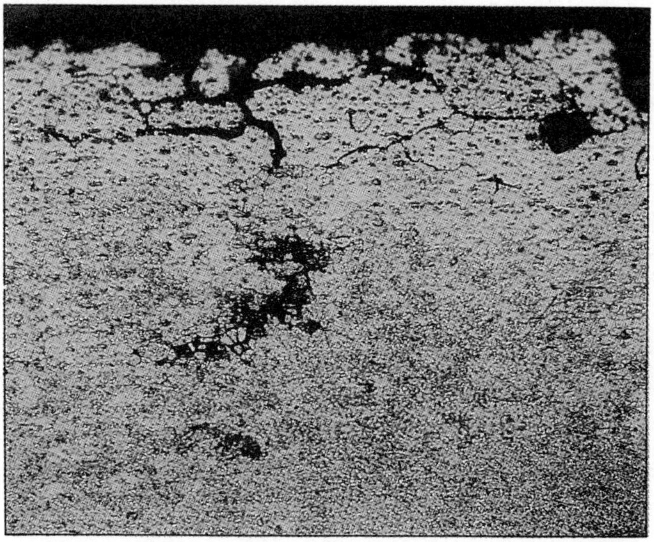

Fig. 4.20 Severe intergranular corrosion attack in the surface as well as body of an aluminium alloy wrought product (as-received) (× 50)

Stress Corrosion Cracking

It results from the simultaneous action of residual or applied tensile stress and chemical action by a corrosive environment. Stress corrosion manifests itself as a crack in the metal not necessarily accompanied by corrosion. This simultaneous influence of corrosion and stress vastly differs from the mechanism of cracking of a metal exposed to any one of them. Time required for the initiation, nucleation and subsequent propagation of stress corrosion cracking that leads to the ultimate failure is dependent on various factors such as material composition, fabrication history, type and intensity of corrosive environment and magnitude of internal, residual or applied stress. Stress corrosion cracking occurs frequently in both aluminium and magnesium alloys as well as in steels. It is recognized only after cracking has occurred in service.

Characteristics

- Stress corrosion cracks are progressive, beginning with a small crack and propagating in a plane generally perpendicular to the operating tensile stress. Cracks often show branching off of secondary cracks from the main crack.

- The resulting fracture has a brittle appearance.
- Normally (not essentially) there are numerous cracks that tend to 'wander through' the metal imparting a coarse as well as layeral woody appearance (Fig. 4.21).

Fig. 4.21 Layeral woody structure, typical of stress corrosion cracking in a high strength Al-zn-Mg alloy extrusion (in-serivice) (× 3)

- A polished section through the metal damaged by stress corrosion cracking usually shows a network of fine cracks (secondary intergranular cracks parallel to the main crack (Fig. 4.22).

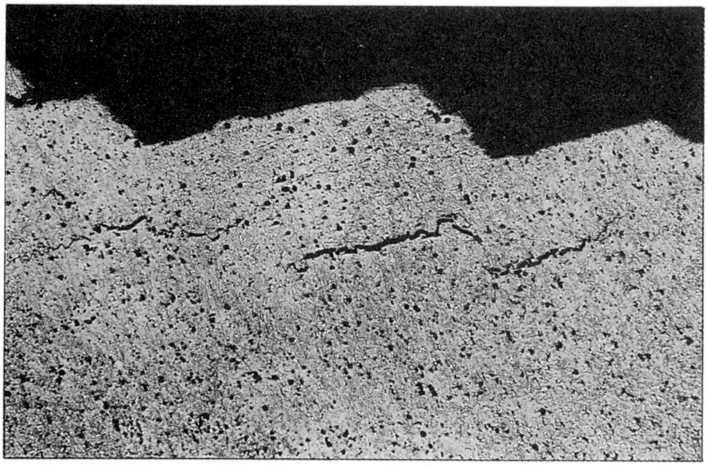

Fig. 4.22 Parallel secondary cracks of a stress corrosion phenomenon in the part seen in Fig. 4.21 (× 200)

- Stress corrosion cracking are generally intergranular although in certain steels and copper alloys it may be either intergranular or transgranular.
- Fracture surface usually contains easily identifiable regions (such as corrosion pits, bright half moon zones, in Fig. 4.27) of crack initiation, slow crack propagation and final rupture.
- Area of slow crack growth often contains corrosion products, stained or otherwise discoloured with respect to the area of final fast fracture.

Exfoliation Corrosion

It is an advanced form of intergranular corrosion and exhibits itself in a lamellar form resulting from rapid lateral attack[9] along grain boundaries or striations within the grains, parallel to the metal surface. This directional attack results in a leafing action aggravated by the voluminous corrosion products accumulated at the grain boundaries just below the surface, forcing the uncorroded strata to split apart (Fig. 4.23).

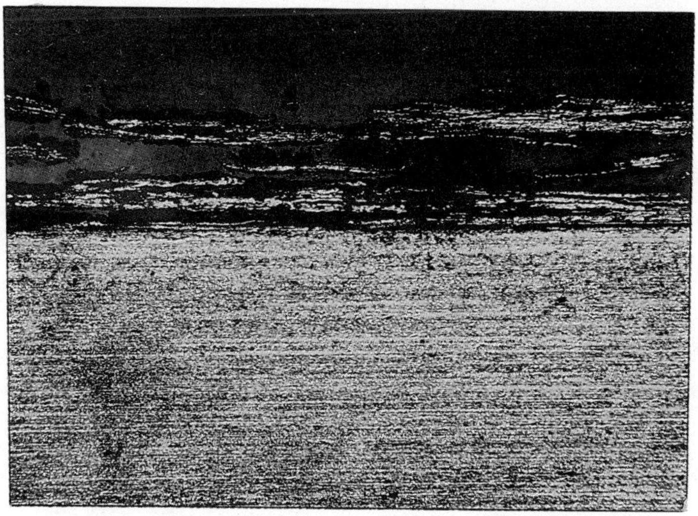

Fig. 4.23 Directional attack and leafing action along the grain fibres of an Al-Zn-Mg-alloys extrusion (in-service) (× 50)

Exfoliation phenomenon is much more pronounced with (marked) directionality of the grain structure. Therefore, it is usually found occurring in extruded sections where the grain thickness is less than in rolled form and in the manufacturing products such as forging, plate and thick sheet where grains are relatively thin compared to their width.

Potential loss of strength as a result of corroding away of the bearing surface and loss in effective cross-section are the key dangers of exfoliation corrosion.

Fretting Corrosion Fatigue

Generation of oxide debris at the interface of two surfaces in contact, when subjected to relative oscillatory motion by virtue of adhesion, indicates the phenomenon of fretting corrosion. The rubbing action removes the protective film on a metallic surface and the small particles that are created then oxidize and act as abrasives inducing fretting pitting. Fretting destroys the dimensional accuracy of closely fitted parts, ruins bearing surfaces (particularly ball and roller bearings) and increases the susceptibility to fatigue failure. When fatigue failure results due to initiation and propagation of fatigue cracks from such fretted areas, it is termed as fretting fatigue.

Characteristics

- Surface discolouration due to oxidation, most commonly noticed in ferrous materials as reddish brown colour and blackish colour in aluminium alloys.
- Formation of debris or welded zones and deep pits at the contacting surfaces (Fig. 4.24).
- Fatigue crack originating at the fretted zone.
- Other fracture features are similar to fatigue failure.

Fig. 4.24 A typical fretting fatigue fracture showing the origin (marked by arrows) of fatigue on the fretted zone in a failed rotating steel part (× 3)

Corrosion Fatigue

It occurs due to the combined action of corrosion and cyclic stress on metal parts. The corrosion fatigue mechanism involves the formation of extrusions and intrusions at the surface, destroying the protective surface layer at these points and allowing electrochemical attack to form microscopically small discontinuities. These discontinuities act as a notch for fatigue fracture to occur under the influence of continuous chemical reaction. Fatigue life of a dynamically loaded parts may be reduced by 50% in a moist or corrosive environment than in dry air.

The process of failure by corrosion fatigue is composed of two stages. In the first stage the combined action of corrosion and cyclic stresses damage the metal by pitting and initiate a crack. Second stage is essentially a fatigue stage in which fatigue crack propagation is accelerated by corrosion and is controlled primarily by stress corrosion effects and the physical properties of the metal.

Characteristics

- Surface of the metal is generally pitted and cracks are initiated at the bottom of the pit (Fig. 4.25).

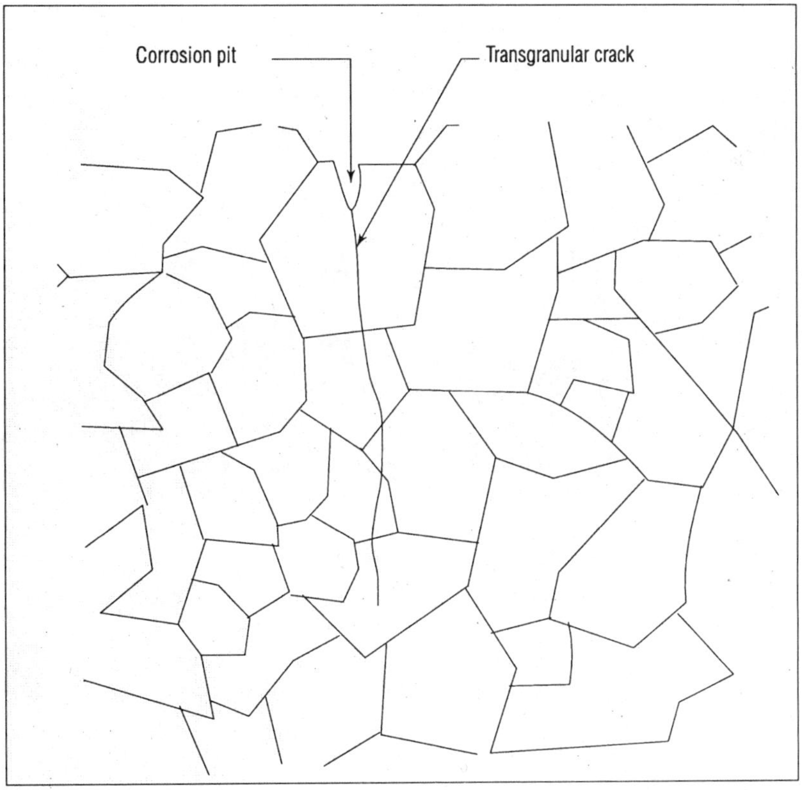

Fig. 4.25 Schematic representation of a corrosion fatigue cracking

- Cracks are predominantly transcrystalline and progress in the direction normal to the operating stress.
- Cracks are intercrystalline in ferritic steels (particularly 13% chromium steel) depending on the stress intensity.
- Several cracks initiate and propagate simultaneously.
- Fracture features are identical to those of fatigue failure with its origin at the corrosion pit. However these features are sometimes obliterated to some extent by corrosion.

Hydrogen Embrittlement

It is a mechanical environmental process of failure that results from the initial presence or absorption of excessive amounts of hydrogen in metals, usually in combination with residual or applied tensile stresses. Hydrogen embrittlement failure occurs primarily in stressed low alloy steels which have been heat treated to 180,000 psi or greater ultimate tensile strength.

Characteristics

- Crack mostly originates internally unless there is a stress raiser on the surface.
- Almost without exception, a single crack that shows no significant branching is produced in hydrogen induced cracking (Fig. 4.26).

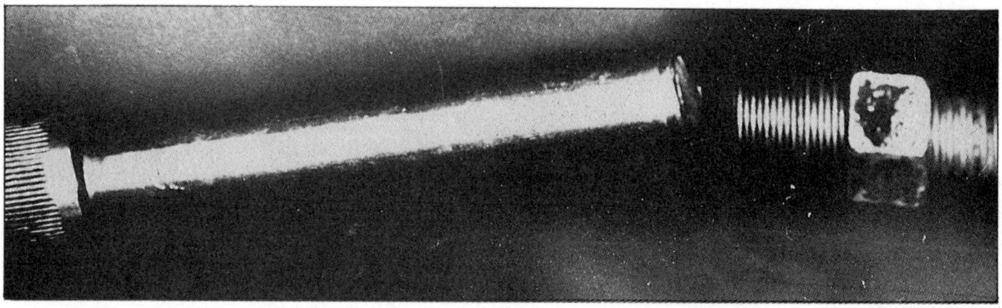

Fig. 4.26 Straight cracking in a hydrogen embrittlement failure of a steel bolt (× 1)

- Cracks can occur either in inter or transcrystalline fashion relative to the prior austenite grains.
- Botanical appearance resembling that of a curved leave (a particular characteristic of micro-quasi-cleavage feature of transcrystalline fracture structure) at high magnification under SEM.
- Presence of random micropores in the brittle facet.
- Flakes and fisheyes which usually surround the non-metallic inclusion (typical evidence of transcrystalline hydrogen embrittlement).

- Intercrystalline subsidiary cracks (yawning, grain boundaries), ductile hair lines (crow feet) and micropores at the grain boundaries (Fig. 4.27).

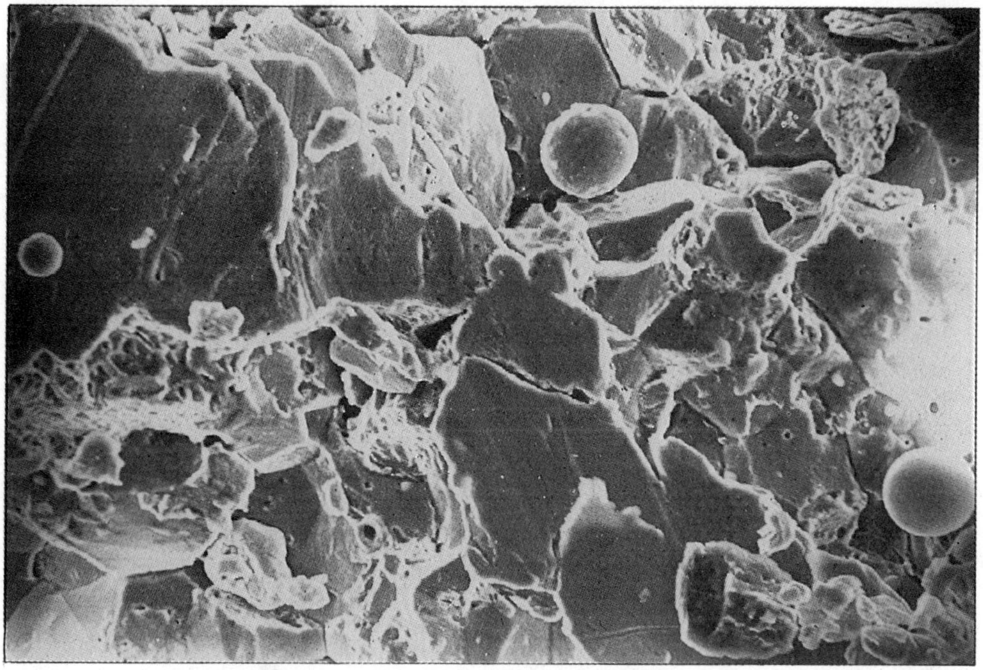

Fig. 4.27 SEM photograph of a hydrogen embrittled fracture showing grain boundary yawning, intergranular cracking, micropore, ductile hairline crack, dimple structures and brittle facets (× 450)

4.5 CREEP FAILURES

Creep is the change in dimension of a metal or alloy under an applied stress. It occurs at a temperature generally above the recrystallization temperature. Creep strain may cause sufficiently large changes in the dimensions of a component to render it useless for further service. The creep strain may lead to fracture. This type of failure is called stress rupture. Creep properties become important in the following areas:

1. The soft metals used at about room temperature such as lead pipes and white metal bearings.
2. Steam and chemical plant operating at 450-550 °C.
3. Gas turbines working at high temperature.

Characteristics

- Easy to recognise by the local deformation and multiplicity of intergranular cracks that are usually present, adjacent to the main crack (Fig. 4.28).

- Section close to the fracture shows at higher magnification multiplicity of creep voids at grain boundaries (Fig. 4.29).
- Cracks originate at triple points.

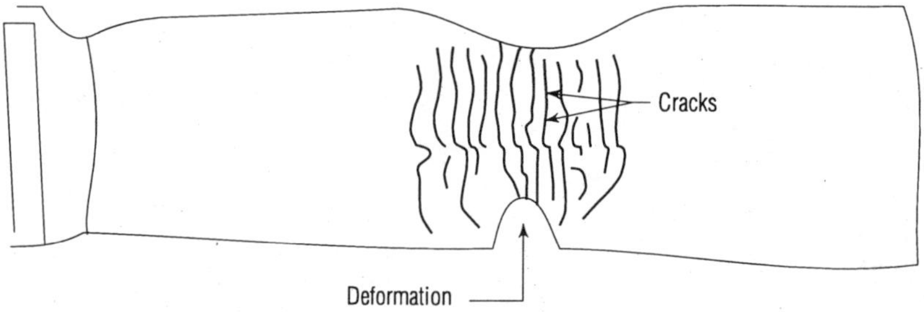

Fig. 4.28 Schematic representation of creep deformation and cracks in a turbine blade

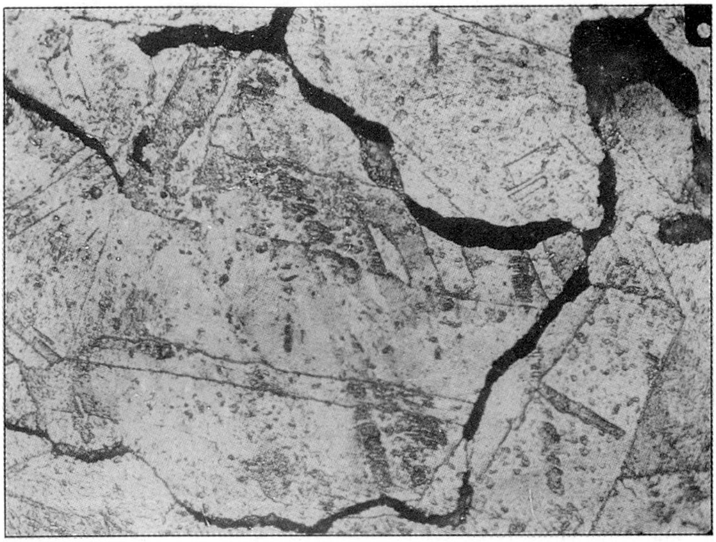

Fig. 4.29 Creep voids at the grain boundaries and cracking at triple points in the microstructure of a creep failed Cu-Cr material (× 400)

4.6 WEAR FAILURES

Wear is a surface phenomenon occurring as a damage to a solid surface. It is caused by the removal or displacement of material from the surface by the mechanical action of a contacting solid/lubricant, liquid or gas.

Unlike fatigue and corrosion, wear is less catastrophic but greatly reduces the operating efficiency. Wear failure depends on the degree of the wear which in turn is governed by the change in load, speed or environmental condition.

The types of wear generally encountered in engineering systems fall into five distinct categories (a) abrasive; (b) adhesive; (c) erosive; (d) corrosive; (e) erosion-corrosion. Of these abrasive and adhesive modes of failures are common in the aircraft industry.

Abrasive Wear

It is a form of displacement of the metal from the surface due to contact with high spots or asperites on the mating surface or due to loose hard particles. In the case of hard particles, they may either be trapped in between the two sliding surfaces resulting in abrasion of one or both of them, or embedded in either of the surfaces abrading the opposing surface. In abrasive wear also, material is removed from the surface by the cutting action of abrasive particles.

Characteristics

- Wear may range from the mild polishing type to the rapid and severe removal of material with accompanying surface roughening.
- Fairly regular grooves in the smooth surface with or without loosely attached metallic debris, producing damages like.

(i) scoring or gouging abrasion identified by the removal of large particles with heavy gouge on the worn surface;
(ii) high stress grinding abrasion by progressive fragmentation of abrasion particles;
(iii) low stress scratching abrasion caused by scratching of the metal surface.

Adhesive Wear

It occurs when two metallic surfaces slide against each other under high localised pressure and due to which microscopic projection/asperites get bonded at the sliding interface. The sliding forces simultaneously break up the bond at the weakest point, resulting in a metal tear from the surface and transference of the same to the other surface.

Adhesive wear is common under dry contact or non-lubricated conditions. Service defects commonly encountered due to this type of wear are surface fatigue, spalling, seizing, scuffing, scoring.

Characteristics

- Exhibits a torn/dragged and furrowed appearance with marking in the direction of sliding.
- Signs of severe scuffing in the zone of sliding contact.

Chapter 5

Techniques of Failure Analysis

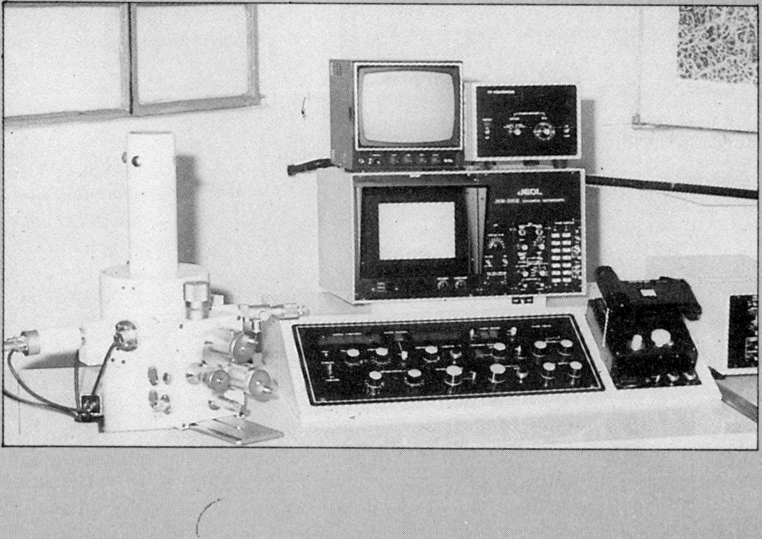

5 Techniques of Failure Analysis

5.1 NON-DESTRUCTIVE TESTING TECHNIQUES

In the event of any failure, whether catastrophic or small, failure analysis has been recognised as an important tool by the metallurgists for finding out the means for corrective action. So much importance is attached to it for the simple reason that it gives a significant output not only in terms of reduced capital investment but also in terms of security to life. This is particularly so in the case of high cost moving equipment such as automobile, aircraft or spacecraft. With the introduction of more and more advanced technologies and newer materials, failure analysis has rapidly grown over the recent past into a separate entity of its own as a science, developing complex and sophisticated technique for better understanding of the failure mechanism. Facture analysis thus constitutes an exceptionally challenging and exciting task.

When conducted in a systematic manner, the direct benefit of failure analysis is to derive the shortcomings in design efficiency, correctness of selection of materials, processing capacity, soundness of raw materials or any other aspects leading to the failure.

To carry out failure analysis objectively it is imperative to have not only the necessary sophisticated equipment but also adequate expertise which is often neglected or not available in small organisations. In addition to a sound background in fracture mechanics, fabrication technology, etc., the investigator should possess an analytical mind and also a knack for carrying out investigation. Besides, he should keep abreast of modern developments in materials and processes, and be thoroughly conversant with the various techniques of failure analysis both in terms of knowledge as well as use from their application point of view. However, despite availability of modern equipment and facilities, the success of the analysis primarily depends upon the investigator who has the ability to sift facts from irrelevant information.

This chapter describes the various techniques of failure analysis. They can be broadly divided into two categories, destructive and non-destructive, and can further be divided as given below.

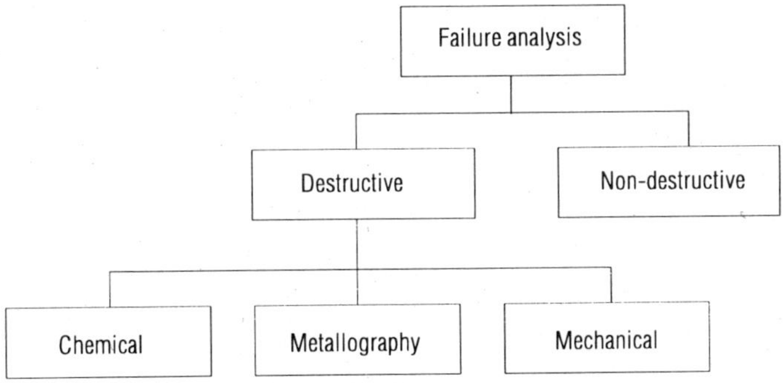

The non-destructive testing technique (NDT) usually adopted in the failure analysis programme are:

1. Visual examination
2. Liquid penetration
   - Dye penetrant
   - Fluorescent penetrant
   - Hot oil chalk test
3. Anodising and etching
4. Magnetic particle inspection
5. Radiography
6. Ultrasonic
7. Eddy current
8. Other techniques (Acoustic Emission Holography, Tomography)

Visual Examination

Visual examination (or otherwise called preliminary investigation) is one of the most important steps for surface studies and constitutes the simplest form of NDT techniques. This is done using a powerful source of light and is aided by a low-power mangifier. Its success depends on the perceptive power of the investigator. This examination, whether unaided or aided, enables him to locate blemishes/imperfections and to study the fracture graphy for assessing the mode of failure.

During visual examination, all significant information regarding the primary location and nature of failure, secondary factors influencing the main event, corrosion or extraneous products, geometry at the location of failure, damages, etc. are recorded and photographed not only for systematic analysis but also for future reference. Today in failure analysis, colour photography techniques are most conveniently used to reveal the actual colour of the contaminants/corrosion deposits or any discolouration for the easy recognition of facts which ultimately relates to the failure. It is sometimes necessary during the preliminary investigation to use the low powered stereomicroscope to magnify evidences for better clarification and understanding. From the visual as well as the close studies, it is possible to ascertain from the fracture surface the mode of failure, i.e. whether ductile, brittle, fatigue, etc. Before attempting to clean the area of the fracture, any relevant deposit in the close vicinity of the fracture needs to be carefully analysed. This is essential for correct interpretation of fracture impressions. Further, it is important to see that while extracting such deposits or during the cleaning process undue damage is not caused to make interpretation difficult or misleading.

Different procedures are adopted for acid solution cleaning of the fracture surfaces depending on the material composition, surface protections, etc. Usually lossely adhering debris is removed using a dry air blast.

Normally camel hair brushes are used to apply suitable solvents for cleaning the surface. Where the deposits are strongly adherent ultrasonic cleaning is resorted to. In the case of failures or accidents involving fire, where the fracture surface gets oxidised, solvents containing special inhibitors are recommended for cleaning purposes in order to reveal better fractographic features. A list of solvents commonly used for cleaning the fracture surface of different materials is given below.

For aluminium and its alloys:

1. Sulphuric acid (conc)	: 15% (by volume)
Chromic acid	: 5% (by volume)
Water	: Remainder
2. Sulphuric acid (conc)	: 15% (by volume)
Sodium dichromate	: 7% (by volume)
Water	: Remainder
3. *Alkaline etch*	
Sodium hydroxide	: 10% solution
(followed by 1 : 1 HNO_3 Dip)	
4. *Corroded parts:*	
Chromic acid	: 1 gm
Phosphoric acid (conc)	: 1 ml
Water	: 100 ml
(at boiling condition)	

For Magnesium and its alloys:

1. Nitric acid	: 5 ml
Water	: 95 ml

For Titanium and its alloys:

1. Nitric acid (conc)	: 47%
Hydrochloric acid (40%)	: 2 %
Water	: Remainder

For Nickel and its alloys :

1. Hydrofluoric acid (40%)	: 125 ml/l
Ferric sulphate (anhydrous)	: 200 gms/l
Water (at 70°C)	: Remainder

For steels (low alloy)

1. Hydrochloric acid (conc)	: 50 ml
Water	: 50 ml
(Desmut by vigorous scrubbing with fibre brush)	

2. Stainless steel

Ferric sulphate (40% solution)	: 4 parts
Hydrofluoric acid	: 1 part
Water (at 150 °C)	: 15 parts

Beside these conventional procedures, occasionally the replica technique is adopted.

After proper cleaning of the samples, the critical areas are carefully examined under bright light to reveal the sources of origin of the failure. Apart from the manner of loading, critical scrutiny reveals the level of the applied stress, the mechanism involved and the relative ductility or brittleness of the material. It also reveals other aspects such as the presence of hardened cases, apparent grain size and internal imperfections. Visual examination also involves detecting the presence of design-induced stress concentrators, imperfections due to material fabrication or machining deficiency which may be the ultimate source for the failure. However, certain types of physical defects such as grinding burns/cracks, segregations, brushes, etc., do not permit themselves to be easily recognised in the as-is condition. These defects need to be highlighted either by macroetching the surface of the failed components or by special print techniques such as sulphur, oxide, lead and phosphorus prints for revealing the extent of distribution of these elements. Application of these techniques during visual examination helps the investigator to assess the risks involved in the failure due to the presence of such hidden imperfections. However, there are many failures which do not involve fracture but are caused due to wear and corrosion. Careful examination with the aid of a low power stereoscopic microscope is helpful in the initial stages for determining the type, nature and extent of wear or corrosion defect. However these have to be confirmed by microscopical tests.

In short, visual examination is required to bring out preliminary but definite information details in the following order.

1. Visually survey the entire component and if necessary related assembly parts to obtain an overall understanding of the function of the component. If required relevant details are photographed.
2. Classify the failure from a macroscopic point of view as ductile, brittle, fatigue, etc.
3. Determine the sources of origin of the failure by tracing back to its starting point.
4. Based on the fracture features, estimate the manner of loading (tension, compression, bending, torsion, etc), the relative stress level and the stress orientation.
5. Ascertain any abnormal features (assembly, mismatch, distortion, bend, etc.).

A variety of NDT methods such as radiography, ultrasonics, magnetic particle inspection, dye penetrant, anodising and etching, and eddy current are conventionally used to determine the manufacturing defects in the failed component, depending on the method of fabrication and type of material. Each of these methods constitutes only a particular domain of testing but in various combinations they are regarded as a powerful system of testing and evaluation of defects. They are primarily employed to detect flaws in the material present at or near the surface as well as in the interior. Since failures are often found to have originated from pre-existing flaws created during service by the action of stress and environment, NDT technique serves as an important tool to reach the root cause without destroying the valuable fracture features. The conventional NDT techniques are categorised broadly into two groups:

(i) those detecting surface and sub-surface flaws, and
(ii) those detecting internal flaws.

The following section gives a brief summary of each of the techniques[17,18,19] along with their areas of effectiveness and limitations.

Liquid Penetrant Inspection

This method is applied to detect discontinuities along the surface in solid and essentially non-porous materials. In this process, liquid penetrants seep into various minute surface openings by capillary action. Indication of flaws is moreover, independent of size, configuration, internal structure or chemical composition of the material. The resolution of the technique depends on the wetting characteristics of the liquid penetrant on the surface and can be as low as 0.1 micron. The basic process involves five steps: (i) surface preparation, (ii) penetration, (iii) removal of excess penetrant, (iv) development, and (v) inspection. Three methods are adopted depending on the type of penetrant.

Dye Penetrant

The technique employs oil or water based penetrating liquid (Ardrox 996) which is applied over the surface of the component under study either by immersion or by means of a brush. Since the dye has low surface tension it easily penetrates into very thin and light cracks. After sufficient soaking, the excess penetrant is removed from the surface and the component is cleaned with a suitable solvent (trichloroethylene or water). A thin layer of developer is applied uniformly over the surface. This absorbs the penetrant from small openings and spreads it over a large area as an indication of defects.

Fluorescent Penetrant

This technique is similar to dye penetrant except for the use of a fluorescent fluid instead of a colour dye and examination under ultraviolet light in a darkened enclosure. The resolution of this technique is superior to that of the dye penetrant. High resolution penetrants under the trade name "Wink Penetrants" like ZL30, ZL60, P133, etc. are available.

Hot Oil Chalk Test

This involves immersing the component in a neutral paraffin base oil at 100 to 140 °C for sufficient time so as to attain completely the temperature of the oil. Due to thermal heat, the component expands and the paraffin penetrates into the defect. Then, the excess oil on the surface is washed with a suitable solvent and a powder developer is sprayed. As the component cools, the paraffin that was entrapped in the defect rises to the surface and wets the chalk thus indicating the location of the defects. This process is generally used for large sized castings of aluminium, magnesium and other light alloys.

For the purpose of interpretation of results, the following general guidelines are given to identify defects in castings.

1. Scattered/ cluster of dots generally indicating porosity, blisters and pits.
2. Solid red line generally indicating cracks, cold shuts and misruns. Depth is indicated by the degree of spread of indication.
3. Dotted red line indicating generally light cracks, cold shuts and misruns.
4. Large blobs generally indicating voids of any type.

The main areas of effective application of dye penetrant inspection are given below.

Product/Process	Effective detection of
All types of materials, castings, weldments, assemblies, ground and machined components.	All discontinuities open to the surface, e.g., cracks, hot tear, grinding cracks, heat-treat cracks, porosity, shrinkage, sponginess, cold shut, etc.

The main advantage of penetrant inspection is its economy and ease of application. The main limitation is that this method is applicable only to detect defects that are open to the surface. Further, it is difficult to evaluate highly porous components.

Anodising and Etching

Anodising is a process of electrolytic conversion of an insulated oxide film on the surface of metals. Chromic acid anodising, in addition to its extensive use for protection against corrosion of aluminium alloys is successfully used for detection of surface discontinuities. It serves as a complimentary method to dye penetrant inspection during the manufacturing process and service failure.

The process involves vapour degreasing, hot alkaline degreasing, cold water swilling and anodising followed by cold water swilling and drying using an air jet. The components are then allowed to rest for a minimun of two hours to allow chromic acid to seep out from the surface cracks. Inspection is carried out under bright light and cracks show up as yellow or brown stains with a distinct darker line defining the actual geometry of the flaw. Since an anodised surface reveals a clear indication of grain pattern on forgings, fine demarcation of folds will show up distinctly as dark lines of different intensity without any stain.

The areas of effective application of anodising inspection are given below:

Product/Process	Effective detection of following faults
Aluminium alloy forgings, stampings, wrought products and castings.	Folds, laps, cracks, hot tears.

The advantage of the anodising method of inspection is that it is superior to the conventional dye penetrant inspection for detection of tight cracks in aluminium alloy forgings that have been camouflaged by reworking and oxide folds. The limitation of this method is that it is slow compared to the dye penetrant inspection and gives a low contrast of revealed defects against the general grey background of the surface. Further, the method is applicable only to certain alloys which can be anodised.

Etching Method

It is a process of removal of a thin metallic layer from metal surface by careful control of chemical reactions. Etching, as a method of inspection, is employed to detect surface defects in aluminium alloys and steels.

The process of etching involves precleaning of the components by descaling, dry/wet blasting, stress relieving, degreasing in trichloroethylene bath and cold rinsing. The components are then subjected to an alkaline etching bath in case of aluminium alloys and to an acid etching bath in case of steels. The process then involves cold water swilling, neutralising and drying and inspection is done under bright light.

The areas of application of etching are given below:

Product/Process	Effective detection of following faults
Aluminium alloys and steel forgings machined and ground components.	Cracks, folds, gross inclusion, segregation, grinding abuse open to the surface.

The main advantage of the etching method of inspection is its simplicity and economy and its ability to reveal surface defects quickly. Care must be taken to limit the etchings to avoid too much loss of material as it may affect dimensions as well as erase any superficial surface defect.

Magnetic Particle Inspection

This technique is used for ferromagnetic materials to detect surface and sub-surface flaws. In this method, the component is magnetised by using a direct current or a half wave rectified alternating current. In any homogenous ferromagnetic material, a magnetic field forms an uninterrupted circuit from pole to pole through the material. However, in the presence of a discontinuity at or near the surface, the magnetic field is deflected and forms a leakage field. Detection of this leakage field forms the basis of this technique.

The component intended for magnetic particle testing is cleaned, magnetised in a magnetic field and finely divided particles of magnetic oxide of iron (Fe_3O_4) suspended in kerosene oil or dry powder of Fe_3O_4 are applied on it. Surface or sub-surface discontinuities if any, attract the magnetic particles and they are held by the leakage field and form an indication of the defect. The fields produced by dc generally penetrate deeper into the part and help in detecting sub-surface discontinuities. On the other hand, fields produced by ac tend to flow at or near the surface of the metal and thus help efficient indication of discontinuties.

Dry powder is generally used on large components with a rough surface finish, whereas the magnetic particle suspension in kerosene is used on almost all types of components. Fluorescent ink is used on components with a fine surface finish and this is more efficient in defect detection.

The current required for magnetising a given component is estimated by using the formula

$$I = \frac{45{,}000\, D}{LN}$$

where, I is the current in amperes, D is the diameter of the component, L is the length and N the number of turns in the magnetic coil.

The areas of effective application of magnetic particle inspection are as follows:

Product/Process	Effective detection of
All ferromagnetic materials—castings, weldment, forging, assemblies, ground and machined components.	Surface/sub-surface cracks, grinding cracks quenching cracks, stringer type non-metallic inclusions, porosity, laps and folds.

The main advantage of this method is that it is reliable for detecting surface cracks, especially, very fine and shallow ones and there is no limitation as to the size or shape of the part being tested. Moreover, this method is effective through thin coatings of paint and other non-magnetic covering such as plating and is simple and quick.

The limitations of this method are that applicability is confined only to ferromagnetic materials and it is not always possible to locate discontinuities which lie deep below the surface.

Radiography

It is based on the principle that penetrating radiations are differentially absorbed by the material through which they pass. The absorption is primarily dependent on the energy spectrum of the radiation and the density and thickness of the material. The radiation commonly used in radiography are: (i) X-rays, (ii) Gamma rays, and (iii) Neutrons.

Modern X-ray machines operate with tungsten, gold or platinum at 50 kVp. to 400 kVp. The sharpness of the image is determined by the size of the target, the ratio of the source-specimen and film-specimen distances. Screens of aluminium, copper and lead are used to improve the definition. Radiographic sensitivity is measured in terms of the minimum percentage of thickness of the component that corresponds to the observed change in the photographic density of the radiograph. The size of the flaw is detected by using a penetrometer.

Radiography by X-ray[18] requires a predetermined intensity of the X-ray beam from an X-ray machine to pass through the component under examination. The beam which emerges out of the component, after undergoing differential absorption, is allowed to fall on an X-ray film. The film is enclosed in a light-tight cassette positioned closest to the component. The information contained in the emergent beam is recorded as a latent image on the X-ray film. The latent image is converted into a visible image by chemical treatment of the film with a developer. Thus the information is permanently recorded. Wherever permanent recording is not required, X-ray fluoroscopy can be used for quick disposition of parts.

The main areas of effective applications are given below.

Product/Process	Effective detection of
Cast ingots, shaped castings, weldments.	Internal discontinuities, e.g. shrinkage, hot tears, cold shut, sponginess, cavities, porosity, inclusions, segregation, cracks (which are not very tight), misruns.
Assemblies	Corrosion pits, misalignments, cracks (not very tight) and entrapment of dense materials.

The greatest advantage of radiography is that a permanent record of examined components can be preserved for future reference. However, it is essentially a slow and expensive method of inspection. Another major limitation is that if the differential absorption is below a certain limit which cannot produce a readable density on the film, the information will not record on the film. Precisely for this reason, planer defects such as tight cracks, folds, laminations and debonds are not detected by radiography.

Radiography by gamma rays is similar to X-ray radiography, except that radiation is obtained by the disintegration of atomic nuclei within radioactive matter. Gamma-ray equipments consist of a gamma-ray source such as Ir 192, Co 60, placed in an adequately shielded container. Selection of a particular isotope for gamma radiography depends on the nature and thickness of the material. The properties to be considered while selecting the isotope are gamma-ray spectrum, half-life, specific emission, curie strength and its cost. Ir 192 is capable of penetrating 35 mm of steel and Co 60 is capable of penetrating 150 mm of steel or its equivalent.

Gamma-rays are not recommended for radiography of aeronautical components except in cases where the component thickness is beyond the range of X-ray penetration, and where poor flaw sensitivity is acceptable.

The advantages of gamma radiography are (i) unlike X-rays no electrical power or cooling system is required; (ii) deeper penetration because of higher energies compared to X-rays; (iii) portable; and (iv) cheaper compared to X-ray equipment.

The disadvantages are (i) the energy of gamma radiation cannot be controlled at will; (ii) sensitivity is comparatively poor; (iii) longer exposure time; (iv) frequent renewal of isotopes in cases of shorter half-lives and (v) special safety precautions are required.

Ultrasonic Inspection

This method of testing[19] consists of transmitting energy in the form of high frequency sound waves, both longitudinal and transverse (frequency 20,000 Hz) into the material under test. Ultrasonic energy is introduced into the specimen by means of a probe equipped with a transmitter and a receiver or by means of a separate transmitter and receiver through a coupling media such as water, glycerine, mineral oil or grease. The probe generates ultrasonic waves when an electrical field is applied to piezoelectric crystals such as quartz, barium titanate, etc. For most of the materials, the testing frequency of ultrasonic waves ranges between 1 and 15 MHz. If the beam is interrupted by a discontinuity such as crack, void etc., the energy is reflected back. This reflected energy is picked up.

There are three types[20] of information display from the pulse echo inspection which is a widely used system in industry.

1. A-Scan, which is a quantitative display of intensity and elapsed data obtained at a single point on the test piece.
2. B-Scan, which is a quantitative display of elapsed time data obtained along a line on the surface of the test piece.
3. C-Scan, which is a semi-quantitative display of echo intensity obtained over an area of the test piece.

A comparison of the three scans is shown in Fig. 5.1.

The first step in ultrasonic testing is to calibrate the equipment to ensure linearity of indications, required level of sensitivity, resolution and probe characteristics. This is done with the help of the calibration blocks. After calibration the next step is to apply a suitable couplant on the cleaned surface of the component and move the probe on the surface in a zig-zag manner, and to record the readings obtained on the cathode ray screen of the equipment.

The main areas of effective applications of ultrasonic testing are given below:

Product/Process	Effective detection of
Billets, forgings, extrusions, plates, sheets, bars, rods, weldings.	Detection of internal discontinuities such as cracks, bursts, laminations, inclusions, voids, debonding.
Assemblies	Stress corrosion cracking, corrosion pits and fatigue cracks.

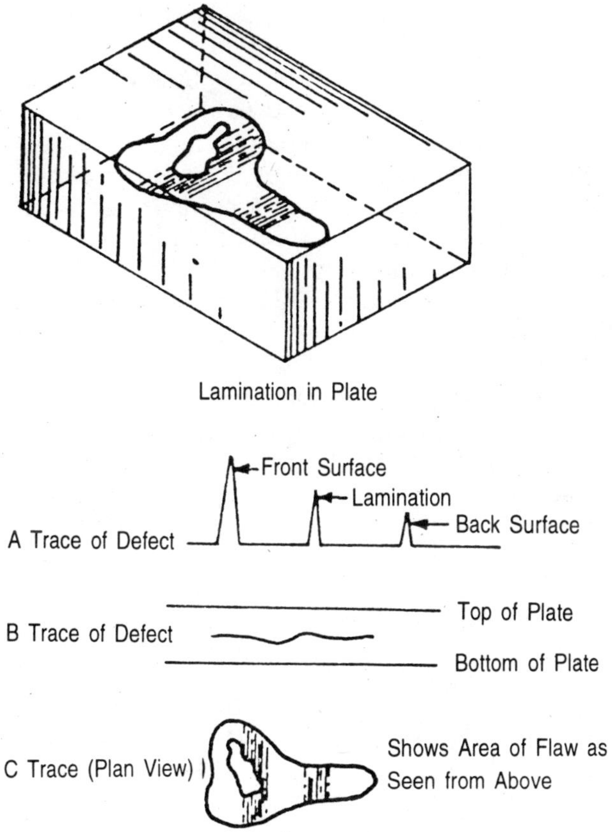

Fig. 5.1 Comparison of A, B and C-scan techniques[20]

The main advantages of ultrasonic testing is the ability to get instantaneous results and a fast inspection rate. Its main limitations are complete dependence upon the competence and integrity of the operator to interpret the result, the difficulty in discriminating between different types of defects and the difficulty in detecting defects near the surface owing to the dead zone of the probe.

Eddy Current Inspection

This is based on the principle of electromagnetic induction. An alternating electric current flowing in a coil causes an alternating magnetic field of the same frequency in the vicinity of the coil. This alternating magnetic field induces eddy currents in an electrically conducting material located near the coil. The eddy currents thus generated, produce their own alternating magnetic field which opposes the existing alternating magnetic field of the coil. This causes a change in the impedance of the coil. This change is related to a variation in physical properties and the presence of surface and sub-surface discontinuities in the material.

The eddy current test system consists of a source of alternating current of known frequency applied to a test coil assembly and an electronic detecting system to sense and indicate variations in the output of the coil assembly. These variations are caused by changes in the electromagnetic field of the coil due to discontinuities, and also due to a change in physical properties effected by variation in composition and treatment. The test system may also include a phase discriminator, filter circuit and magnetic saturation devices as required for a particular application.

Eddy current methods are normally used for the detection of surface and sub-surface defects only. This method is complementary to ultrasonic testing which is used for detecting defects close to the surface. It is also complementary to the dye penetrant inspection which cannot reveal sub-surface defects. The equipment is calibrated with appropriate standard/s depending on the purpose of inspection.

The areas of effective application of eddy current inspection are as follows:

Product/ Process	Effective detection of
All conducting materials	Surface/sub-surface cracks, laps, seams, voids and inclusions. Dimensional measurements such as plating/coating thickness. Determination of physical properties such as conductivity and hardness. Sorting of metal mix-up.

The advantages of eddy current inspection are its versatility, high speed of inspection, non-requirement of couplant between the test coil and specimen, and ability to detect cracks filled with non-metallic contaminants.

This method cannot be used on non-conducting materials. Local variations in conductivity and permeability of an acceptable nature may interfere with accurate crack detection. The measurement of metal coating thickness is difficult unless a substantial difference in conductivity exists between the coating and the substrate.

Other NDT Techniques

Acoustic Emission

This is a phenomenon arising from the energy released when a solid material undergoes plastic deformation. Part of this energy is converted into elastic waves which propagate through the material and can be detected at its surface using high sensitivity sensors. The most important industrial applications are:

1. Monitoring of flaw during hydrostatic proof test of pressure containers.

2. Continuous surveillance of known or suspected flaw areas in an operating pressure system.
3. Formation of crack during weld cooling.
4. Growth of fatigue cracks under various loading conditions.

Holography

It is a photographic process which allows the formation of a three-dimensional record of an object. When an object is illuminated, it scatters and reflects light. This electromagnetic disturbance carrying information about the object is propagated in space. The information consists of (i) the amplitude of disturbance; and (ii) the phase of disturbance. Holography records the amplitude and phase of the total wave leaving the object surface and this can be reproduced subsequently. Thus the optical effect of the object can be reproduced. Acoustic holographic records are produced similarly. Applications of holography include:

1. Displacement measurement to study stress distribution.
2. Crack detection on surface of machined components.
3. Growth of cracks due to fatigue/stress corrosion.
4. Study of composites and laminates.
5. Pneumatic type inspection.
6. Vibration analysis.

Computed Tomography

This is a relatively new X-ray method that retrieves information normally lost in standard X-ray imaging. Since no single conventional testing method provides a total solution to the industry's requirements for non-destructive evaluation in terms of dimensioning internal structure, mapping density distribution, or precisely locating, orienting and sizing flaws, only X-ray computed tomography is considered to have wide-scale application.

In X-ray computed tomography, a collimated X-ray beam passing through the test object is measured by an array of detectors located on the other side of the object. The object is rotated until measurements are made from 180°. Measured projections are manipulated by a computer to produce a two dimensional map of the X-ray attenuation coefficients in the irradiated cross-section.

The method has several advantages over conventional radiography. The most important is its superior large area and low contrast image resolution which is a measure of the ability to detect differences in density. Due to the increased contrast resolution and improved 3-D spatial resolution CT is an ideal NDE technique to detect, locate and size details that may be planer or volumetric.

In addition, images are produced in a quantitative, ready-to-use digital format. They provide detailed physical information such as size, density and composition to aid in evaluating defects. Methods are being developed to use this information to predict failure modes or system performance under varied operating loads.

5.2 METALLOGRAPHIC TECHNIQUES

Microscopic examination constitutes an important aspect of failure analysis. Certain valuable information regarding the origin of the crack or fracture, which can only be gathered by proper analysis of surface or sub-surface related phenomenon through microscopic examination, is essential for correlating the various factors leading to failure. In other words, it aids the emergence of a correct picture of the failure by accumulating corroborative evidence during microscopic examination. It becomes more important when the component failure assumes criticality in an accident. Moreover, apart from his dependence on the established techniques discussed below, the analyst is required not only to identify the critical failure but also to pinpoint its cause and to reconstruct the sequence of events. However, it is needless to mention that the correct choice of specimens, locations and their preparation during microscopic examination assumes greater importance in the event of a catastrophic failure.

In this task of examination, the analyst is assisted by certain tools of investigations which help him to work at the fracture surfaces more carefully and at a greater depth. This chapter describes the process of microscopic examination with the specific aim of answering questions like:

- What goes on?
- How to go about it?
- What instruments to use?

Specific examples have been included in the description for a better understanding of the concepts.

Macroscopic Examination

Fractographic studies involve general evaluation of the fracture surface including any deterioration adjoining the fracture zone that the component is likely to have suffered during service prior to failure. This process of segregation of possible contributing factors will enable the investigator to study the observations in proper perspective. A general observation includes a careful study of the following aspects.[21]

1. The type of fracture and the presence of any secondary cracks.
2. The origin of the fracture.
3. The presence of external deposits or discolouration on the fracture surface.
4. The presence of any evidence to show plastic deformation at the fracture zone.
5. The presence of any wear marks in the vicinity of the fracture or elsewhere, indicative of fouling during service.
6. The presence of any dimensional changes or distortions in the component.
7. The presence of any corrosion surrounding the fracture.
8. The presence of any mismatch in the component assembly.

It needs to be kept in mind that mere analysis of the mode of failure is not the answer to the cause of the failure. It is necessary to go beyond fracture identification and find out why it happened. Only then will it be possible to suggest corrective measures for prevention of a given failure. Hence, for remedial action, an overall understanding of all aspects of the problem is a prerequisite. The first step adopted is the use of an optical stereomicroscope. This is a common tool readily available for macroscopic examination of fracture characteristics at very low magnification. Reasonably large specimens can be handled and the surface needs no preparation. The aim of such a study is to get a general idea of various aspects of the fracture surface and the origin of the fracture before further examination. In many cases the origin is readily seen with the naked eye or with the help of a hand lens. Figs 5.2 and 5.3 give examples of macro-photographs of fracture surfaces. In Fig. 5.2, the fracture origin is located with the help of radial markings. The radial markings are the orthogonals of the fracture front as it travels, thus identifying the origin of the fracture. Fig. 5.3 is a typical fatigue failure identified by the presence of beach marks that appear like contour lines. In this case, the origin is easily recognised by virtue of convergence of shell markings either to single or multi-point location. In general, when the bulk of the material fails because of a flaw, it becomes the origin of failure (either inherent or developed during service).

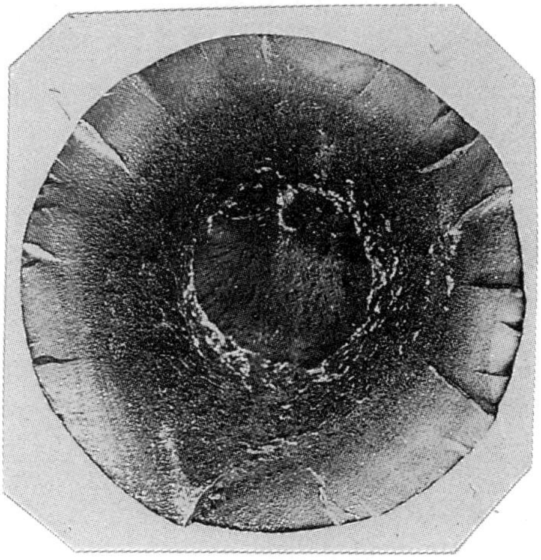

Fig. 5.2 Macrograph of a brittle fracture surface. Radial markings indicate (multiple) origins of fracture at the outer periphery

Fig. 5.3 Macrograph of a typical fatigue fracture surface with beach marks originating from multiple points, indicated by arrows

A pure ductile fracture of a component will not have an origin like that of a fatigue fracture. Once the origin is located, it is studied in greater detail at higher magnifications to derive further information for complete analysis of fracture mode, magnitude and nature of stress etc. This also helps in assessing the following:[21]

1. Length of an existing crack.
2. Presence of inclusions at the origin.
3. Presence of beach marks and determination of their spacing.
4. Length of a fatigue crack before it became critical.
5. Presence of any corrosion products and other important evidences directly related to the fracture.

Metallography

Optical Microscopy

Use of the optical microscope is one step further in the analysis of the problem. Micro examination reveals information about the structural quality of the material used in the failed component. The microstructure can yield information regarding the cleanliness, heat treatment and even chemistry of the material, besides the condition in the sub-surface structure, crack path, deformation characteristics, grain-flow contour, grain size, etc. Therefore, the role and techniques for preparation of metallographic specimens to obtain accurate information, are briefly discussed as follows.

Selection of Specimens

A careful choice of specimen in terms of location and direction is a prerequisite to extract not only the maximum amount of information about the defect, but also to detect the exact cause of the failure mechanism from the structure point of view.

Quite often a single microstructure may be enough to satisfy the investigator but sometimes more sections may be necessary depending on the complexity of the problem. Since it might be necessary to section a component before a metallographic specimen is prepared, great care has to be exercised so that valuable evidence is not lost or undue heat is not generated during sectioning causing lowering of material property. When necessary, the fracture surface should be preserved by an anti-corrosive coating. Transverse sections of cracks reveal crack paths easily and are useful in identifying sub-surface defects. In all cases of microstructural observation, the surface under study has to be prepared properly.

Specimen Preparation for Optical Metallography

Polishing The preparation consists of mechanically (and/or electrolytically) polishing the surface under consideration. Larger specimens can be handled as such, provided retaining the edge is not mandatory, and smaller ones are embedded in a resin mount. The surface is then polished successively with finer particles of emery and finally with diamond or alumina paste. This preparation is to be done in a systematic manner to remove the original coarse scratches from the surface a stepwise manner and bring about a minor finish without causing undue deformation on the surface. Automatic polisher takes the human element out of this partially and polish the surface free of mechanically induced deformation. A final electrolytic polishing is resorted to in cases where the material is very soft or when some special features are to be relatively observed.

Etching

Etching is a chemical process by which the microstructure is revealed. This process consists of preferential attack of the etchant on the metallographically polished surface. The preferential attack is brought about by differences in composition, orientation of grains and other local variations in the specimen. This process reveals the presence of grain boundaries and the different phases in the structure. However, in case of identification and evaluation of non-metallic inclusions, it is beneficial to observe the structure without etching. The process of etching can be selected in such a way as to achieve a special effect on phase characterisation.

Optical Microscope

The optical microscope is a simple instrument which enables observation of flat surfaces at reasonable magnification of 1000×. It consists of an objective lens and an eye-piece. The two-stage magnification produces an image which can be projected on a screen. Unlike the biological microscope, this is a reflected light microscope microscope with the objective and the illumination system on the same side of the object. The contrast that is obtained is mainly from the differences in the reflectivities from various regions. Another technique that goes with the optical microscope is the use of polarised light[21] for viewing. This is very useful when studying materials that are optically anisotropic and specially for identification of non-metallic inclusions.

The main disadvantages of the microscope are its limited resolution power and very poor depth of focus. While it is very handy in studying the microstructure of flat sections it cannot be used to study a fracture surface.

Special Techniques

Surface Finish Another area where the optical microscope could be of help in identifying surface unevenness caused during machining. In order to evaluate the unevenness, the specimen is cut at an angle to accentuate the undulation. It is also good practice to give a coating on the surface before sectioning. This serves the dual purpose of preserving the surface and increasing the contrast by virtue of well defined projection of the surface which stands out better as an interface.

Flow Line Study of flow lines or grain flow pattern has assumed importance in the failure analysis programme. Usually flow lines refer to the direction along which the metal flows during metal formation. It is common knowledge that forging with an improper flow line pattern such as asymmetrical grain flow or kink is inferior in quality. These unfavourable grain flow patterns (Figs. 5.4), that exist in a critical part often lead to unexpected failure. The failure may be either due to reduced fatigue life or due to applied stress along the grain fibres. Therefore, it is good practice to look for the flow lines. This is accomplished by grinding the surface under study and etching deep. Design analysis reveals that discontinuities in the flow pattern constitute sources of weakness.

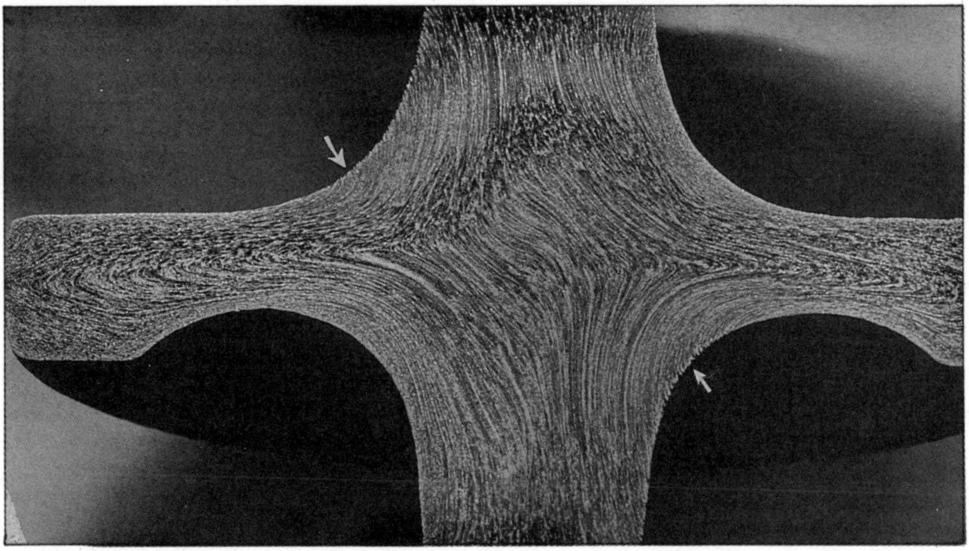

Fig. 5.4 Macrograph of an asymmetrical grain flow pattern in a highly stressed gear part. Note the flow lines terminating in the radius zone (indicated by arrow)—an unfavourable condition for reduced fatigue life (× 1)

Inclusion Identification of inclusions in the material used for critical components is an important prerequisite for quality conformance. The existence of inclusions in the highly loaded parts has quite often been identified as the source of catastrophic failures.

Inclusions are the extraneous particulate material present in metals and alloys. These may have clearly defined chemical composition and crystal structures such as simple oxides, carbides, sulphides and silicates, or they may be ill defined such as glassy slags, fluxes etc. In general, inclusions are detrimental to the property of the material. Since they have different properties compared to the matrix, they behave differently during subsequent fabrication operations and may get spread out during rolling. These inclusions[22] can also act as stress raisers in the matrix since they break the continuity of the matrix (Table 5.1).

Inclusions are best observed in the unetched condition after metallographic polishing. However, this will not differentiate the various kinds of inclusions if they exist in the combined form. In such a case preferential etching is adopted in order to reveal an inclusion of one kind. A quantitative estimate of the inclusion population is an important parameter from the acceptance limit point of view for aircraft quality steel. Sophisticated instruments like quantitative metallograph aids are used for determining the population of inclusions of different sizes. As the presence of inclusions is inevitable, standards have been prescribed based on their number and size. Depending on the criticality of the component cleaner materials are specified. Identification of inclusion for its chemical content (which is necessary for establishment of type and character of the inclusion, if primarily responsible for failure) can be done easily with the help of an electron probe or an extraction replica.

It is well established that cleanliness of material is vital for steel components used in critical applications, especially in dynamic machines. Inclusions do have an adverse effect on material property, particularly fatigue, toughness and corrosion. The characteristic size and shape of four major inclusions in steel are shown in Table 5.1, along with their effects.

Table 5.1 Characteristic shape and distribution of major inclusions and their effect[22]

Type of inclusion	Characteristic shape and distribution		Effect
Sulphide —MnS —FeS		At X100 (unetched) Shows dark grey or brownish yellow in streaks.	MnS is least harmful. It improves machinability and prevents ill effects of FeS. FeS causes brittleness due to red shortness and poor weldability.
Alumina		Dark grey clusters.	Causes pitting or spalling
Oxide	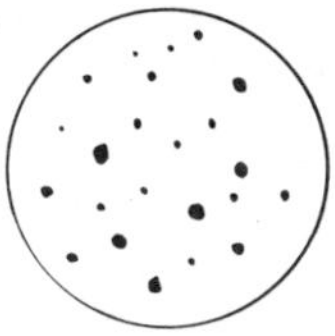	Medium grey or dark grey globular constituent.	Causes notch effect.
Silica	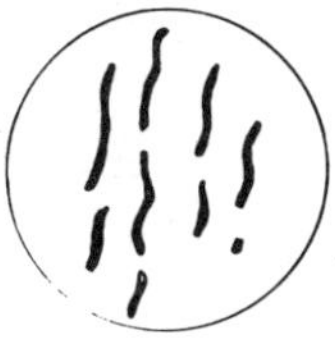	Slate grey in wavy streaks.	Causes notch effect.

Microfractography

In fracture analysis there are unique features that characterise a given type of failure. Based on this, it is possible to identify unambiguously the type of failure suffered by a service component by observing special features of the fracture. However, it is not always true as in the case where it involves two different types of failure mechanisms, resulting in an almost similar fracture appearance. Therefore, observation of surfaces in greater detail at higher magnifications is necessary before a failure is attributed to a given mechanism. For examination of fractures at higher magnification two widely used instruments are the transmission electron microscope (TEM) and the scanning electron microscope (SEM).

Transmission Electron Microscope[21]

The conventional light microscope has the following two major limiations when used in studying fracture surfaces:

- very limited depth of focus, and
- poor resolution.

The electron microscope, because of the small wavelength of the electron beam used for imaging (typically 0.05 Å, for 50 kV accelerating voltage) and the small value of beam divergence from the optic axis (typically 10^{-2} radians), has excellent resolution and a very good depth of field. The only disadvantage is the special technique for the specimen preparation that is necessary before a fracture surface can be observed under the microscope. Since it is not possible to put the specimen under the microscope, only the impression or the replica of the surface is used for the study. This has considerable advantages. Replicas can be made from fractured components at the site of the accident itself. Plastic replicas have very good dimensional stability and they offer a method of recording surface or metallographic evidences for possible study later. The method is very simple and does not destroy the surface. However, great care has to be taken to get an identical representation of the features on the fracture surface.

Replica Technique Replication is a way of transferring the surface features onto a replica which could be used for studies under a TEM. Two kinds of replication techniques are known, *viz,* single stage replica and double stage replica. The double stage replica technique is discussed here in detail (Fig. 5.5).

A cellulose acetate sheet (normally 75 μm thick) softened in acetone is pressed firmly on the fracture surface, which has been properly cleaned, till the sheet sets in. This is allowed to dry under dust free conditions. When dry, the sheet comes off carrying with it the negative impression of the fracture surface. This plastic replica is mounted in an evaporator and a thin layer of carbon (~ 500 Å) is uniformly deposited on the impression side of the sheet. It is also advantageous to shadow the surface with a heavy element like chromium, gold, etc., in order to obtain better contrast and accentuate shallow features. The evaporation is done in vacuum better than 10^{-5} torr. The coated replica is cut into small bits and the plastic backing is gradually dissolved in acetone, leaving behind a positive carbon (chromium shadowed) replica. This is fished onto a copper grid and used for examination.

Great care is necessary to make the replica an exact representation of the original surface. Several features not present on the original fracture surface may get introduced during the TEM observation.

TEM Observation The transmission electron microscope can be considered as a counterpart of an optical microscope in which a beam of electrons serves as the illumination

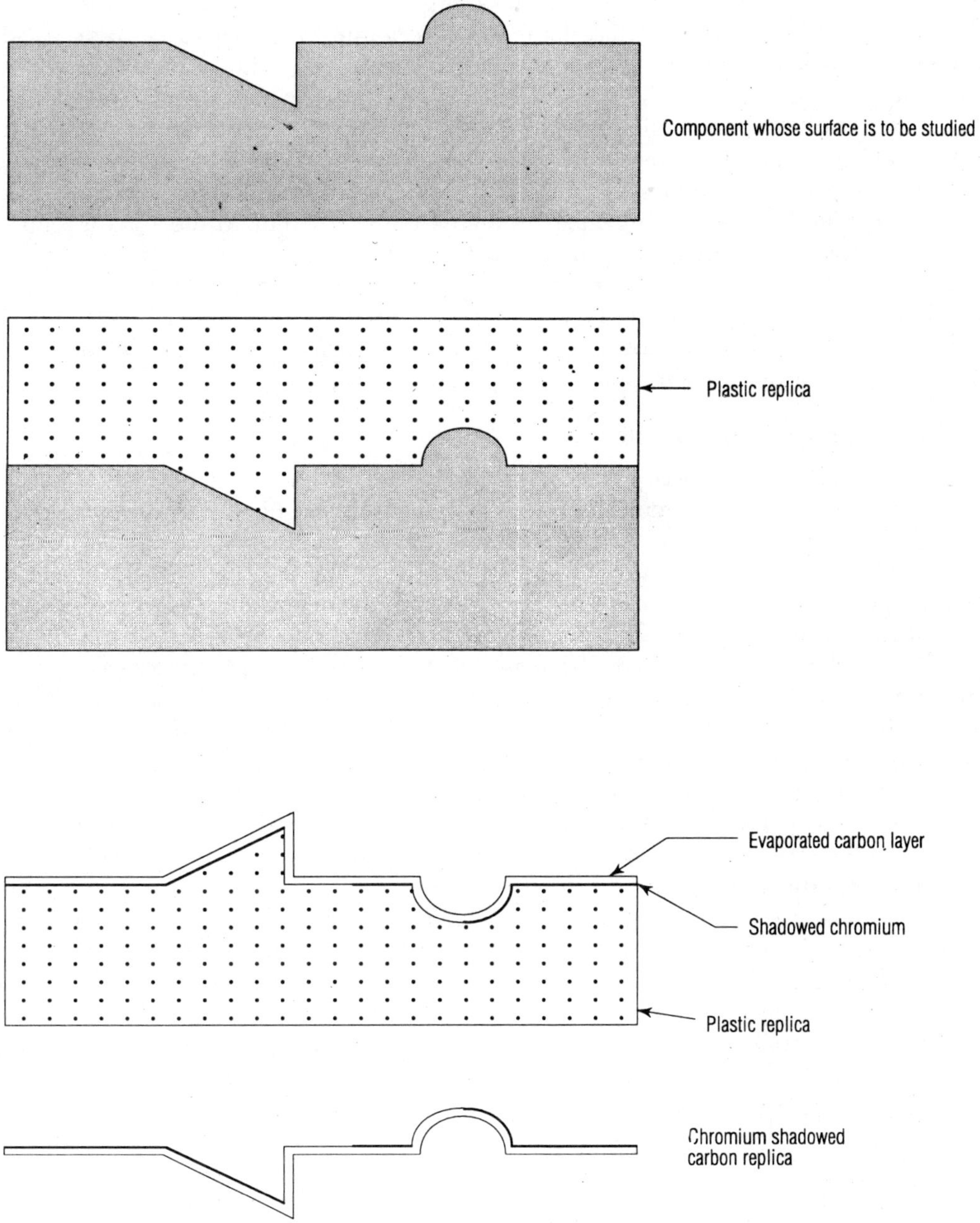

Fig. 5.5 Double stage replication technique for fracture surface observation[21]

system, electromagnetic lenses replace glass lenses and a fluorescent screen takes the place of a ground glass plate. It is customary to operate the microscope at an accelera ting voltage of ~50 kV for replica observations and employ minimum beam intensity to avoid damage to the replica. Since the replica is mounted on the copper grid, it should be borne in mind that part of the replica is hidden behind the copper and is not available for observation. Sometimes this could mask some important detail. Also, the replica can be tilted during viewing and it is possible to take stereographic pairs of photographs if one is interested in calculating measurements perpendicular to the fracture surface. Another easy way of height measurements is to incorporate spheres of known diameters in the replica before shadowing and measure the length of shadows in the image.

While it is relatively easy to replicate, it is far more difficult to associate once findings on the replica with the actual location on the fracture surface. Numbered grids come in handy and its use partially mitigates this problem.

Scanning Electron Microscope[21]

The scanning electron microscope has been of great help in the study of fracture surfaces. It allows a direct study of fracture surfaces of specimens small enough to be accommodated in the specimen chamber of the SEM.

The SEM is different from any microscope in that there is no imaging lens; and the actual image is built-up of a large number of picture points on the viewing cathode ray tube (CRT). The intensity of each picture point is governed by the strength of signal emitted from the specimen when the primary beam of electrons impinges on it. This beam is moved in a systematic manner, scanning a specific area of the specimen surface and so is the picture point in the CRT. Thus the image is built-up of a number of picture points each representing a corresponding point on the specimen surface. A variety of signals emanate from the specimen *viz.* secondary electrons, back scattered electrons, characteristic X-ray, etc. The most common mode of imaging is the secondary electron imaging, i.e. the strength of the secondary electron signal from each object point is used for modulating the intensity of the picture point in the CRT. The image contrast is obtained because of differences in the inclinations of the facets to the primary beam, thus resulting in the varying yield of secondary electrons for collection. Figure 5.6 shows a typical scanning electron micrograph of a hydrogen induced fracture surface. The striking three dimensional form of the fractograph is readily seen.

The great potential in analyzing the chemical contents of any area of the surface is capitalized in any failure investigation. The recently introduced energy dispersive spectrometer allows the identification of elements in the least time.

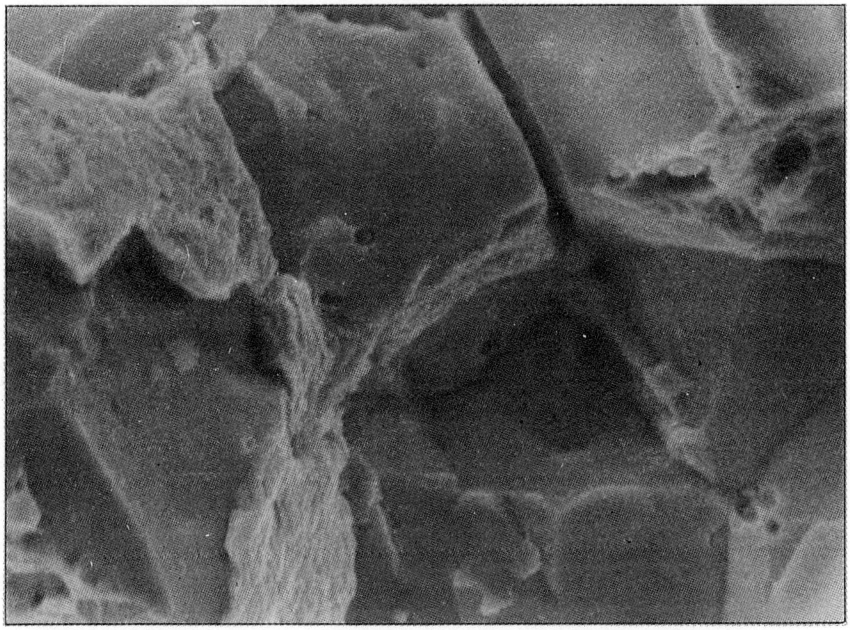

Fig. 5.6 A three dimensional feature of hydrogen induced fracture with intergranular facets, standing out conspicuously

Chapter 6

Concept and Mechanism of Failure

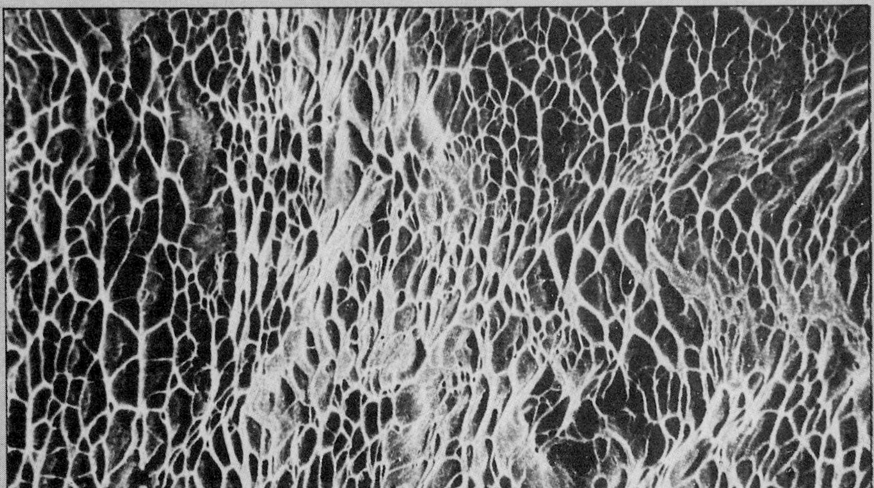

6 *Concept and Mechanism of Failure*

6.1 FATIGUE

The most common cause of damage to engineering components is fatigue, because of dynamic rather than static loading, particularly in aeronautical applications. It is implied that the maximum load or stress to cause fatigue failure is much less than the static breaking load. This type of fracture usually occurs in the components that are subjected to repeated applications of alternating stress.

Fatigue is thus an important and complex phenomenon as it accounts for 80% of all structural failures. It was first discovered in 1800 in the broken surface of a locomotive axle and a solid 125-year research has gone into it. Therefore, it is reasonable to ask why fatigue failures still occur with such regularity and in such a wide variety of industrial products. The predominant reason is that complex metal structures may contain hidden or unrecognised areas of stress concentration and defects. A majority of component fatigue failures originate at some form of stress raiser.

The tensile stress, whether applied or residual, is the major component of fatigue failure and is generally much lower than would be required for static fracture. After some critical number of stress cycles, a crack will form in the region of highest stress concentration. With continued cycling, the crack will grow in length and in a direction perpendicular to the applied tensile stress. The initiation and propagation of a fatigue failure is a very complicated process, both occurring as a plastic deformation. Unlike ductile fracture, fatigue cracking usually does not show any deformation or necking in the region of failure. Therefore, no prior indication of the onset of fatigue cracking can be detected visually. Normally fine cracks are difficult to be detected until they grow to macroscopic sizes leading to partial or complete rupture.

The basic method of presenting engineering fatigue data is usually in the form of S-N curves (i.e. fatigue curves with stress against the number of cycles to failure). Figure 6.1 illustrates a typical S-N curve from rotating beam tests.

It is seen that the S-N curve is concerned chiefly with fatigue failure at a higher number of cycles (i.e., $N > 10^5$ cycles). Under these conditions the stress on a gross scale is elastic. At higher stresses, the fatigue life is progressively decreased. For low cycle fatigue which is concerned with $N < 10^4$ cycles, tests are conducted with controlled cycles of elastic plus plastic strain instead of controlled stress cycles[23]. As strain amplitude increases, N cannot be sustained at higher values. Hence it drops.

Low cycle fatigue failure is taken into consideration in the design of nuclear pressure vessels, steam turbines and other types of power machinery, whereas high cycle fatigue is concerned with dynamic structures. Low cycle fatigue often occurs where repeated stresses are of thermal origin, due to thermal stresses produced by thermal expansion of the material. Hence, this type of fatigue is due to strain rather than stress cycling. The usual method of presenting low cycle fatigue data is by plotting the plastic strain versus number of cycles on a log-log scale in the form of a straight line. This relationship is known as Coffin Manson Law and is represented by the equation

$$\Delta\varepsilon p N^b = \text{Constant}.$$

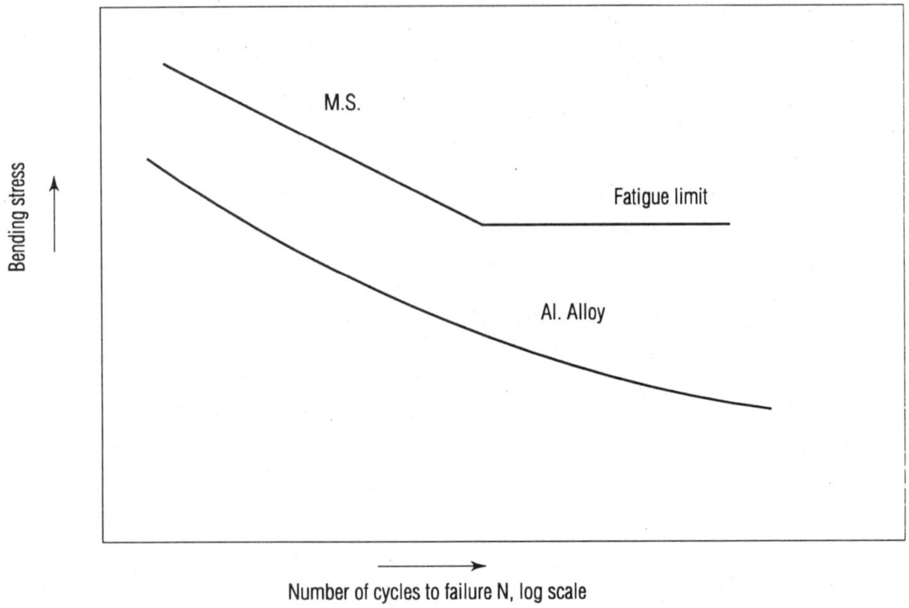

Fig. 6.1 Illustration of relationship between stress amplitude and life in cycles

However, due to the difficulties in extrapolating the results obtained from laboratory tests to full size structural components, fatigue tests are now-a-days conducted on actual full size components such as aircraft wing by subjecting these to simulated vibration.

Metallurgical Aspects of Fatigue Failure

Four Steps of Fatigue

A fatigue failure occurs in four steps, namely,

1. micro crack nucleation.
2. micro crack propagation (Stage I),
3. macro crack propagation (stage II), and
4. final rupture.

Role of Stress Raisers in Fatigue

A stress raiser is an irregularity from a geometrical point of view either in shape/section or surface causing a peak in stress at or near the irregularity and which is higher than the average stress in the surrounding areas. Stress raisers can be mechanical or metallurgical. Mechanical stress raisers are actual discontinuities or separation in product (*viz.* fold, inclusion, lamination, crack, etc.). Metallurgical stress raisers arise from the localised differences in properties such as soft and hard spots, decarburisation, grain flow variation, chemical segregation, banding, etc.

Source/Origin of Stress Raisers

Since a large part of fatigue failure analysis involves detection and identification of stress raisers at the fatigue initiation site, broadly four categories of sources are identified.

1. Resulting from design weaknesses or defects (i.e. holes, corners, key ways, threads).
2. Carried over from metal refining processes (i.e. pipe, inclusion, segregation).
3. Caused during the fabrication process (i.e. seams, laps, tears, cracks).
4. Induced as a result of operation/environment (i.e. corrosion, irradiation, foreign objects and damages).

Fatigue Crack under Alternating Loading

Microscopic and macroscopic aspects of fatigue crack nucleation and growth.

The microscopic aspect[23] of fatigue involves the phenomenon of crack nucleation and growth with the help of theories of micro-plasticity, whereas, the macroscopic aspect deals with the growth of a pre-existing macro-crack or a crack like defect with the help of macro-plasticity. A free surface normally constitutes the preferential area for the initiation of a fatigue crack. Alternating stresses produce slip bands. The slip deformation mechanism both in fatigue and fracture under monotonic deformation is common as far as occurrence on the same slip system is concerned. But they differ widely in their deformation. In monotonic stressing the slip is widespread covering all the grain, whereas in fatigue some grains will show slip lines without any evidence of slip in others. However, the type of surface deformation varies considerably in their orientation and surface distribution depending on the inhomogeneities of the underlying structure.

Figure 6.2 shows two types of surface deformation[24] namely, *extrusions* (material projects from the surface) and *intrusions* (where shallow channels are formed).

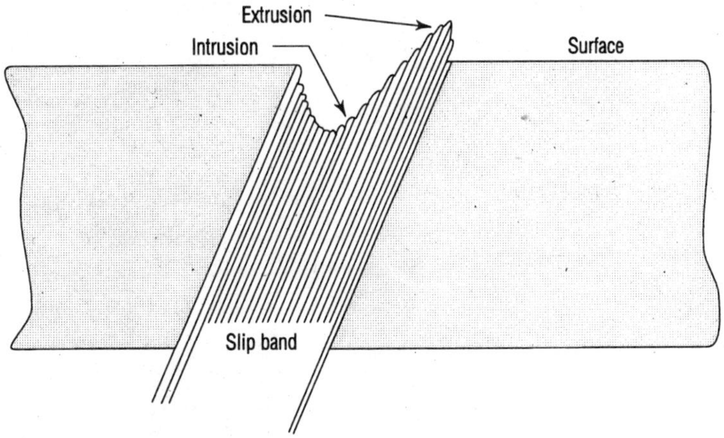

Fig. 6.2 Surface deformation processes of extrusions and intrusions in metallic materials subjected to alternating loading (for additional reading ref. J.C. Gross Kreutz, ASTM STP 495, 1971)

The slip lines are formed during the first few thousand cycles of stresses and the successive cycles produce additional slip bands. There are generally several slip bands which are more persistent than the rest and which will remain visible, while others are polished away. These persistent slip bands constitute the embryonic stage for fatigue cracking.

SEM micrographs of intrusions and extrusions have indicated that the process is associated with the formation of micro-notches and micro-cracks. Once formed, the fatigue crack tends to propagate at an angle of approximately 45° to the tensile stress (GrossKreutz identified the process[24] as stage I of the crack propagation). The surface where the latter takes a direction perpendicular to the maximum applied stress is known as the stage II of crack propagation (Fig. 6.3). The 45° crack is easily identified as the narrow initiation zone on the fracture surface after failure in most of the alloys, except in high strength nickel and cobalt based alloys[24] where large areas of fracture surface themselves consist of this type of crack. The rate of crack propagation in stage I is very low and is of the order of a few atom diameters per cycle compared to rates of a few microns per cycle in stage II.

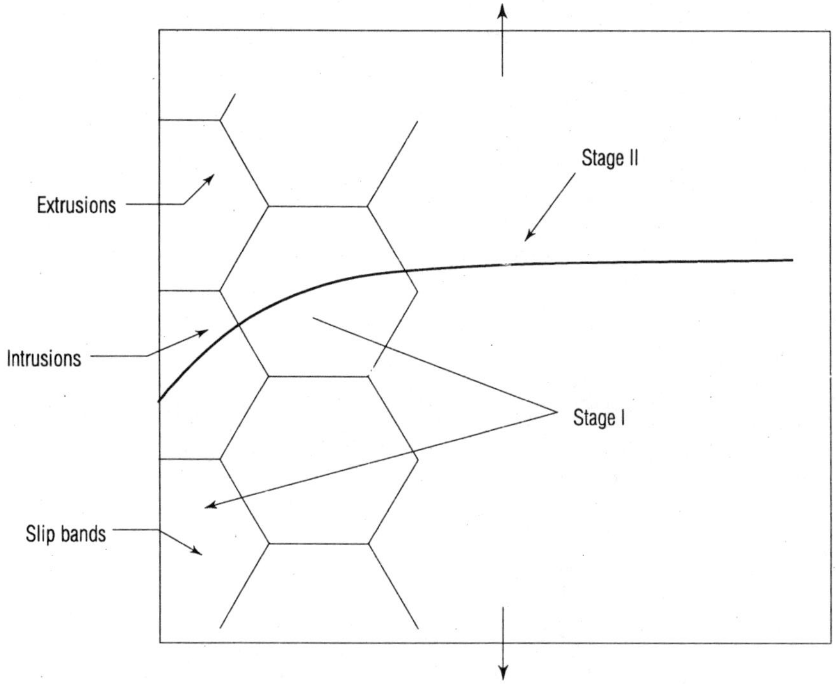

Fig. 6.3 Representing fatigue crack propagation in Stage I and Stage II (for additional reading ref. C. Laird, ASTM STP 415-1967)

An important microscopic feature which is unique to fatigue deformation is the formation of ridges and grooves on both the intrusion and extrusion slip bands. A mechanism of formation at intrusion and extrusion, as suggested by Cottrell and Hall is shown schematically in Fig. 6.4.

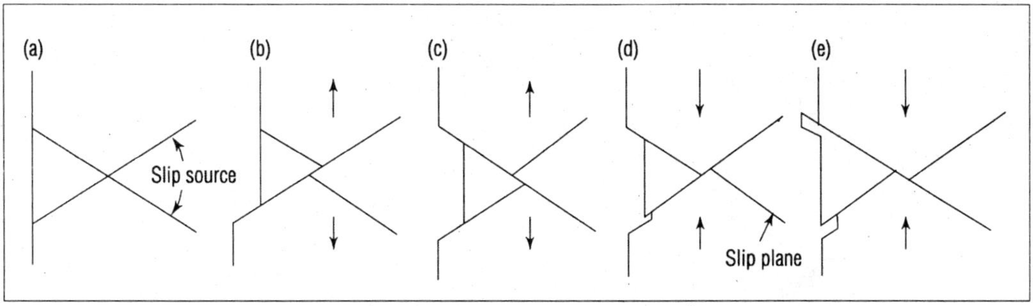

Fig. 6.4 Illustration for the mechanism of formation of slip band, extrusion and intrusion

During the tensile part of the stress cycle, both slip systems operate sequentially to produce surface stress. When the stress becomes compressive, intrusion forms by slip on the first system and extrusion forms when the other slip system operates.

It is known that a fatigue failure can also occur in a perfectly smooth polished specimen without any apparent presence of stress concentration. But the crack initiation process is considerably influenced by irregularities on the surface, such as, inclusions at the free surface, foreign particles at the sub-surface or cracks formed by shot peening, scratches and surface damage caused by fretting or other mechanical means.

Fatigue Crack Propagation

After crack initiation in Stage I, Stage II starts with the propagation of crack at right angles to the direction of tensile stress till fracture takes place due to overloading (Stage III of crack propagation). During Stage II, the crack advances a certain distance with each load cycle and in the process a series of striations are produced on the fracture surface[24]. The crack propagation in Stage II shows a characteristic propagation pattern at microscopic level and is indicated on the fatigue fracture surface. They reveal the extent of fatigue by the crack front from the transition to the final catastrophic fracture. The variations in the beach mark pattern (which are lying parallel to the fatigue striations) form due to changes in loading and/or in the crack tip environment.

Extensive research has been carried out to determine the laws of fatigue crack propagation for the Stage II growth. Reliable crack propagation relationship generally permits the implementation of a safe design philosophy, which predict that the presence of cracks or crack like defects in engineering structures is inevitable. It aims at determining the safe load and crack length which will preclude failure in a conventionally estimated service life. Among the various empirical relationships[23] the following equation is widely used.

$$da/dN = A\,(\Delta K)^P$$

where da/dN is the crack propagation rate, a is the crack length, A and P are constants, $K = y\sigma\sqrt{\pi a}$, where Y is the geometrical correction factor and ΔK is the stress intensity range. Figure 6.5 shows a schematic representation of relationship[23] between crack growth rate and ΔK for a structural alloy tested in non-aggressive environments.

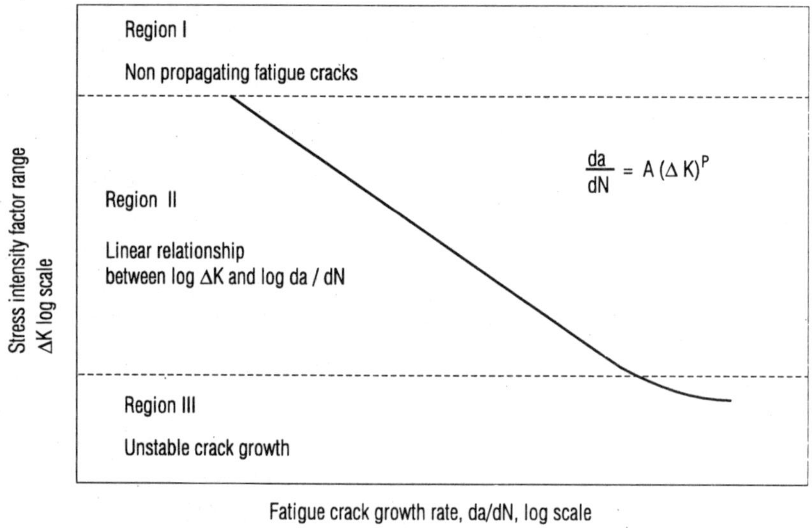

Fig. 6.5 Schematic illustration for fatigue crack growth behaviour

Fatigue cracks propagate commonly in a ductile manner. Each load cycle produ- ces an intensive plastic deformation around the crack tip and the cracking process takes place predominantly in a transcrystalline manner. In the case of polycrystalline microstructures, the crack front is sub-divided locally on to a number of separate planes[24] (Fig. 6.6), resulting in the formation of the microscopically visible structure of many parallel crack paths.

The profiles of the individual striations depend on the material and the loading condition. In absence of definite mode of formation of individual markings on the striation profile, crack propagation model as proposed by C. Laird[24] is generally acceptable.

According to the model, the formation of individual striation occurs as a result of successive opening and closing of the crack flanks. In the process, the crack advances during opening, whereas in closing, the plastically deformed material at the crack tip gets compressed. As a result a fold is produced which subsequently stands out in the form of a striation. The variation in striation profile depends on the intensity of loading and the ease of slip formation at the crack tip.

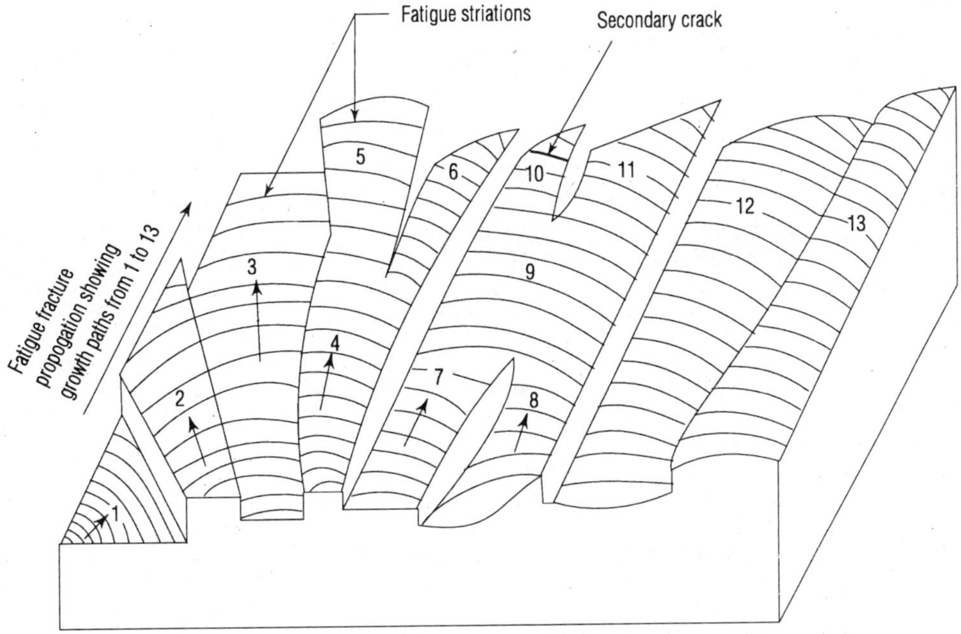

Fig. 6.6 Schematic representation of fatigue growth path in different planes (For additional reading ref. C.D. Beachem Trans ASM 60, Philadelphia, USA 1967 S 325)

6.2 STRESS CORROSION

The stress corrosion phenomenon occurs frequently in both aluminium and magnesium as well as in steels. This includes simultaneous processes involving the corrosion induced initiation and the growth of cracks in the presence of a tensile stress which exists above some critical level on the surface of the part. The mechanisms of crack growth involve the synergistic action of the corrosive environment and the sustained stress acting on an exposed surface and the failure mode is known as stress corrosion cracking (SCC). The simultaneous occurrence of three distinct elements influence the stress corrosion phenomenon. These are:

(i) the existence of a sustained surface tensile stress of sufficient magnitude,
(ii) the presence of an aggressive environment that can access the surface, and
(iii) a material which is susceptible to localised corrosion along a specific path.

Stress corrosion is a particularly insidious form of corrosion, since it does not involve any visual evidence of surface corrosion such as discolouration or the build up of recognisable corrosion products. Stress corrosion crack growth generally occurs in a slow, time dependent manner. Stress corrosion cracking is basically a localised form of corrosion. It affects either grain boundaries or specific crystallographic planes, resulting either in intergranular decohesion, transgranular cracking or in a combination of the two. The fracture surface of the transgranular cracking may show up brittle, cleavage—like markings as well as areas of dimpled features of ductile, fracture. The crack that follows the fracture which is independent of microstructural path, occurs approximately perpendicular to the direction of the tensile stress with extensive branching effect. The. combined stresses involved in stress corrosion cracking may be due to residual stress from manufacture (forging), fabrication (machining) or heat treatment, assembly stresses (interference fit, tightening, etc.) or due to loads imposed during flight or ground operation.

It has been experimentally observed that additional stresses are developed at the crack tip by the build up of corrosion products during the initiation of stress corrosion cracking. These stresses can sustain the crack growth beyond the region of influence of the stresses responsible for SCC. Besides, irrespective of effects by the corrosive nature of the environment and the stress, the alloy composition and its microstructure greatly control stress corrosion cracking phenomenon.

Initiation and Propagation of Stress Corrosion Cracks

In stress corrosion cracking the site of initiation is always sub-microscopic in nature and is influenced by local difference in alloy composition, concentration of corrodent and stress factor. Pre-existing damage or flaw such as a scratch, crack, etc. acts as an area for initiation of stress corrosion cracking. It can also start at the site created by other forms of corrosion such as pitting, fretting, intergranular corrosion which may contribute to the initiation process as stress concentrations.

Once the stress corrosion cracking starts it will continue to grow, until and unless the sustained tensile stress has fallen below the critical value or when the corrosive environment is not active. The crack may permanently be arrested if residual stress are released by the initial crack advance. If the stresses are assembly related such as interference fit, torque or service operations, the cracks may continue to grow until the component can no longer sustain them. The speed of crack growth for a given material environment combination is dependent on the crack tip stress state. The crack tip stress intensity can sometimes indicate the crack velocity. Numerous laboratory studies have shown that for a given system the velocity of stress corrosion crack growth will remain constant over a range of stress intensities, but as the stress intensity is decreased, a point will come when the crack velocity will also start decreasing.

If stress-corrosion crack velocity versus crack-tip stress intensity are plotted on a logarithmic scale, two major regions of stress corrosion cracking (namely, stress dependent cracking as 'Region I' and plateau as 'Region II') are observed. Also, a second stress dependent cracking (referred to as 'Region III') occurs at higher intensity, normally difficult to establish. In the region of slow crack growth a value of stress intensity KISCC can often be established, below which crack growth proceeds at such a slow rate that the crack appears to be at rest. This is referred to as the threshold stress intensity for stress corrosion cracking.

Mechanism of Stress Corrosion Cracking

The mechanism of stress corrosion cracking generally varies in different situations. Among the several theories[15] advanced to explain the mechanism of stress corrosion cracking, the electro-chemical and stress-sorption theories are considered important.

Electro-chemical Theory

It explains that galvanic cells that are set up between metal grains and anodic paths are established by heterogenous phases, resulting in dissolution of the material at the crack tip. In the case of intergranular cracking, the grain boundaries are likely to be more active with respect to corrosion than the adjacent grain. This is perhaps due to the segregation of alloying elements or depletion of alloying elements through percipitation of second phases. The anodic paths are often associated with heterogenous phases such as $CuA1_2$ which will precipitate along grain boundaries in 2000 series aluminium alloys during heat treatment and produce copper depleted paths adjacent to the grain boundaries. When such an alloy is stressed under tension and exposed to a corrosive environment, it opens up a crack due to the combined effect of the ensuing localised electro-chemical dissolution of the metal and localised plastic deformation. With sustained tensile stress in action protective films that form at the tip of the crack break up, causing fresh anodic material to be exposed to the corrosive medium. During its cyclic process the stress corrosion crack continues to propagate. The above theory is supported by the existence of a measurable potential in the metal at grain boundaries which is negative (or active) with respect to the potential of the grains. Furthermore, this theory has been extended to include metals that do not form intermetallic precipitates but produce phase changes or segregation of alloying elements or impurities during the process of plastic deformation of the metal at the crack tip, resulting in composition gradient which sets up galvanic cells.

Stress-sorption Theory

It relates to the fact that stress corrosion cracking generally proceeds by weakening of the cohesive bonds between surface metal atoms due to adsorption of damaging substances from the environment.[15] According to this theory, adsorption of any kind that reduces surface energy should favour crack formation. Due to reduction in the surface energy the possibility of forming a crack is increased under tensile stress.

Susceptibility of Metals to Stress Corrosion Cracking

Susceptibility of a given metal to stress corrosion cracking in a specific environment is governed by its overall, and local alloy composition and its metallurgical structure. The microstructural variables include grain size, grain shape and the nature of precipitates. Effects of composition include the presence of alloying elements and impurities either in low concentration or even in trace elements.

6.3 HYDROGEN EMBRITTLEMENT

The hydrogen embrittlement phenomenon constitutes one of the greatest areas of concern for metallurgists particularly those dealing with dynamic loaded high strength constructional materials. This is partly because of its unpredictable nature of occurrence and partly due to its complex mechanism. In recent years no other failure mechanism has evoked so much interest as has hydrogen embrittlement owing to the use of advanced materials.

Hydrogen embrittlement can be broadly defined as the reduction in the ductility of the metal due to presence or absorption of hydrogen. Generally all kinds of damages that hydrogen brings about in metallic materials are termed as hydrogen embrittlement. The phenomenon of hydrogen embrittlement was discovered at the turn of the century by Heyn and others in steels, but it was not until 1926 that any serious thought was given to the problem. The fact that hydrogen interacts with metal and alloys making them unfit for their intended use is the basis of the study on unified theory.

Atomic hydrogen is a cathodic product of many electrochemical reactions. It is formed during corrosion as well as the plating or pickling processes. Hydrogen whether liberated as gas or atomic hydrogen is absorbed by the metal depending on the surface chemistry of the metal. Substances like phosphorus, arsenic, antimony, sulphur, selenium, tellurium, iodine can act as a cathode poison impeding the combination of pairs of hydrogen atoms to form gas molecules and favouring the entry of atomic hydrogen into the metal. Hydrogen dissolves endothermically in iron and is soluted during metallurgical processing at high temperature and is retained as a supersaturated solution in the lattice unless the cooling rate is too high to allow it to diffuse out. Due to its small size and mass atomic hydrogen has very high diffusibility in most metals and therefore can penetrate appreciably into the metal surface and migrate rapidly to favourable sites where it may remain in solution, precipitate as molecule hydrogen (forming small pressurised cavities, cracks or large blisters) or it may react with the base metal/alloying elements to form hydride.

Classification of Hydrogen Damage

There are four broad categories[25] of hydrogen damage, *viz.*

1. Hydrogen embrittlement.
2. Hydride embrittlement.
3. Solid solution hardening.
4. Creation of internal defects.

and can further be sub-divided into various damaging processes as shown in Fig. 6.7.

Hydrogen Embrittlement

It results mostly from the pervasive and damaging effect of hydrogen in materials and is marked by the substantial decrease in the energy absorption ability of the material.

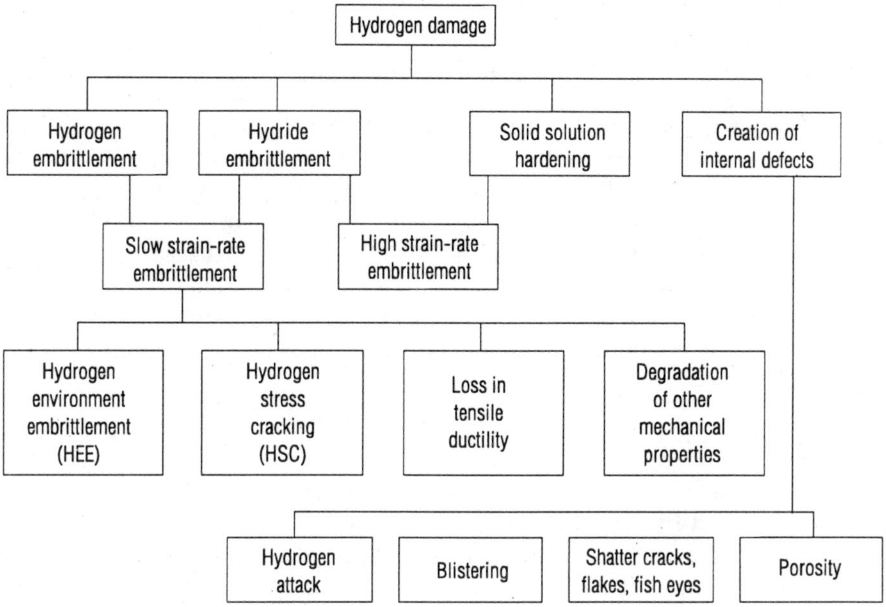

Fig. 6.7 Various classifications of hydrogen damage[25]

The phenomenon which is most pronounced at slow strain rates is termed either as Slow Strain Rate Embrittlement (SSRE) or Hydrogen Induced Brittle Cracking (HIBC)/ Hydrogen Assisted Cracking (HAC). For the sake of convenience, multi-hydrogen embrittlement processes are grouped into four phenomena and each of them are briefly discussed below.

Hydrogen Stress Cracking Also known as internal hydrogen embrittlement, it is a cracking process which involves brittle delayed failure of normally ductile materials charged with hydrogen. Mostly high strength materials such as steels, titanium alloys and aluminium alloys are prone to such a phenomenon.

Hydrogen Environment Embrittlement It also referred to as external hydrogen embrittlement, it is the embrittlement phenomenon that occurs in plain carbon, HSLA and alloy steels, metastable stainless steel and alloys based on nickel, titanium, uranium and niobium, when tensile loaded in contact with gaseous hydrogen.

Loss in tensile ductility is produced by hydrogen in most materials. In cases where a drastic embrittlement effect is not observed (i.e. low strength steels, austenitic steels, aluminium alloys, etc.) the decrease in percentage elongation or loss in percentage reduction in area in tensile tests, indicate the probable existence of hydrogen embrittlement.

Degradation of other Mechanical Properties Hydrogen affects the plastic flow behaviour of metal and alloys, thereby resulting in increased or decreased yield strength, serrated yielding and lowered or raised work hardening rates. Presence of hydrogen is reported to have degraded fatigue properties of many materials. Similarly, the high temperature creep properties of steels and titanium alloys are reduced by hydrogen[25].

Hydrogen Attack On exposure to high pressure hydrogen at high temperature for a long duration, hydrogen forms methane bubbles within the material while reacting with carbon of the steel and causes formation of fissures due to expansion and joining of such bubbles. Due to decarburisation near the bubbles, a carbon concentration gradient is set up which causes the dissolution of neighbouring carbide particles[25]. If confined to the metal surface, hydrogen attack causes decarburisation and surface weakening. Hydrogen attack is reported to have been observed in plain carbon steel, low alloy steels and even some stainless steels operating above 200 °C. It is said to be a major cause of failure of steam generator components.

Hydride Embrittlement

At low concentrations of hydrogen (i.e. below the solid solubility limit), the embrittlement is caused by stress assisted hydride formation. When the solid solubility decreases, hydrogen leads to precipitation of brittle hydrides on the slip planes and cause severe embrittlement. This kind of embrittlement is enhanced by increased strain rates, and increased temperatures in presence of notches[26]. Another feature of hydride embrittlement is to cause ductile-to-brittle transition in toughness with decrease in the temperature of testing.

Creation of Internal Defects

Blisters, shatter cracks, flakes, fish eyes and porosity are some of the common defects produced by hydrogen when present in structural materials in sufficient quantity.

Blisters Very large blisters or bubbles can form as a result of the generation of atomic hydrogen which enters the metal only to condense as molecular hydrogen at internal microvoids or at lines of oxide or sulphide inclusions. High pressure is built up at such places as atomic hydrogen continues to diffuse in, ultimately forming a blister. Blisters grow and finally rupture.

Shatter Cracks, Flakes, Fish Eyes and Microperforations Flakes and shatter cracks are internal fissures generally observed in large forgings. Fish eyes are bright patches of embrittlement regions visible on the fracture surfaces. Hydrogen picked up during melting segregates and precipitates at voids and discontinuities, producing these defects.

Porosity An excessive liberation of hydrogen gas during cooling of the molten metal like aluminium or magnisium whose solubility of hydrogen is known to decrease with temperature causes prosity. The extent of porosity depends on the hydrogen contents of the melt, its cooling rate, crystallisation behaviour, external pressure and equilibrium partial pressure of hydrogen[26].

Solid Solution Hardening

In certain metals[26] like niobium and tantalum, hydrogen causes solid solution hardening and embrittlement at concentration much below the levels at which hydrides precipitate. At low concentrations of hydrogen βTi exhibits cold shortness at room temperature.

Sources of Hydrogen

Hydrogen is present in almost all environments and metals easily pick up hydrogen to get damaged in their service life. Table 6.1 gives the possible sources of hydrogen.

Table 6.1 Various sources of hydrogen pick up[24]

Sources	Cause of Hydrogen Evolution
Casting welding	Due to thermal dissociation and ionisation of water. Hydrogen solubility in iron increases from 0.1 ppm at room temperature to 30 ppm in the molten state.
Pickling	$Fe + H_2SO_4 \rightarrow FeSO_4 + H^+$
Plating	For galvanic nickel plating hydrogen yield is caused as per reaction. $Ni + 2e + 2H^+ + 2OH \rightarrow Ni + 2H^+ + 2OH^-$
Gas - curburisation	$3Fe + CH_4 \rightarrow Fe_3C + 4H$

Theories of Hydrogen Embrittlement

It is well-known that temporary immersion of iron in hydrochloric or sulphuric acid brings about an extraordinary decrease in its toughness. The effect is only temporary, because with the lapse of time the metal slowly regains its original toughness and strength. In other words, the damage is reversible. Half an hour's immersion in hydrochloric acid will make a piece of steel (containing say 0.6% C) so brittle that no amount of exposure to air or heat will ever completely restore it. The damage is thus irreversible. On hardened and tempered steels, the decrease in toughness produced by immersion in acid is greater and more rapid than with the same steel in a soft condition, indicating the influence of the level of strength.

Till date, five major mechanisms have been proposed to explain hydrogen embrittlement.

1. Pressure Theory

Zapffe and Sims proposed that molecular hydrogen can precipitate into internal voids in the lattice, thereby building up pressure sufficient to cause premature failure[27]. De Kazinczy and Tetelman modified the theory with concept of hydrogen diffusion[28,29] for maintaining the internal pressure within a propagating crack. Later the concept of transport of hydrogen to cracks or voids was included to maintain high pressures needed to cause continued crack propagation. However, it is not always true to assume that hydrogen simply acts as a source of internal pressure. Since the void enlarges due to straining, it results in a decrease in hydrogen pressure. For further damage to occur, continued precipitation of hydrogen to the voids is necessary. If the rate of straining is greater than the restoration of necessary pressure, embrittlement should decrease. The effect should be similar if the test temperature is reduced, because the rate of hydrogen delivery into the voids and gas pressure inside the void would both decrease with decrease in temperature.

However, the theory is only partially applicable because is does not explain hydrogen embrittlement which occurs in low hydrogen gas. These conditions make the creation of high hydrogen impossible.

2. *Surface Absorption Theory*

Petch and Stables proposed that the hydrogen already in a metal lattice diffuses to the crack tip and lowers the fracture stress[30] by being adsorbed on the crack surface. The presence of a pre-existing crack is assumed or micro-cracks can form as a result of dislocation pile up at grain boundaries and hydrogen is adsorbed by the internal surfaces of these crack to lower their surface energy and thus the stress required for crack propagation. The drawback of this theory is that other elements (e.g. oxygen) are also strongly adsorbed on iron but do not embrittle them.

3. *Decohesion Theory*

Troiano originally proposed that hydrogen diffuses to highly stressed region[31] in the metal and concentrates there. In stressed zones hydrogen accumulates at microcracks or voids and upon reaching a critical concentration, embrittles the region and causes the crack to extend to just beyond the hydrogen concentrated volume. The process repeats after a time necessary to accumulate the critical hydrogen concentration at the new site. Troiano postulated that stress induced diffusion of hydrogen to regions of high triaxial stress within the lattice resulted in a reduction of the cohesive or fracture strength of the lattice (i.e. cleavage is made easier).

Later on Troiano proposed that hydrogen influences the bond breaking at the tip of the crack by lowering the cohesive strength of the material. Oriani and Josephic also proposed[32] that hydrogen from any source greatly increases the decohesioning of the lattice atoms by decreasing the local maximum cohesive force per unit area, leading to crack growth.

In high strength steels the role of hydrogen is to influence decohesion by overcoming inter atomic cohesive force in the immediate vicinity of the crack tip and within the plastic enclave of about 10 μm diameter.

This theory may well explain failures of the cleavage or intergranular type, but it fails to account for the dimple rupture and blistering observed in a hydrogen environment.

4. *Slip-softening Theory*

Beachem proposed a theory[33] in which a model of hydrogen assisted cracking was used in explaining the microscopic plasticity exhibited by almost all materials that failed in the presence of hydrogen. Hydrogen dissolved in the lattice just ahead of the crack tip enhances deformation there. Hydrogen induced delayed plasticity has been observed in various steels and in 7075 type high strength aluminium alloys, supporting this hypothesis. This model was later modified to include a counter effect, namely that this plastic deformation aids the entry of hydrogen into the critical volume of the material. It is an established fact that hydrogen can enhance dislocation motion. Enhanced screw dislocation motion by double-kink nucleation, dislocation injection at crack tips[34] and promotion of shear instabilities[35] can be caused by hydrogen.

However the drawback in this theory is that hydrogen induced hardening has been observed in several materials and the strong chemisorption of hydrogen at the crack surfaces tends to prevent dislocation emission. Hence the slip softening model cannot be accepted as a general mechanism for hydrogen embrittlement.

5. *Hydride Embrittlement Theory*

In hydride forming metals like Ti, Zr, V and Nb hydrogen embrittlement is attributed to stress induced hydride formation[36] at the crack tip and cracking of the brittle hydrides. According to Westlake the hydrostatic stress field at the strained crack tip is required for hydride formation[37]. Recent studies[38] on Vanadium suggest that even the hydrostatic component of stress may not be essential for hydride formation. In iron and other non-hydride forming elements even on exposure to high pressure hydrogen, no hydride is detected. However, theoretical calculations have shown that a Fermi-Dirac atmosphere of hydrogen could be formed at strained crack tips in iron. This atmosphere, extending over 100 sites per atom plane will have a hydrogen concentration of 10^{-8} which is very close to the room temperature solubility limit of 3×10^{-8}H atoms/atom of iron in α-Fe. For all practical purposes, such zones are hydrides and may initiate cracking. It is well-known that elements like Ti and Mn added to iron can stabilize complex alloy hydrides which are brittle and may act as crack nucleii. Cluster chemistry studies[39] have shown that impurities like As, Sb and Sn can form complexes with hydrogen causing lengthening of the atomic bonds and hydrogen embrittlement.

All the above theories require transport of hydrogen either from the bulk material or from the outside atmosphere to regions of stress concentrations.

6. *Unified Theory of Hydrogen Embrittlement*

It is believed that none of the above mechanisms or models is totally applicable to all materials, as these theories do not deal with the essential atomistic mechanisms by which hydrogen embrittlement degrades the metal.

A new unified theory of hydrogen embrittlement[25] proposes two basic tenets.

1. An essential atomic interaction between hydrogen and the matrix atoms exists and it is the basis of all the various degradation phenomena observed.

2. A critical minimum hydrogen concentration is required to initiate the degrading interaction.

Practical Solutions to Hydrogen Embrittlement Problem

Having discussed various concepts and mechanisms, and since many of the alloys used in airframe, engine and pipeline construction are affected by hydrogen embrittlement including steels, aluminium, titanium alloys and nickel base alloys, it will be of great interest to include here some of the metallurgical variables that become critical in assisting the phenomenon and discuss them with guidelines with a view to resist hydrogen embrittlement during alloy design.

The important variables are alloy chemistry, microstructure and strength level. These variables are interrelated and the variable of microstructure can be subdivided in terms of grain size, crystal structure, precipitate type morphology and texture. Metals with body centered cubic structure are generally more prone to hydrogen damage than those with face centered cubic and hexagonal crystal structures. FCC alloys (like austenitic steels, nickel alloys and aluminium alloys) as well as HCP alloys (titanium and zirconium) are also found susceptible. However, FCC metals like gold, silver and copper are nearly immune to hydrogen embrittlement.

Alloy Chemistry

Depending upon the type of solid solution alloying elements are added to steels. Out of these elements Cr, Mo, Si, V, Nb act as substitutional and C, N, P as common interstitial elements. On the other hand As, Sb, Bi and S are treated as trace elements.

However, effects of elements like Mn, Ni, Cr and Ti are sometimes dual, i.e. increasing the hydrogen susceptibility of steels under one condition and decreasing it under another condition. Si addition to steel is beneficial owing to its ability to resist epsilon carbide from acting as a cathodic site for hydrogen discharge and its capacity to act as a hydrogen trap and its effect on the diffusivity of hydrogen.

Mn is found to be deleterious in the presence of hydrogen in high strength (4340 type) steels[40]. Mn is believed to make the crack tip more cathodic in an aqueous environment.

In aluminium alloys the total amount of dissolved solutes is required to be restricted to ensure better hydrogen performance. High strength 7000 series alloys perform better with higher Zn, reduced Cr, Fe and Si and lower total impurities.

In titanium alloys, aluminium below 5% and oxygen below 1000 ppm ensure better immunity to hydrogen. High Mo + V is also beneficial.

As, Sb, Sn and S are generally harmful to the hydrogen embrittlement resistance of alloys.

Microstructure

Hydrogen susceptibility[25] of common micro-constituents in steel vide Table 6.2 indicates the dominant role of microstructure in the hydrogen embrittlement of alloys.

Table 6.2

Steel			
Composition:	Carbon (%) 0.26, Chromium (%) 1.02, Molybdenum(%) 0.30	Nickel (%) Iron	0.25 Remainder
Hydrogen: Electrolytic charging for 48 hours			

Structure	Reduction of area (%)	
	Uncharged	Charged
Martensite	47.4	0
550 °C tempered martensite	63.1	3.5
Fine bainite	59.9	30.8
Coarse bainite	38.5	2.8
Fine pearlite	65.8	52.3
Coarse pearlite	62.4	36.6
Spheroidized structure	76.0	71.6

Steels

Strength level is an important variable affecting several types of steel and the susceptibility to hydrogen embrittlement increases as the alloy strength level increases. However, this relationship holds good for high strength steels since steels with yield strength below about 700 MPa are usually not markedly affected by hydrogen embrittlement.

It is clear from Table 6.2 that of all the microconstituents, martensite lends most susceptibility to hydrogen embrittlement and the most spheroidized structure gives least susceptibility. In general, a fine grain size in materials favours superior resistance to hydrogen embrittlement and this is explained by the fact that a small grain size provides more traps for hydrogen due to the increased boundary area, thereby reducing the quantity of diffusible hydrogen.

In austenitic stainless steels carbide precipitation increases the susceptibility. Susceptibility is also increased due to presence of Fe-Ni-Al or Ni-Nb compound precipitates in the martensitic structure of pH grade steels[25]. However, austenitic stainless steel composition which favours austenite stability generally provides increased resistance to hydrogen embrittlement (i.e. high nickel contents and chromium levels either below 15% or above 20%). Presence of γ' N_{13} (Al-Ti) precipitates in Ni-based super alloys contributes critically to the hydrogen induced ductility losses[25].

In aluminium alloys, peak aged structure obtained through precipitation of coherent clusters of precipitates, exhibits greater susceptibility to environment assisted fracture due to slip planarity (caused by small, ordered, coherent precipitates). Overaged structure reduces susceptibility due to formation of a wavy slip. Thermo-mechanical treatment[25] which involves ageing prior to working (producing a structure close to the peak aged condition) followed by warm working and further ageing causes a reduction in susceptibility by avoiding formation of a planer slip. It is reported that hydrogen embrittlement contributes stress corrosion cracking in the case of 7000 series. Studies in Al-Zn-Mg. alloy (7075 and 7475) have shown that hydrogen bubbles may form on grain boundaries in alloys exposed to moist atmosphere, but interestingly they are found to occur in overaged alloys. The source of hydrogen in these studies is believed to be the reaction between aluminium and water, which at 70 °C produces a duplex film consisting of aluminium oxyhydrate (ALOOH) and beryerite ($AL(OH)_3$) as per reaction

$$\text{Anode AL} + \begin{matrix}(3H_2O)\\(2H_2O)\end{matrix} \rightarrow AL(OH)_3 + 3H^{-}{}_{-3e}$$

$$\text{Cathode } 3H_2O + 3e \rightarrow 3/2\ H_2 + 3OH^{-}$$

Thus hydrogen produced by the reaction can penetrate the metal and in the presence of a stress can lead to a brittle intergranular fracture. This brittle intergranular fracture can occur with no evidence of concurrent anodic dissolution of the metal, suggesting that true stress corrosion cracking is not responsible. The embrittlement effect is reversible and hydrogen evolution may be detected together with a recovery of the normal ductile fracture mode.

Inclusion

Inclusions are very often recognised as the origin of hydrogen induced cracks. For example, MnS is the cause of blistering in pipeline steels. Elongated inclusions are more harmful than rounded ones, by virtue of their higher concentrations. The elongated shape of MnS inclusions can be made more rounded by adding such elements as Ca or rare earths to the melt which make them harder, so that their original round shape is retained during the rolling operations. A small particle size is desirable and this can be achieved through addition of Ti, Hf or Zr to the melt in case of MnS. Addition of Ti or rare earths also induces early precipitation of sulphides in the melt, preventing their segregation in a row where ever the structure is sensitive.

Chapter 7

In-Process Failure: Case Studies

7 *In-Process Failure: Case Studies*

7.1 CRACKING DURING FORGING OF EXTRUDED ALUMINIUM ALLOY BAR STOCK MATERIAL DUE TO SEGREGATION

Summary and Recommendations

During the pre-production stages of forging an initial batch of 50 mm dia aluminium copper alloy extruded bar stock material was found cracking randomly. It was a rather unusual phenomenon because the quality of the forging stock was checked prior to acceptance of the raw material for production.

An in-depth analysis was carried out to probe into the metallurgical factors that caused random cracks, using optical as well as scanning electron microscopes and elemental analysis facility. Micro-examination revealed the presence of round segregated particles along the intergranular crack paths besides gas voids and oxide films and the same was confirmed by the scanning electron microscope. Microprobe analysis of the segregation identified them as copper rich particles. On the basis of results, a conclusion was reached that the unusual cracking of the forged blanks during different stages of forging had obviously resulted from the use of poor quality raw material, with segregation as the prime contributing factor. The reason for segregation was suspected to be the use of improperly made master alloys.

It was recommended to adopt improved melting technique and use proper master alloys (aluminium-copper alloy in particular) for avoiding segregation and other harmful defects such as oxide films and gas voids which could have additionally favoured the cracking. Also, a need for a close and careful microstructural analysis for detection of not-so-easily recognisable segregation particles (particularly under low magnification examination) was emphasised, prior to acceptance.

Background

Applications

A batch quantity of 50 mm dia aluminium 4% copper extruded rods was used as forging stock for production of different types of forgings. Quality of the raw material was microstructurally checked and strength properties in heat-treated conditions were evaluated prior to acceptance for forging.

Circumstances Leading to Failure

For the initial quantity of 50 parts of forgings made out of the same batch, a majority of them showed visible cracks and random burstings, both in the pre-finished and finished forging stages (Fig. 7.1).

Fig. 7.1 Cracks and random bursting in pre and finished forging stages (× 1/2)

Specifications

The materials used for indigenously produced forging stock was an Indian alloy of HE 15A specification, which is similar to the British BSL L77 alloy.

Selection of Specimens

Three sections across the cracked regions of the parts (one each from upsetting, pre-forging and final forging stages) were selected for low-powered optical (macro and micro) and high powered scanning electron microscopic examination, along with a cut specimen from the supplied raw material, for detailed and comparative studies.

Visual Examination of General Physical Features

Examination of the failed parts showed several fold type discontinuities/cracks besides radial bursts (Fig. 7.1).

Testing Procedure and Results

Macro examination of the cross sections of the extruded bar as well as the preforged failed sample, under the low-powered stereoscopic microscope, showed the presence of oxide-films confined to the coarse grain envelope (Fig. 7.2).

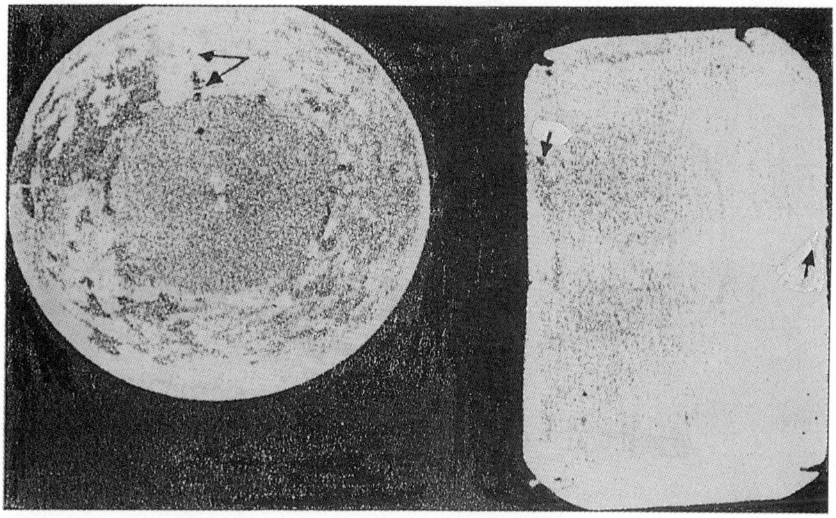

Fig. 7.2 Oxide film confined to coarse grain envelope (× 1)

Metallography

Microstructural Analysis Micro-examination of the sections across the defect revealed intergranular cracks tracing the path of round segregated particles and oxide film discontinuties (Fig. 7.3). In addition, the structure showed pockets of gas voids in chain

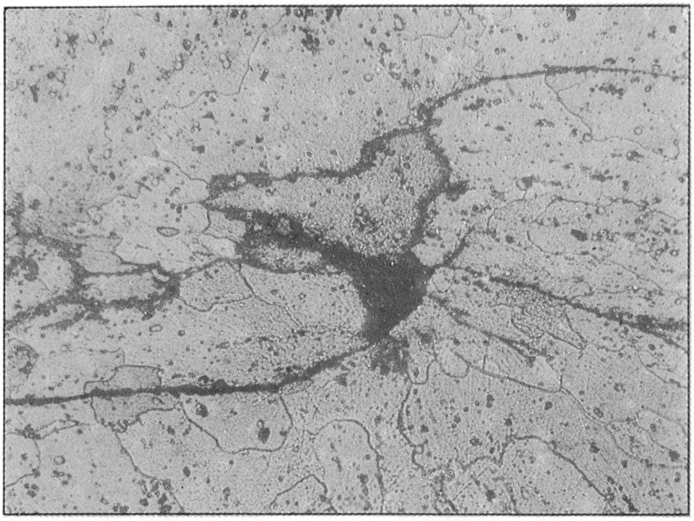

Fig. 7.3 Black segregated particles along the path of intergranular cracks (× 250)

form which appeared as crack-like discontinuties (Fig. 7.4). Further, micro-examination of the raw material also confirmed the presence of gas voids mostly in the chain form and zones of segregation.

Fig. 7.4 Gas voids in chain form in the forging (× 200)

To identify the nature of the segregated particles, and also to differentiate them from other microstructural constituents that are normally present in the heat treated aluminium 4% copper alloy structure (i.e. Fe - Mn_3, Si - $A1_{12}$, Cu - $A1_2$, Cu_2 Mg_8, Si_6 - $A1_5$), a differential etching test with 10% phosphoric acid was carried out on the micro-specimens. It was observed that while normal constituents were easily identified due to changes in their respective colour, the remaining particles were unchanged, suggesting the presence of segregation. Further, on scanning electron microscopic examination the segregated particles concentrating along the path of cracking were found to be dark-coloured in appearance against the white round particles of Cu-$A1_2$ (Fig. 7.5). However, to confirm this, an etched specimen of a satisfactory heat-treated aluminium 4% copper alloy extruded material (which was free from any segregation on micro-examination) showed no dark-coloured particles on SEM examination (Fig. 7.6).

Chemical/Micro-probe Analysis

Chemical analysis of the forging stock material was found conforming to L77 specifications (Table 7.1). Further, elemental analysis using a microprobe identified the segregations as copper-based particles (Fig. 7.7).

Discussion

Macro and microstructural analysis clearly proved that unusual cracking/radial bursting of the forgings was associated with the basic poor quality of the raw material, which

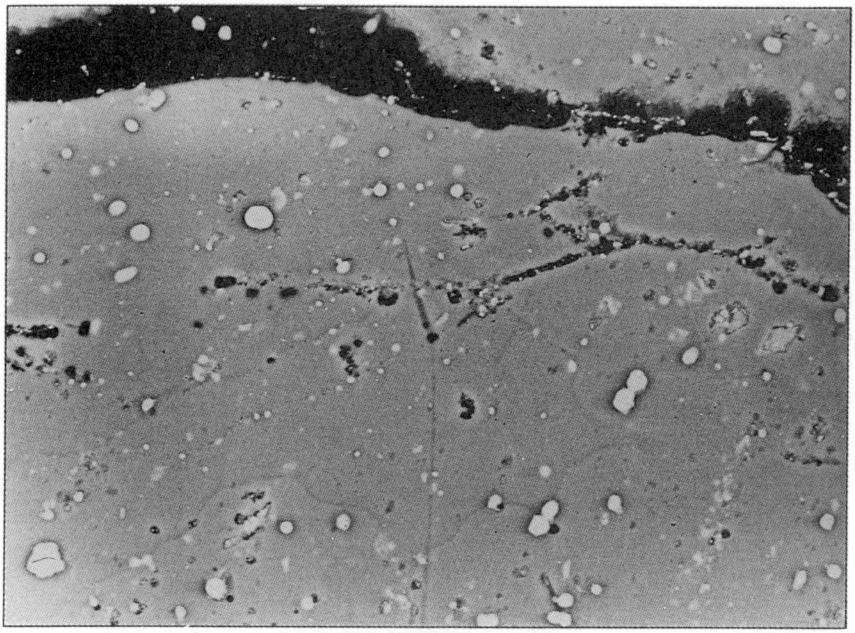

Fig. 7.5 Dark coloured particles as segregation along the path of cracking against white round particles of $CuA1_2$ under the scanning electron microscope (× 450)

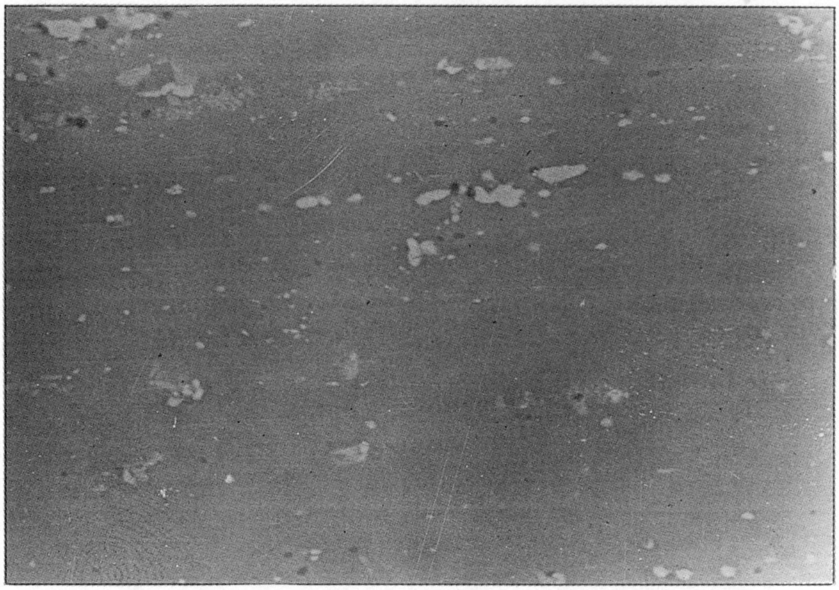

Fig. 7.6 The microstructure of a satisfactory heat-treated aluminium-4% copper alloy with no evidence of segregation, under the scanning electron microscope (× 450)

Table 7.1 Chemical specification (Wt.%) for the indigenised aluminium 4% copper alloyfrom which forgings were made and BSL 77, to which it is similar

Element	L77		Indigenised (failed part)
	Min.	Max	
Copper	3.9	5.0	4.15
Magnesium	0.2	0.8	0.40
Silicon	0.5	0.9	0.65
Manganese	0.4	1.2	0.53
Titanium + zirconium	-	0.2	0.015
Iron	-	0.5	0.34
Nickel	-	0.2	0.04
Zinc	-	0.2	0.12
Lead	-	0.05	0.001
Tin	-	0.05	0.002
Chromium	-	0.2	0.09
Al	Remainder		

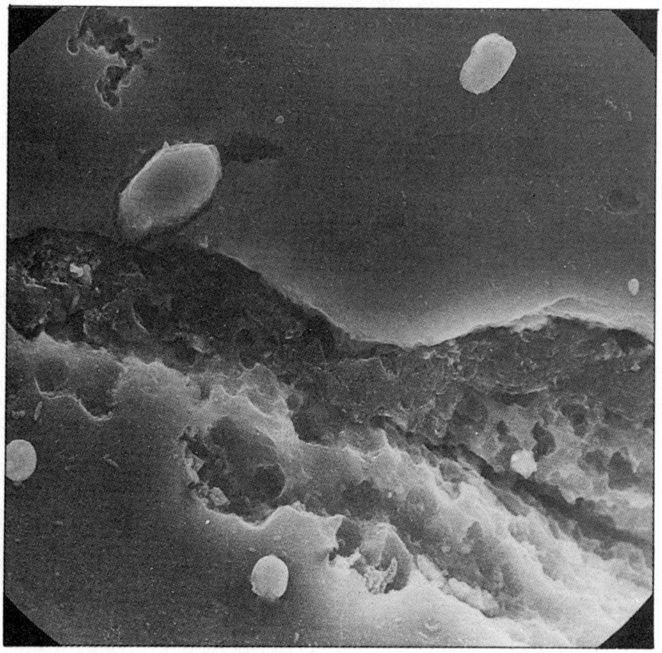

Fig. 7.7 Segregated zones of copper base particles identified by elemental analysis using microprobe facility (× 1000)

inherited harmful segregation of copper-rich particles, oxide film and gas voids from improper melting practices. The major source of weakness for failure was found to be the high concentration zones of segregation.

Since the raw material was cleared prior to forging, the investigator had to carry out a detailed microstructural examination, including scanning electron microscope and microprobe analysis, to determine the nature and type of segregation. This may have probably escaped the attention of the examiner at the time of raw material acceptance.

It is probable that stray zones of segregated copper-rich particles could not be differentiated from the normal constituents under low powered optical micro-examination (at ×100). Hence, the need for an in-depth examination was not thought of during the routine acceptance test.

Conclusion and Recommendations

Most Probable Cause

The basic cause of random cracking/bursting of the pre-finished and finished forged parts was the segregation of copper-rich particles which were present in the material. In fact, these segregated particles provided the matrix with sites for inherent brittleness during forging operations. Apart from this, presence of oxide film, gas voids in chain form and coarse grain envelope had additionally favoured the process of cracking.

Remedial Action

To prevent recurrence of such a failure, the supplier was advised to improve the melting techniques and the quality of master alloys so that formation of harmful segregation, oxide films, gas voids and other defects were avoided. In addition, the shop was instructed to carry out proper quality control check by ultrasonic test on the extruded forging stock, and by macro/micro-examination test of slices cut from the front and back ends of the supplied extruded rods.

How Failure Could have been Avoided

Due to the above findings, samples from the left over quantity of the batch were microstructurally examined and found unacceptable for further production purposes. This unusual problem of failure on mass scale could have been avoided had the initial acceptance examination taken note of the existence of these harmful defects and been aware of its role during the heavy deformation processing.

7.2 FAILURE OF ALUMINIUM ALLOY FORGED VALVE BODY DUE TO HEAT DETERIORATION

Summary and Recommendations

A batch of aluminium 4% copper alloy forged valve body developed dark brown line indications all over the surface on anodising after heat treatment.

Micro-examination of a section across the defects revealed severe heat deterioration effects in the form of grain boundary cracking/melting throughout the structure. It indicated that the furnace temperature was higher than the actual solutionising temperature. Since overheating would normally reflect structural deterioration in the form of either blisters or cracks on the surface and not in the form of dark line discontinuties, the cause of the unusual line defects was attributed to the increased current density developed during anodising, in the presence of melted constituents/ grain boundary cracks.

Background

Applications

A batch quantity of 25 valve body parts (Fig. 7.8) was forged out of aluminium-copper alloy extruded bar and heat-treated. Subsequently, these were subjected to anodising. These parts are used for critical applications in the aircraft.

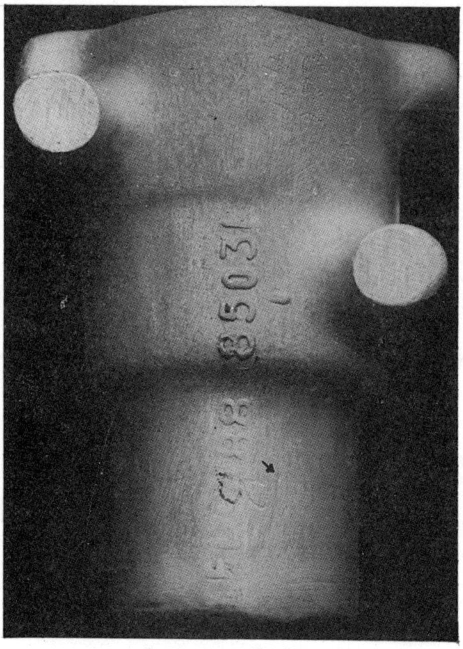

Fig. 7.8 Dark brown lines on the surface of the part (× 1)

Circumtances Leading to Failure

All the parts showed numerous line discontinuties on the surface on inspection after anodising.

Specifications

The material used for fabrication of the parts was an Indian alloy to IS 24345-HE 15 specification for aluminium 4% copper alloy as forging stock.

Selection of Samples

One defective part was selected for low power stereoscopic as well optical microscopic examination along with a sample sliced from the supplied extruded bar used for forging.

Visual Examination of General Physical Features

Examination of the anodised parts showed numerous dark brown (parallel) line indications all over the surface of the body, in varying magnitudes (Fig. 7.8).

Testing Procedures and Results

Macro Fractography

Examination of the line defects at a magnification of × 3.5 revealed them as deep furrow type linear surface tears (Fig. 7.9).

Fig. 7.9 Linear surface tears at close up magnification (× 2)

Metallography

Microstructural Analysis Micro-examination of the sections taken across the defects revealed numerous grain boundary cracks emanating from the surface (Fig. 7.10). At higher magnification, numerous zones of grain boundary cracks and eutectic melting (Fig. 7.11) were observed, indicating severe heat deterioration effect. However, examination of the microstructure of the forging stock was found free from any structural inhomogenities or defects.

Fig. 7.10 Grain boundary cracking in the microstructure at the defect zone (× 50)

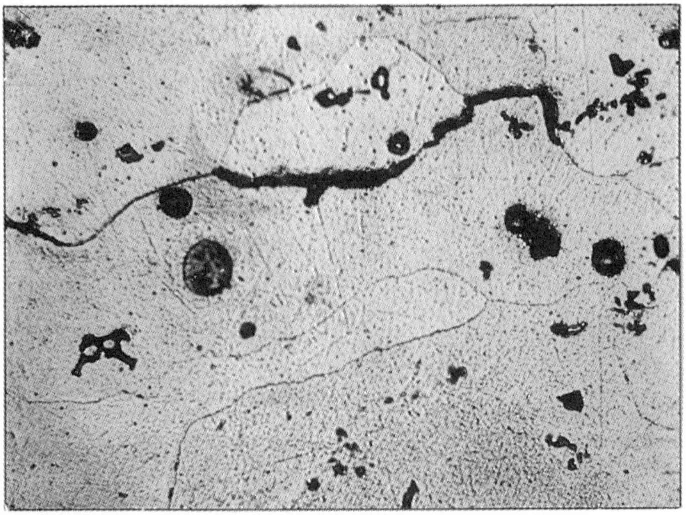

Fig. 7.11 Severe heat deterioration effects in the core structure (× 400)

Chemical Analysis

Chemical analysis of the forged valve body was found conforming to IS 24345 - HE 15 specification (Table 7.2).

Table 7.2 Chemical specification (Wt.%) for the Indian aluminium- copper alloy forging stock (IS 24345-HE-15) from which valve body was forged and actual composition of the failed part

Element	Min.	Max.	Failed part
Copper	3.90	5.00	4.15
Magnesium	0.20	0.80	0.52
Manganese	0.40	1.20	0.65
Silicon	0.50	0.90	0.48
Iron	-	0.50	0.39
Nickel	-	0.20	0.16
Chromium	-	0.10	0.07
Zinc	-	0.20	0.12
Titanium + zirconium	-	0.20	0.13

Discussion

Macro and microstructural analysis clearly established that the formation of unusual defect indications in a linear form on the surface of the forgings was a result of severe heat deterioration caused by undue overshooting of the solutionising temperature during the batch heat-treatment.

It is well-known that heat-treatment of aluminium alloy containing 4% copper is very sensitive to temperature, i.e. solutionising temperature lies very close to the melting zone of the alloy and any slight rise in temperature, exceeding the specified limit, tends to overheat the material. Line defect indications on anodising appear due to the extra sensitiveness of the overheated zones (in this case in the form of melting/cracks) to the anodising treatment by virtue of an increase in current density. Any increase in current density for copper bearing aluminium alloy above 2% tends to blacken the surface, particularly when they are overheated. This phenomenon of surface blackening during anodising of overheated structure is reported vide case study 7.4.

Conclusion and Recommendations

Most Probable Cause The basic cause of the linear surface defect indication is overheating of the parts during heat-treatment.

Remedial Action To prevent recurrence of such failure the heat treatment shop was advised to take effective steps in controlling solutionising temperature within specified limits of 500 ±5 °C and periodic calibration of the furnace temperature was recommended before loading the critical parts for heat-treatment.

7.3 FAILURE OF ALUMINIUM ALLOY FORGED BELL CRANKS DUE TO LINEAR SURFACE DISCONTINUITIES

Summary and Recommendations

A batch of forged aluminium copper-magnesium-nickel alloy bell crank showed linear discontinuities along the fillet radius on anodising treatment.

Macro and micro-examination of the defective area revealed the presence of a 'push through defect' in the radius zone, as a consequence of emergence of re-entrant type grain flow pattern in the form of surface discontinuties. The abnormal grain flow pattern was caused by the excess metal which failed to escape out of the already filled-in die cavity and re-entered through the inner radius towards the flash line.

It was recommended to introduce an intermediate sequence of operations with a suitable pre-finish forging die set in order to offset the re-entry of excess metal.

Background

Applications

A batch quantity of 30 nos. of bell crank parts was made out of aluminium alloy extruded bar stock material by close die forging and was heat-treated and anodised. These parts are subjected to critical stresses during the flight operations.

Circumstances Leading to Failure

All the forged parts on subsequent anodising showed linear discontinuities in single at location of the concave fillet radii portions (Fig. 7.12).

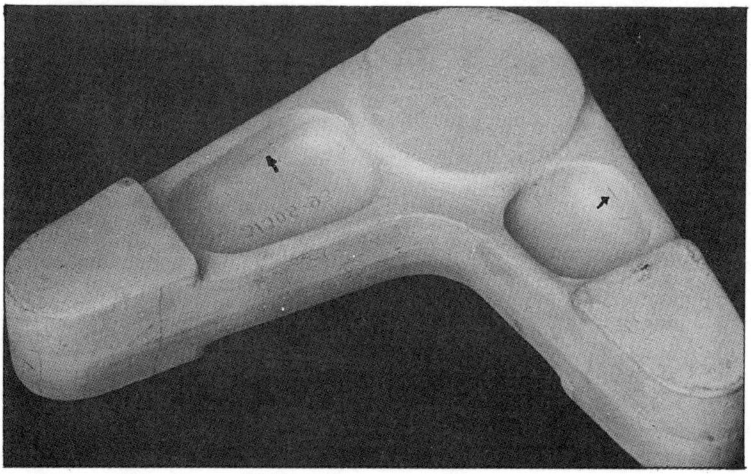

Fig. 7.12 Linear discontinuity along the fillet radius (× 1)

Specifications

The material used for fabrication of the parts was a French alloy to AU 2GN specification. The specification calls for aluminium-copper-magnesium-nickel alloy extruded bar as forging stock.

Selection of Samples

One defective part was randomly selected for the macro as well as microscopic examination.

Visual Examination of General Physical Features

On visual examination, the as-forged bell crank parts showed linear discontinuities in single along the fillet radii at two locations (marked in Fig. 7.12).

Testing Procedures and Results

Metallography

Macro-examination Macro-examination of sections across the linear defects, on etching with 5% sodium hydroxide solution, revealed re-entrant type grain flow pattern in the fillet region (Fig. 7.13). The defect is typically 'push through' in nature as normally

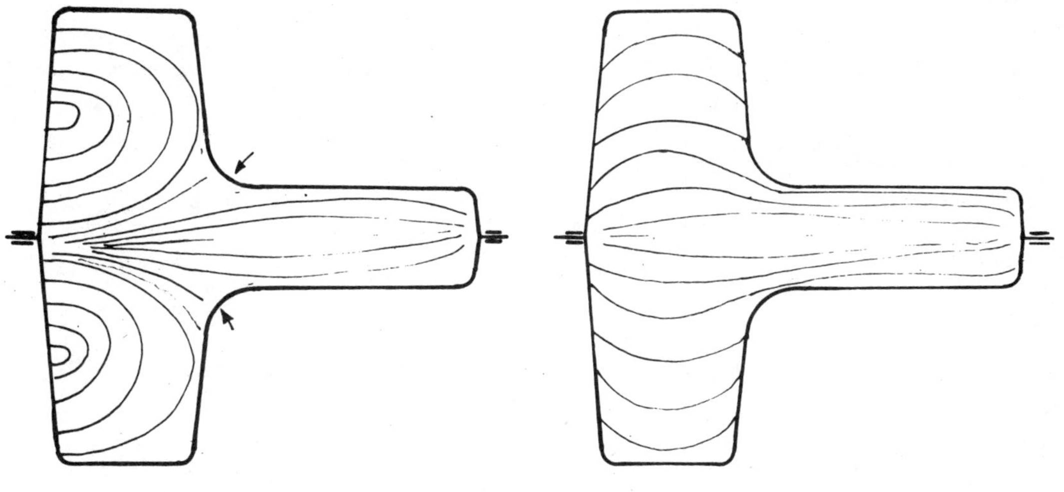

Fig. 7.13 Grain flow pattern as 'push through' defect in the failed part (left) along with desired type of grain flow pattern

encountered in forging defects. For the purpose of comparison, a desirable grain flow pattern in the section is also illustrated in Fig. 7.13.

Microstructural Analysis Micro-examination of the same section used for macro-examination revealed a 'U' shaped cavity (Fig. 7.14) in the defective zone, indicating the presence of a fold/lap to a maximum depth of 0.50 mm in the section examined.

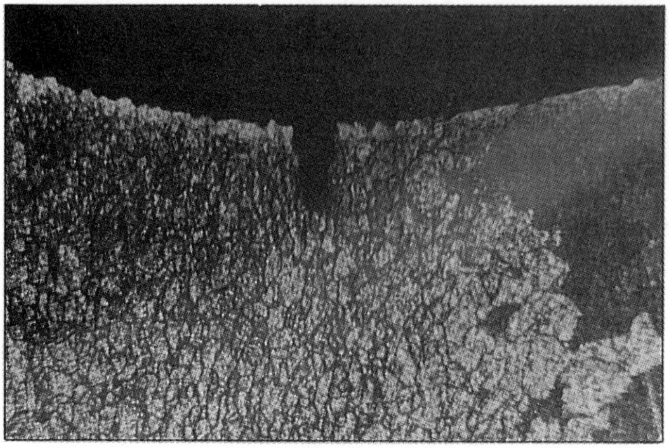

Fig. 7.14 The presence of a fold in the form of 'U' shaped cavity in the defective zone (× 12)

Chemical Analysis

Chemical analysis of the defective part was found conforming to AU2GN specifications (Table 7.3).

Table 7.3 Chemical specification (Wt. %) for the French aluminium-copper-magnesium-nickel alloy (AU2GN) from which bell crank was made and actual composition of the failed part

Element	Min.	Max.	Failed part
Copper	2.00	2.60	2.25
Magnesium	1.20	1.80	1.60
Silicon	0.10	0.25	0.15
Iron	0.90	1.40	1.15
Nickel	0.90	1.40	1.20
Titanium	0.05	0.15	0.12

Discussion

Macro and micro structural analysis clearly established the fact that superficial indication of linear discontinuities along the fillet radius zones had occurred due to development of a 'push through' defect during forging operations.

On close scrutiny of the grain flow pattern of the forging (Fig. 7.13), as well as that of a desirable one (Fig. 7.13) it was evident that grain flow fibre ends of the defective part were found terminating at the radius, instead of running along the contour of the profile in the section change-over. Even in the absence of indication of discontinuities, parts thus forged with such abnormal grain flow pattern are very susceptible to development of fatigue crack owing to (1) depletion of localised strength in the critically loaded areas (as a result of loss of grain orientation and deformation texture) and (ii) less resistance to fatigue load (for mis-orientation of the grain fibre).

The 'push through' defect in this case is developed by the excess metal (that is ejected out of the already filled die cavity of the ribs) which have found a way through the inner radius towards the flash line along with excess metal from the web, causing oxide surfaces to get embedded in the radius and disturbing the desired grain flow pattern.

Conclusion and Recommendations

Most Probable Cause

The basic cause that led to the formation of linear discontinuities along the fillet radii, was due to folds/ laps resulting from the occurrence of a push through defect.

Remedial Action

In light of the nature of the defect present in the critical zone of the part, all the forgings were recommended to be rejected. To overcome the problem, a suitable pre-finish forging die set was recommended to be introduced to offset the re-entry of the excess metal. Or, in the absence of this intermediate sequence of operations excess metal from either tip of the ribs might be ground off to a pre-determined depth prior to finish forging.

7.4 DEVELOPMENT OF BLACK PATCH TYPE DEFECT IN ALUMINIUM ALLOY END COVER FORGINGS DURING ANODISING

Summary and recommendations

Almost half of a batch quantity of aluminium-copper end cover forgings showed identical black patches on the surface during chrome-acid anodising. Suspecting a process lapse, the defective parts were sand-blasted and etched before re-anodising but the same phenomenon persisted. Micro-examination of the defective zone revealed presence of a discontinuity (black oxide layer or grain boundary thickening) on the surface close to the black patch. To confirm the cause for the development of the black patch, a simulation test was carried out by heat treating the parts to different degrees of heat deterioration (slight to severe), prior to anodising treatment. The experiment revealed that the part containing grain boundary cracking was susceptible to development of black patches. It was concluded that the cause of black patches was from the local burning effect due to rise in current density in the presence of an overheated structure.

Background

Circumstances Leading to Failure

Seventy of a batch quantity of 150 aluminium-copper alloy end cover forgings showed black patches on anodising after heat treatment.

Specifications

The material used for the end cover forging was a British alloy to BS L77 specification.

Selection of Specimens

Two defective forgings were selected for micro-examination and three parts of a satisfactory batch (unanodised) of similar forgings were taken for simulation tests.

Visual Examination of General Physical Features

Examination of the forgings revealed identical black patches, either single or numerous but discontinuous, confined mostly to the base area of the forging (Fig. 7.15).

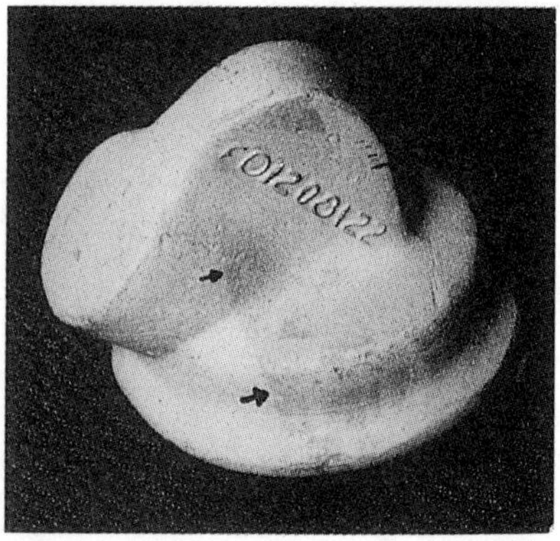

Fig. 7.15 Black patches (× 1)

Testing Procedure and Results

Metallography

Microstructural Analysis Micro-examination of the sections across the defect zone revealed a thin black oxide layer (Fig. 7.16) ingressed beneath the surface (possibly during forging), forming a distinct discontinuity up to a maximum depth of 0.02 mm. However, no anodised layer was noticed in the zone containing the oxide layer. Microstructure of another defective part showed grain boundary thickening (Fig. 7.17) in places very close to the surface in the black patch area.

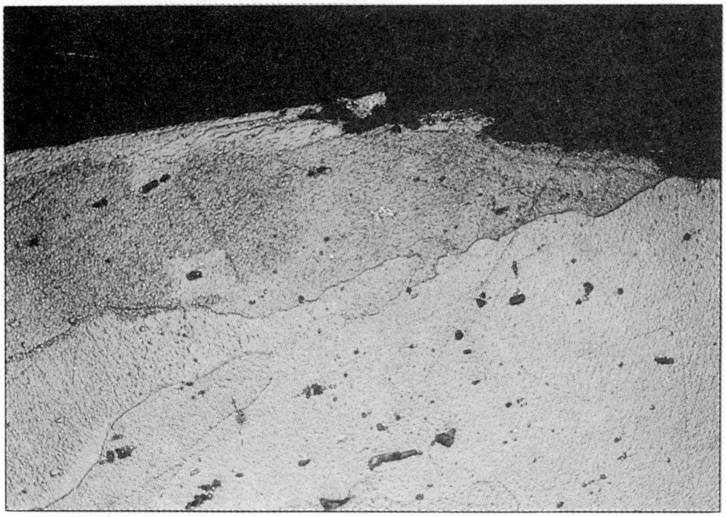

Fig. 7.16 Black oxide layer beneath the surface (× 400)

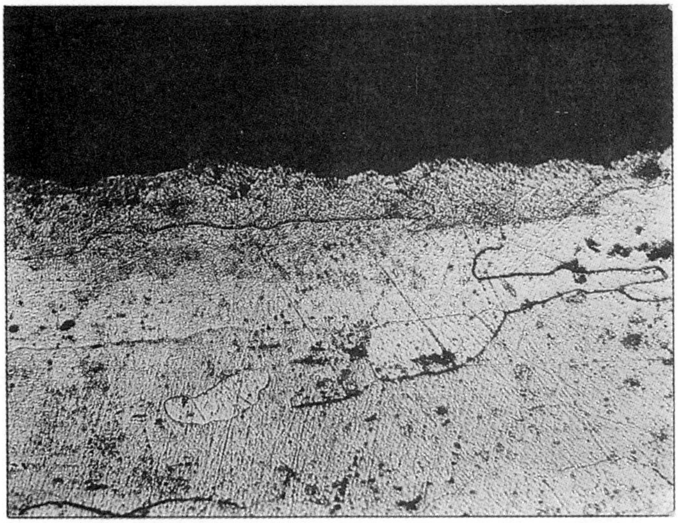

Fig. 7.17 Grain boundary thickening (× 400)

Chemical Analysis Chemical composition of the forging material was confirmed to be of L77 alloy specification (Table 7.4).

Table 7.4 Chemical specification (wt. %) for the British Aluminium-Copper alloy forging stock from which end cover was made and actual composition of the failed part

Element	Min.	Max.	Failed part
Copper	3.90	5.00	4.50
Magnesium	0.20	0.80	0.60
Manganese	0.40	1.20	0.75
Silicon	0.50	0.90	0.62
Iron	-	0.50	0.38
Nickel	-	0.20	0.15
Chromium	-	0.10	0.08
Zinc	-	0.20	0.12
Titanium and zirconium	-	0.20	0.14

Simulation Test

Since the defective parts showed re-appearance of black patches after they were sand blasted and etched before re-anodising, a simulation test was carried out to confirm the reason for black patches. Three parts of L77 forging were subjected to high temperature solutionising treatment in order to develop structures with severe grain boundary cracking/eutectic melting (Fig. 7.18), rounding off of constituents (Fig. 7.19) and normal

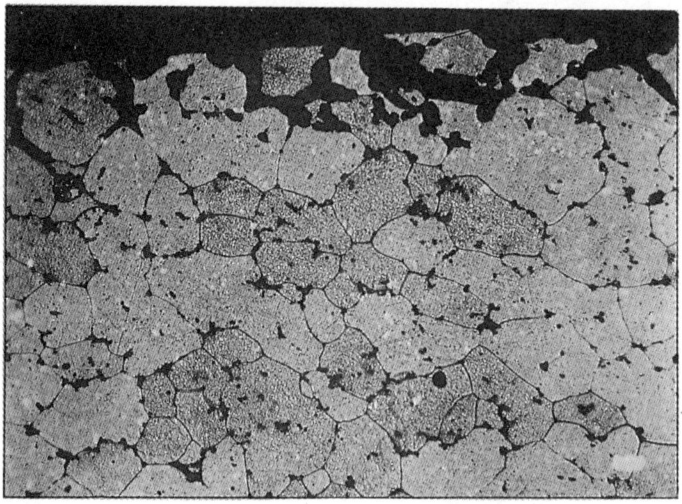

Fig. 7.18 Grain boundary cracking and eutectic melting (× 400)

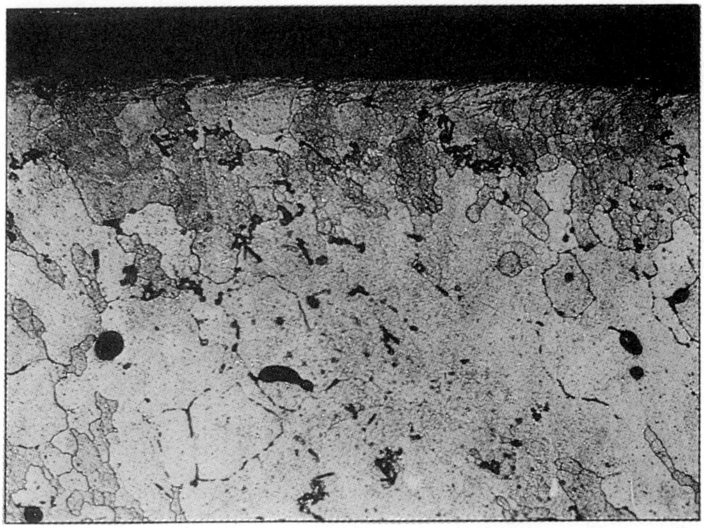

Fig. 7.19 Rounding off of constituents (× 400)

matrix without any overheating effects, respectively. These parts were anodised after etching in the same bath used for the above defective batch quantity. The severely overheated part showed uniform black patch throughtout the entire surface, whereas, the other two were found free from any blackening effect (Fig. 7.20).

Fig. 7.20 Blackening effects all over in the severely heat treated part (left) and no perceptible effect in the slighty overheated part (right) (× 1)

Discussion

Micro-examination of the zone containing the black patches and simulation test established the fact that the blackening effect was due to the local burning caused by the presence of oxide layer and grain boundary thickening in the forgings during anodising. The simulation test also confirmed that the part having severe heat deterioration, in the form of grain boundary thickening or cracking was more susceptible to burning than those with less overheating (i.e., rounding off of constituents) effects. Since almost half of the batch quantity was not showing any black patches, it was surmised that the heat treatment of the batch was quite satisfactory. The development of grain boundary thickening, as observed in the microstructure (Fig. 7.17), could be due to the fact that the base surface areas of the forging were locally overheated. It possibly occurred during critical deformation in the forging operation itself and before heat treatment. It is important to note that grain boundary thickening was confined only to the surface of the black patch area and its core structure was free from any abnormality. It is also well-known that aluminium-copper alloys containing more than 2% of copper have a greater tendency to burning during the high current density anodising treatment.

Conclusion and Recommendations

Most Probable Cause The basic cause for blackening of the surface of the forging was from a local burning effect as a result of rise in current density during chromic acid anodising, in the presence of discontinuities such as grain boundary thickening or cracking/oxide layer.

7.5 SURFACE BLISTERS/LAYERAL CORROSION IN ALUMINIUM-COPPER ALLOY EXTRUDED BAR

Summary and Recommendations

Several pieces of heat treated aluminium –4% copper alloy extruded bar in T6 condition, showed surface blisters during inspection. Visual, macro- and micro-examinations of the affected areas established the cause of surface blisters due to occurrence of exfoliation corrosion phenomenon during storage. Although temporarily protected against corrosion, exfoliation corrosion developed from the unfavourable metallurgical factors such as heat treatment condition and grain structure.

It was recommended to use this alloy in modified temper condition (T851) instead of T6, in which exfoliation corrosion susceptibility is minimal.

Background

Applications

A number of 90 mm dia aluminium copper alloy extruded bars were stored in fully heat treated conditions (T6), for fabrication of finished parts for use in helicopters. The bars were temporarily protected against corrosion with RCOM lacquer.

Circumstances Leading to Failure

Several bars were showing surface blisters and corrosion products (Fig. 7.21), on inspection after a period of storage time.

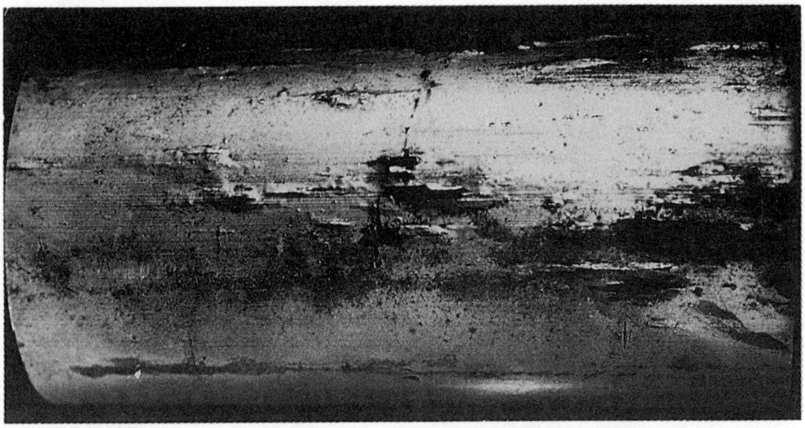

Fig. 7.21 Surface blisters and corrosion products (× 1)

Specifications

The material used for the extrusion was an aluminium-4% copper type British alloy to BS L 65 specification.

Selection of Specimens

A small cut portion from one of the affected extrusion lengths with defects was collected for micro-examination, to establish the nature and mechanism of the defect.

Visual Examination of General Physical Features

Examination of the defective zones visually (and also under a low power stereoscopic microscope) revealed occasional blisters , accompanied with voluminous white corrosion products (Fig. 7.21).

Testing Procedure and Results

Metallography

Microstructural Analysis Micro-examination of the cross-section in the longitudinal direction revealed surface corrosion to a depth of 0.5 mm from the section, resulting from a rapid layeral attack along the grain boundaries. The layeral separations were found parallel to the metal surface with a characteristic leafing action (Fig. 7.22). These features positively indicated the presence of a corrosion mechanism similar to that of an exfoliation type. Microstructure was otherwise found to be satisfactorily heat treated.

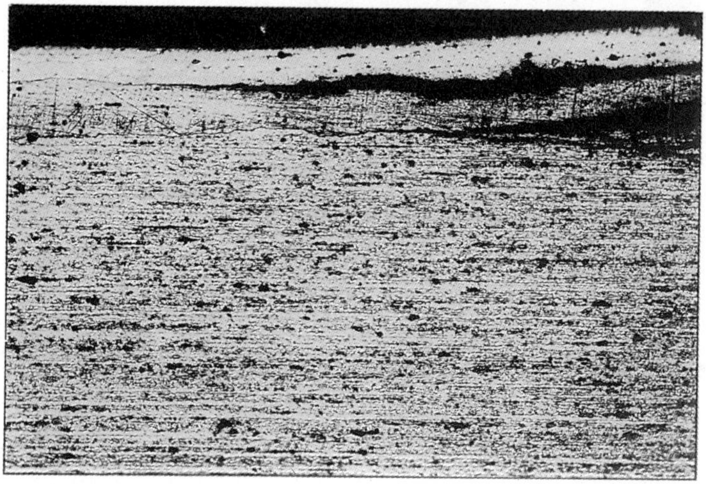

Fig. 7.22 Layeral corrosion and leafing action (× 50)

Chemical Analysis

Chemical analysis of the extruded bar was found conforming to BS L 65 specification (Table 7.5).

Table 7.5 Chemical specification (Wt %) for the British aluminium-copper alloy (BSL 65) from which extruded bar was made and actual composition of the failed part

Element	Min.	Max.	Failed part
Copper	3.90	5.00	4.20
Magnesium	0.20	0.80	0.55
Manganese	0.40	1.20	0.65
Silicon	0.50	0.90	0.72
Iron	-	0.50	0.41
Nickel	-	0.20	0.15
Chromium	-	0.10	0.05
Zinc	-	0.20	0.10
Titanium+zirconium	-	0.20	0.12

Mechanical Properties

Hardness Hardness checked on the micro specimen gave an average value of 136 BHN.

Discussion

Macro and micro-examination clearly indicated the fact that the aluminium – 4% copper alloy extruded rod had suffered an exfoliation corrosion attack during storage. Exfoliation in this case was exhibited in the form of a lamellar corrosion, resulting from a rapid lateral attack along the grain boundaries or striations within the grains. This was additionally confirmed by the presence of voluminous corrosion products which caused the uncorroded strata to be split apart. From a close observation of the micrograph (Fig. 7.22) it was found that exfoliation corrosion progressed through the elongated grain structure with grains relatively thin compared to their width.

Conclusion and Recommendations

Most Probable Cause The basic cause for the surface blisters was attributed to exfoliation corrosion attack to which this alloy product is susceptible in heat treated condition (T6).

Remedial Action As the corrosion was found confined only to 0.5 mm surface layer, the bar was recommended to be used only for static applications after turning off 1 mm layer from the surface.

It is well-known that the A1-Cu, besides A1-Zn, Mg-Cu and Al-Mg wrought alloy products are particularly sensitive to the development of exfoliation corrosion. So, it was suggested to control the degree of susceptibility of the alloy by giving a special metallurgical treatment, i.e. T 851 temper condition.

7.6 HYDROGEN EMBRITTLEMENT CRACKING OF STEEL PISTON RODS DURING CHROME-PLATING

Summary and Recommendations

A number of chromium/molybdenum steel piston rods were found cracked during chrome-plating and some after the baking treatment. A detailed investigation was carried out to locate the source of cracking with the aid of the optical as well as the scanning electron microscope. Gas analysis was also conducted.

Fractographic examination of the cracked surface indicated a coarse and crystallined (brittle) fracture. Scanning electron microscopic examination confirmed that the cracking was due to hydrogen embrittlement, based on characteristic fracture features such as micro-quasi-cleavage, transgranular mode of cracks and a rare botanical curved leaf like structure. Instrumental gas analysis indicated a high gas content of hydrogen in the cracked part.

As the part was heavily machined in the heat treated condition and then plated, it was recommended to minimise the residual stresses by stress relieving treatment prior to plating, and to carry out the baking treatment immediately after plating.

Background

Applications

Piston rods which were heavily machined out of a fully heat treated, low alloy chromium-molybdenum steel bar stock and plated against corrosion, are used in the complex transmission system of helicopters. These parts are subjected to cyclic loading in service and hence, are critical in terms of safety against failure. The machined parts were not stress-relieved prior to plating.

Circumstances Leading to Failure

A few piston rods from different production batches were found cracked in the eye end portion, after the chrome plating and baking treatment. One of them cracked in the plating operation itself instantly broke on slight tapping (Fig. 7.23).

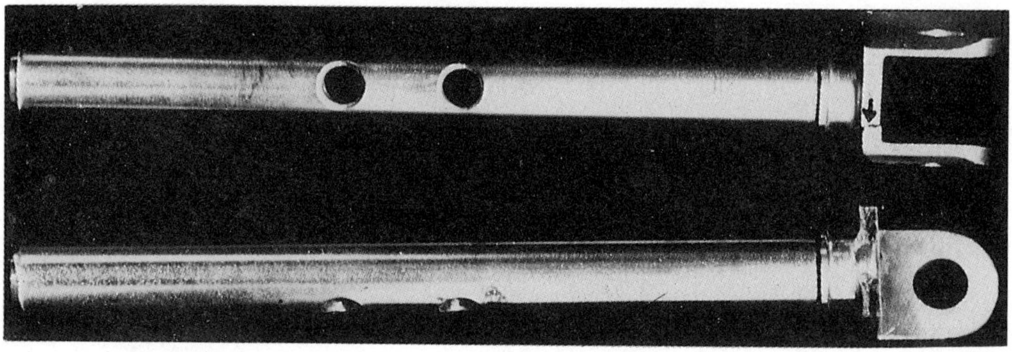

Fig. 7.23 Cracks in the eye end portion of the piston rods (× 1)

Specifications

The steel used for fabrication of the piston rods was a French alloy to 30 CD 12 specification.

Selection of Specimens

Three cracked piston rods were taken and out of them the one already broken was transversely sectioned for examination under the scanning electron microscope. Two more samples were sectioned from the remaining portion of the fracture for microstructural and gas analysis studies.

Visual Examination of General Physical Features

Examination of the piston rods showed identically straight cracks along the outer radii, i.e. 0.5 mm away from one of the two forked eye ends (Fig. 7.23).

Testing Procedure and Results

Surface Examination

Visual On close examination, the rods were found chrome plated up to the outer base surface of the forked eye and, with the inside portion masked prior to plating. Further, numerous tiny gas blisters were noticed all over the chrome plated surface.

Macrofractography Fractography examination of the broken surfaces of the eye ends showed identical fractures, i.e. coarse, irregular and crystalline features, typical of a brittle inter-crystalline failure (Fig. 7.24).

Fig. 7.24 Coarse crystalline features in the fracture after the cracks were opened out (× 6)

Scanning Electron Microscopy Fractography Examination of the fracture by the scanning electron microscope at × 450 magnification revealed a curved botanical leaf (Figs 7.25 and 7.26) like structure (a rare form that occasionally characterises the hydrogen embrittlement phenomenon) and transgranular cracks in the quasi cleavage matrix of the predominantly intergranular (Fig. 7.26) fracture. These fracture features were a clear indication of failure due to hydrogen embrittlement.

Metallography

Microstructural Analysis Micro-examination of a section across the fracture revealed intergranular cracking in the tempered martensitic structure with occasional lifting up of crystals (Fig. 7.27). The feature of lifting up of crystals was indicative of a sudden intergranular mode of failure.

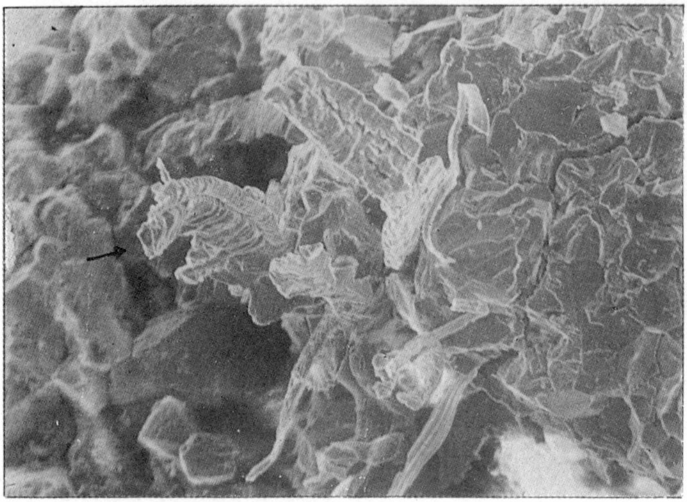

Fig. 7.25 Curved botanical leaf-like structure under scanning electron microscope (× 450)

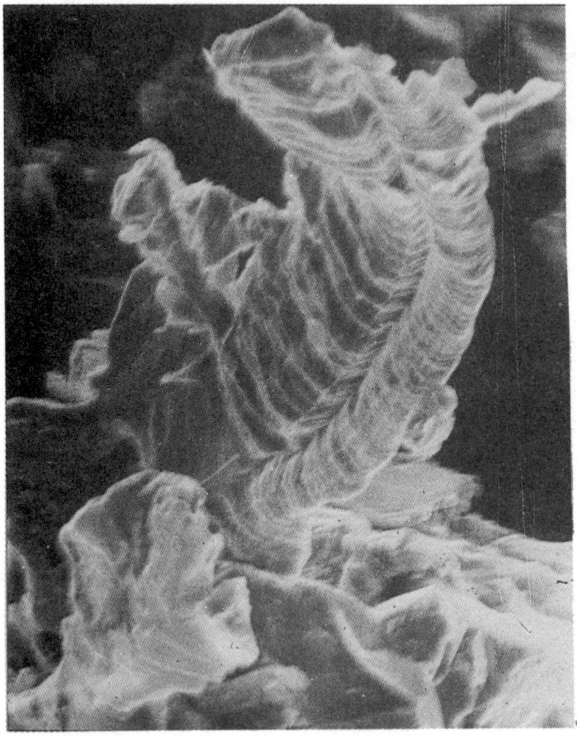

Fig. 7.26 Enlarged view of the botanical leaf as in Fig. 7.25 (× 1200)

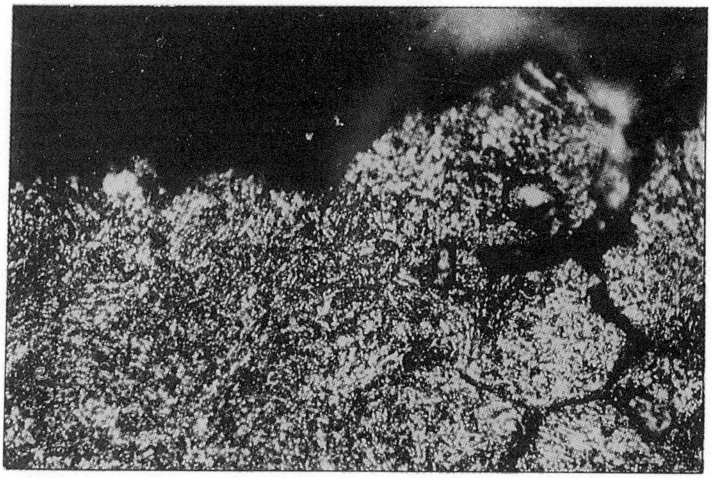

Fig. 7.27 Intergranular cracking with lifting of crystals (× 100)

Chemical Analysis

Chemical composition of the piston rod was found conforming to 30 CD 12 specification (Table 7.6). Further, instrumental gas analysis indicated a hydrogen gas content of the order of 18 ppm in the cracked part.

Table 7.6 Chemical specification (Wt.%) for the French chromium molybdenum steel (30 CD 12) from which piston rod was made and actual chemical composition of the failed part

Element	Min.	Max.	Failed part
Carbon	0.28	0.35	0.31
Manganese	0.40	0.70	0.55
Silicon	0.10	0.40	0.22
Sulphur	-	0.020	0.016
Phosphorus	-	0.025	0.020
Chromium	2.80	3.30	3.10
Molybdenum	0.30	0.50	0.42

Mechanical Properties

Hardness A hardness survey using Vickers pyramid hardness testing machine was carried out on the micro-specimen and the average hardness value was found to be on the higher side of the specification value (400 against 380 VPN max).

Discussion

Fractographic (stereoscopic and scanning electron microscopic) as well as microstructural analysis clearly established that the brittle cracking of the piston rods occurred due to the hydrogen embrittlement phenomenon during chrome-plating process. Several factors acted in combination to enhance the hydrogen gas absorption and caused instant cracking in extreme cases. One of the factors was the presence of high residual stresses in the part, particularly in the location of complex shaped thin forged eye end radius zones due to removal of a heavy mass by machining in the heat treated condition (as ascertained from fabrication route cards/documents). It is a well-known fact that the residual stress increases the rate of hydrogen gas absorption during plating. Presence of numerous gas bubbles/blisters all over the surface of the affected parts, prior to baking, indicated a high range of gas absorption. As time lapses between plating and baking for different batches (as confirmed from the records) were found widely varying, the baking operation was not effective in removing the hydrogen gas. Besides, the hydrogen content of the order of 18 ppm present in the part was sufficient to cause hydrogen embrittlement cracking without any external stresses. It is also pertinent to note that high tensile strength material is susceptible to hydrogen embrittlement. In the present case, hardness of the cracked piston rod was found to have exceeded the maximum limit of the specification (i.e. equivalent strengthwise UTS was 127 against 125 kg/mm^2 max. as per specification).

Conclusion and Recommendations

Most Probable Cause The basic cause for brittle cracking along the forked eye end radius portion was attributed to hydrogen embrittlement phenomenon.

Remedial Action It was recommended to fabricate the parts from annealed/normalised bar stock, and subject them to a proper stress relieving operation after heat treatment and prior to the baking treatment. It was further recommended to ensure that the time delay between plating and baking is kept minimum so as to avoid hydrogen gas permeation.

7.7 HYDROGEN EMBRITTLEMENT CRACKING IN STEEL FORGINGS

Summary and Recommendations

A batch of low alloy steel end frame forgings developed random tiny cracks, mostly the fillet radius portion, on heat treatment. Suspecting a lapse in heat treatment, a second heat treatment was carried out on another batch which revealed identical cracks on magnafluxing. In order to find out the reason, a detailed probe was made with the help of optical as well as scanning electron microscope and instrumental gas analysis.

Microstructural analysis was carried out both on the failed part, as well as on disc of the rolled bar from which the part was made and this was bored to 50 mm dia hole to develop sub-surface cracks. The sub-surface cracks were found to be zig-zag and discontinuous as well as intergranular in nature. On scanning electron microscopic examination of the cracked surface, a mixed mode of fracture involving ductile dimpled structure and brittle flat facets were observed along with the hairline cracks, micropores and rod-shaped manganese sulphide inclusions. All these characteristics indicated a hydrogen embrittlement phenomenon. Moreover on gas analysis, hydrogen content of the order of 22 ppm was found to be present in the material, which was sufficient to cause hydrogen embrittlement without any external stresses.

It was recommended that the hydrogen gas content in the raw material stage must be checked prior to fabrication of the parts . Also, acid pickling, used for the purpose of descaling of the hot rolled bars, needs to be carefully controlled to prevent excessive gas absorption.

Background

Applications

A batch quantity of 120 end frame parts forged and machined out of 200 mm dia medium carbon chromium steel rolled bars, was heat treated as per specification requirements. These parts are used for critical applications in helicopters and hence utmost care is necessary to ensure quality of products right from the melting stage onwards.

Circumstances Leading to Failure

The initial batch of 50 heat treated parts showed several tiny radial cracks on magnafluxing after cadmium plating (Fig. 7.28).

Specifications

The material used for the fabrication of the part was a Russian alloy to 38 XA specification.

Selection of Samples

One cracked sample was selected for low-powered optical and high powered scanning electron microscopic examination, along with 20 mm thick disc sliced from the supplied (imported) rolled bar used for forging.

Visual Examination of General Physical Features

Examination of the magnetic particle inspected parts showed several tiny radial cracks at random, mostly confined to the base edge and fillet radius zones (Fig. 7.28).

Fig. 7.28 Several tiny crack like indications, mostly in the base fillet radius zones, on magnafluxing after chrome plating (× 1)

Testing Procedure and Results

Scanning Electron Microscopy Fractography

One of the tiny cracks was carefully opened out (since depth of the crack is negligible, it was difficult to retain the fracture zone intact) and the fracture zone was examined under the electron microscope. This showed predominantly a dimple structure of a ductile failure, and no clue as to the nature and type of the crack could be established. Suspecting that the hairline cracks could be due to hydrogen embrittlement phenomenon, two sliced discs out of the rolled bar, one in the annealed (as supplied) and the other in the heat treated condition were subjected to a 50 mm dia boring in the centre, similar to the one in the finished end frame part. Both the discs, on magnafluxing, showed tiny discontinuous sub-surface crack like indications on the bore surface (Fig. 7.29). One of the sub-surface cracks was so carefully opened out, that the very small zone of cracking could be retained undamaged for scanning electron microscopic examination. The fracture clearly revealed dimpled structures associated with brittle (cleavage) flat facets, ductile hairline cracks, micropores (Fig. 7.30) and rod-shaped manganese sulphide inclusion (Fig. 7.31). All these fracture characteristics were typical of a hydrogen embrittlement phenomenon.

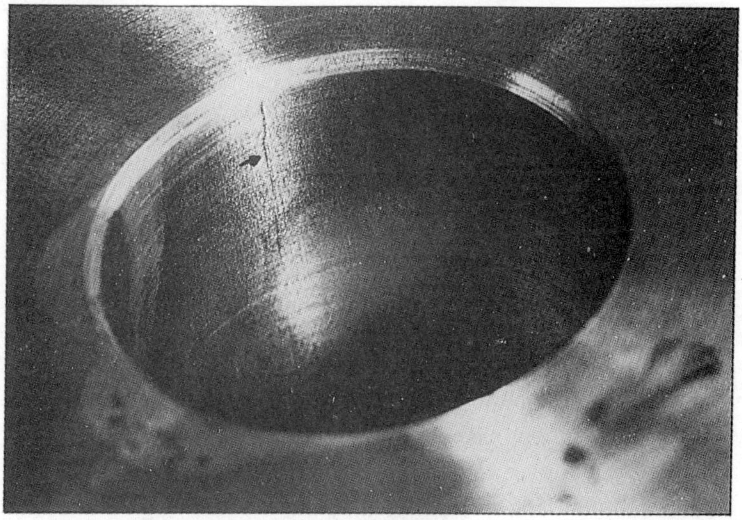

Fig. 7.29 The presence of sub-surface cracks as indicated by magnafluxing in the 50 mm dia machined hole surface of the rolled bar stock used for fabrication of the part (× 1)

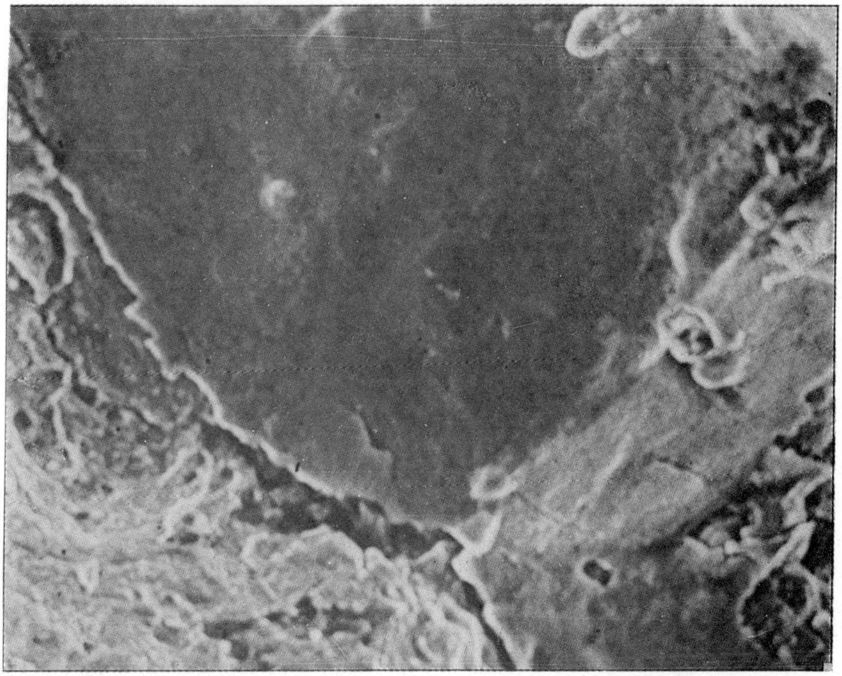

Fig. 7.30 A mixture of heterogenous structures including dimples, brittle facets, micropores and hair line crack under scanning electron microscope (× 450)

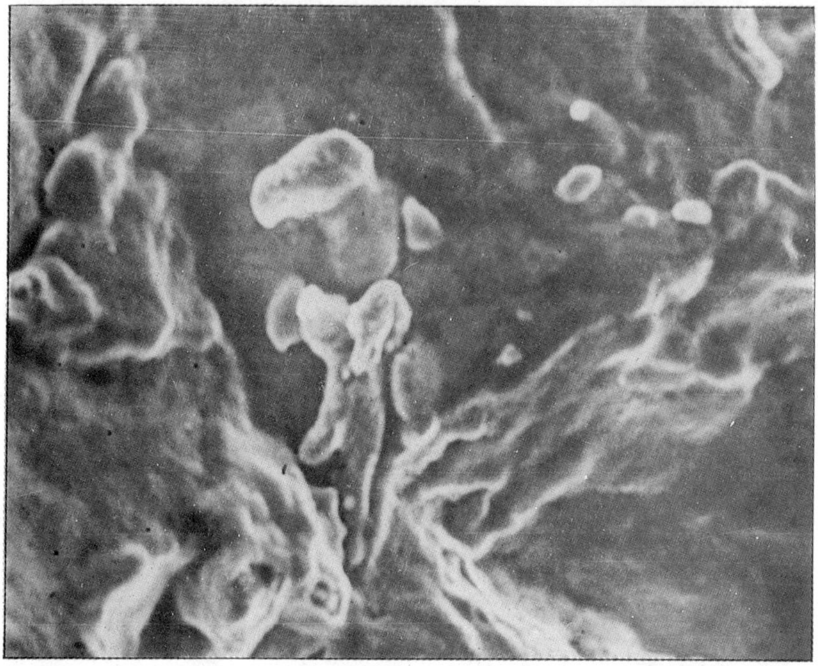

Fig. 7.31 Presence of magnanese sulphide inclusions in rod shaped form in the zone of tiny cracks ($\times 450$)

Metallography

Microstructural Analysis Macro-examination of the sections taken across the defect indications in both the discs revealed sub-surface discontinuous hairline (zig-zag) cracks (Fig. 7.32), which were intergranular in nature without any evidence of branching . Microstructure of the defective part at higher magnification also showed discontinuous intergranular cracks with typical manganese-sulphide inclusions along the path of cracking (Fig. 7.33).

Chemical Analysis

The chemical composition of the end frame material was found conforming to specification (Table 7.7). Further, the cracked end frame part was subjected to instrumental gas analysis and the hydrogen gas content was found to be of the order of 22 ppm.

Mechanical Properties

Impact Toughness Charpy impact properties determined in locations both near and far from the defective zone, in the longitudinal and transverse directions, were found generally low (i.e. 4.5 and 3.5 kg M/cm^2) against specification requirement of 5 kgm/cm^2.

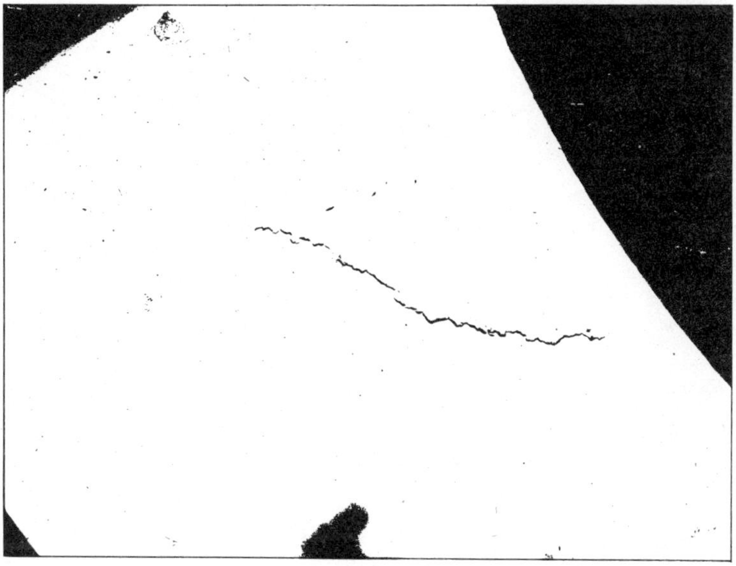

Fig. 7.32 The presence of sub-surface cracks in the zone across the crack indications (as polished) (× 12)

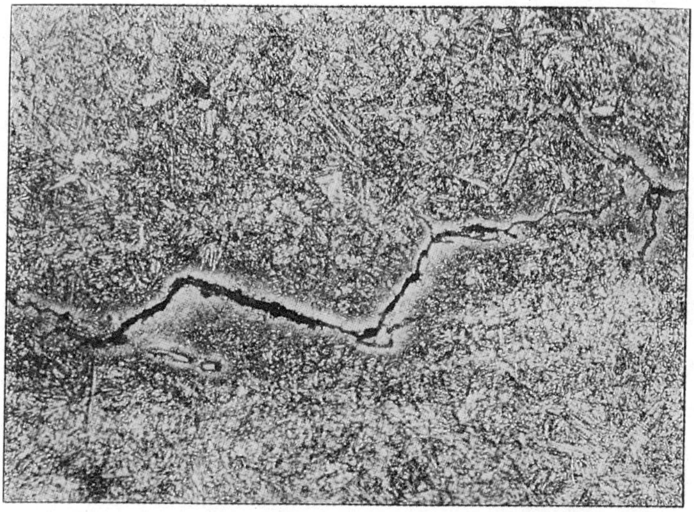

Fig. 7.33 Intergranular cracks with manganese (grey) and oxide (dark) inclusions in the crack path of a section across the radial cracks of the heat-treated part (× 100)

Table 7.7 Chemical specification (Wt. %) for the Russian medium carbon chromium steel (38 × A) from which end frame forgings were made and actual composition of the failed part

Element	Min.	Max.	Failed part
Carbon	0.34	0.42	0.38
Manganese	0.50	0.80	0.60
Silicon	0.17	0.37	0.22
Sulphur	-	0.03	0.02
Phosphorus	-	0.03	0.02
Nickel	-	0.40	0.25
Chromium	0.80	1.1	0.98

Discussion

The scanning electron fractography clearly established that random occurrence of the tiny cracks was due to hydrogen embrittlement phenomenon. This is also supported by the presence of high hydrogen content in the material prior to heat treatment.

Initially from the nature and location of the cracks (mostly in the fillet radius zones), it appeared to be a suspected case of faulty heat treatment. However, when subsequent heat treatment of the second batch showed identical cracks in the same location (despite satisfactory chemistry, microstructure and hardness as per specifications), the random cracking was strongly suspected to be associated with some hydrogen induced phenomenon. Since the tiny cracks of the part could not be successfully opened up, the sub-surface crack of the heat treated disc (as mentioned in section) was forced open and the fracture was examined under the SEM which revealed characteristic hydrogen embrittlement features (Figs 7.30 and 7.31). It may be pertinent to note that a 50 mm dia hole was bored in the disc to develop residual stresses of the order which was sufficient to cause sub-surface cracking. It is also known that under high stresses, hydrogen gas pressure increases rapidly causing flakes/fissures. Moreover, a gas content of the order of 22 ppm present in the raw material was enough for high strength steel to cause hydrogen embrittlement without any external stresses. Further, the low impact properties measured on the failed part also indicated the brittle nature of the material caused by hydrogen embrittlement.

Conclusion and Recommendations

Most Probable Cause

The basic reason for development of tiny cracks was due to hydrogen embrittlement phenomenon.

Remedial Action In the absence of records on manufacturing history of the supplied raw material, the major source of hydrogen absorption was suspected to have come from improper pickling operations of the hot rolled bars. It was suggested to control the acid pickling operation strictly within the specified process parameters in future supplies. Also, quality control was advised to check the level of hydrogen gas in the rolled bars before acceptance for forging.

7.8 FATIGUE CRACKING OF STEEL LINK RODS OF A CABLE ASSEMBLY DURING SWAGING

Summary and Recommendations

Two steeel link rods of a cable assembly were reported cracked at the inspection hole during swaging.

Visual and macro examination conducted on these parts revealed a 2 mm long crack emanating from either side of the hole edge in an identical manner. Fractographic examination of the cracked surfaces indicated the failure was due to low cycle fatigue phenomenon. Based on the observations, fatigue cracking of the rods was attributed to concentration of high tensile stresses at the sharply cut hole edge, during repeated impact load by swaging around the zone in the close proximity of the hole.

It was recommended to carry out swaging at a safe distance from the hole edge, as it acted as a sensitive notch by virtue of its closeness to the highly worked region.

Background

Application

A cable assembly was swaged at the rear end portion of the steel link rods in the assembled condition, with a band of wire strands into the rod bore (Fig. 7.34).

Circumstances Leading to Failure

Both the link rods of the cable assembly showed tiny cracks on either side of the inspection hole on completion of the swaging operations (Fig. 7.35). Development of cracks appeared to be unusual during the manufacturing stage.

Specifications

The material used for fabrication of the link rod was a British alloy to BSS 96 specification (a low nickel-chromium-molybdenum alloy steel).

Selection of Specimen

Both the identical cracks were carefully opened out and one portion was selected for detailed examination under the stereoscopic and optical microscopes.

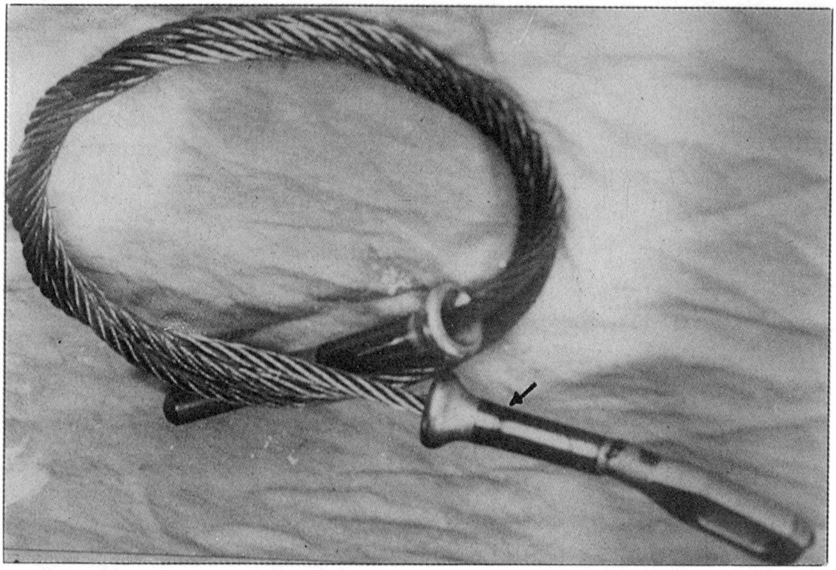

Fig. 7.34 The swaged cable assembly with the steel link rod and its location of crack (× 1)

Fig. 7.35 Cracks on either side of the inspection hole (× 10)

Visual Examination of General Physical Features

Visual examination of both the failed link rods (using magnifying lens) showed feeble indication of tiny cracks emanating from either side of the inspection hole edge (Fig. 7.35). Further the hole edges were found very sharply cut, without any deburring and elongated towards the swaged portion.

Testing Procedure and Results

Non-destructive Evaluation

Magnetic Particle Based on the above indications, both the link rods were subjected to magnetic particle inspection. The test showed distinct crack like indications (Fig. 7.35) at the hole edge.

Surface Examination

Visual Examination of the magnetic particle inspected rod confirmed the cracks and were found extending to a length of 2 mm in a circumferential direction (Fig. 7.35). On close observation, the swaging was found to have been carried out to within 1 mm from the hole edge.

Macrofractography Fractographic examination of the cracked regions, at × 25 magnification, under the stereoscopic microscope revealed a distinct smooth zone with widely spaced but faint beach markings (Fig. 7.36). This is typical of failure under a low

Fig. 7.36 Fatigue zones with trace beach marks (× 6)

cycle fatigue phenomenon. The fatigue cracks were found to have originated at the sharply cut and undeburred hole edge. The fatigue zones on either side of the hole edge were found identical and diametrically opposite to each other, covering 1/15th of the cross-sectional area. The remainder showed over-stressed fracture characteristics, which occurred during forced opening of the crack.

Micro-examination of the section taken across the fatigue zone, under the optical microscope at × 100 magnification, revealed satisfactory hardened and tempered martensitic structures. The material was found to be cleaned with acceptable inclusion ratings.

Crack Origin/Paths The crack had its origin at the sharp edge of the inspection hole and was found to be transgranular in nature, further confirming the characteristics of a fatigue failure.

Chemical Analysis

Chemical compositon of the cracked link rod was found conforming to S 96 specification (Table 7.8).

Table 7.8 Chemical specification (Wt%) for the British nickel-chromium-molybdenum alloy steel (BSS 96) from which link rod was made and actual composition of the failed part

Element	Min.	Max.	Failed part
Carbon	0.27	0.35	0.32
Manganese	0.45	0.70	0.52
Silicon	0.10	0.35	0.23
Sulphur	-	0.020	0.015
Phosphorus	-	0.025	0.018
Nickel	2.30	2.80	2.51
Chromium	0.50	0.80	0.72
Molybdenum	0.45	0.65	0.54

Mechanical Properties

Hardness This was measured on the micro-specimen using the Vickers pyramid hardness testing machine. The average value of 360 VPN was found meeting the specification requirements.

Discussion

Fractographic and microstructural analysis of the cracked zone clearly indicated that development of the tiny cracks on either side of the inspection hole had occurred due to fatigue phenomenon. However, this appeared to be rather unusual from the view point that fatigue seldomly occurs during the manufacturing stage and is generally encountered in the presence of an alternate tensile (cyclic) load application during service. In this case, it seemed that the repeated concentration of high tensile stresses at the sharply cut hole edge (by virture of imposition of high compression loads during swaging in the close proximity of the hole) led to the premature high load low cycle fatigue cracking.

Conclusion and Recommendations

Most Probable Cause

The basic cause that led to the premature cracking of the link rod in the inspection hole area, was the development of a low cycle fatigue during swaging.

Remedial Action In view of the development of fatigue cracks, the limit of swaging was recommended to be at least 3 mm off the hole edge. This is because the hole edge serves as a very sensitive notch by virtue of its closeness to the highly worked region. Further it was suggested to debur the hole edges to avoid unnecessary stress concentration effects.

7.9 SURFACE CRACKING IN THE NEAR WELD ZONE OF A TITANIUM ALLOY FRONT BY-PASS DUCT OF AN ENGINE DURING TYPE TESTING

Summary and Recommendations

A front by-pass duct, fabricated out of welding of 1 mm thick commercially pure titanium alloy sheet, failed during pressure cyclic test with a crack in the near weld zone. The crack was twice repaired as prescribed following failure in the same location. Subsequently a modified patch plate welding was carried out in the defective area, but identical cracks reappeared on pressure testing.

A detailed investigation, involving stereoscopic as well as scanning electron microscopic examination, was carried out in the defective area which revealed a crack of zig-zag pattern and brittle nature. Micro-examination showed entrapment of an oxide layer in the inhomogeneous alpha phase structure. On physical examination, the surface contour indicated a sharp geometrical mismatch of the sheet joint at the defective weld location.

The basic cause of repeated brittle cracking was attributed to a combination of pre-existing notch effect in the weldment (i.e. due to mismatch) and inhomogeneous coarse brittle weld structure.

Hence it was suggested to minimise the mismatch of the sheet joint and use better welding techniques to improve the microstructural homogeneity.

Background

Applications

The front by-pass duct component was circumferentially welded out of 1mm thick commercially pure titanium alloy sheet by the orbital welding technique (Fig. 7.37). The welded component was stress-relieved after satisfactory X-ray examination. But when

Fig. 7.37 The by-pass duct assembly (× 1/10)

pressure tested at 50 psi, a transverse crack was revealed on fluorescent dye penetrant check. When the crack was welded and tested, two transverse cracks showed up in the same location. The same repair technique was repeated but the defect opened up at 90 psi pressure test. A modified repair method (permitted as per instruction manual) was adopted using 'Patch plate' welding in the cut-out zone (50 mm × 25 mm) of the defective area (Fig. 7.38).

Circumstances Leading to Failure

During the final type testing, the modifed weld area, after it had passed cyclic pressure test to 500 cycles, developed surface cracks near the weld.

Specifications

Material used for the duct was a commercially pure titanium alloy to a British BSSTA1 specification.

Selection of Specimens

The front by-pass duct was carefully cut open around the defective zone. Sections were taken through the defect for micro-examination. One of the cracks was opened out for

Fig. 7.38 Weld patch and pronounced undulation on the skin, away from the weld area (× 1/2)

fracture studies under the stereoscope as well as scanning electron microscopes. The fracture was ultrasonically cleaned prior to scanning electron microscopic examination.

Visual Examination of General Physical Features

The defect location was found to be on the outer skin, near the flange along the circumferential weld bead, adjacent to a 50 × 25 mm patch plate (Figs 7.39 and 7.40). The crack was found to be a 'zig-zag' type surface rupture running to an approximate length of 14 mm (Fig. 7.40). The defect location showed a high degree of mismatch as recorded by the plastic cast impression of the fracture profile (Fig. 7.41) and the skin away from the weld area had pronounced undulations (Fig. 7.38). Also the weld bead was found non-uniform.

Testing Procedure and Results

Non-destructive Evaluation

Radiography The defective weld area was examined with X-rays which detected the presence of a very fine zig-zag crack, but no abnormalities were observed within the weld bead around the zone of cracking.

Liquid Penetrant Further, dye check on both the outside and inside surfaces of the defective area showed the crack confined only to the outer surface (i.e. not through and through).

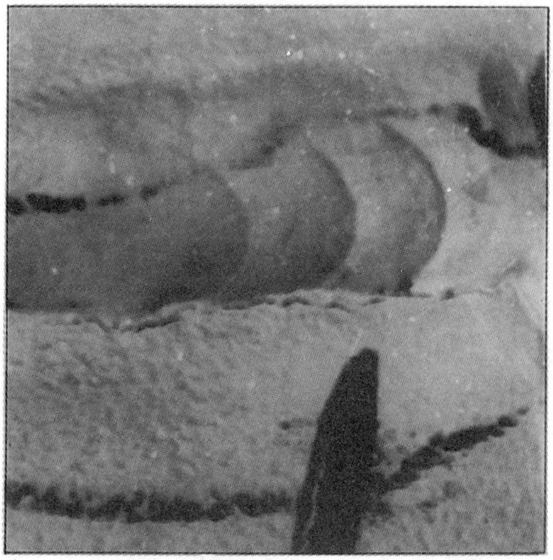

Fig. 7.39 Zig-zag type crack/rupture (× 1)

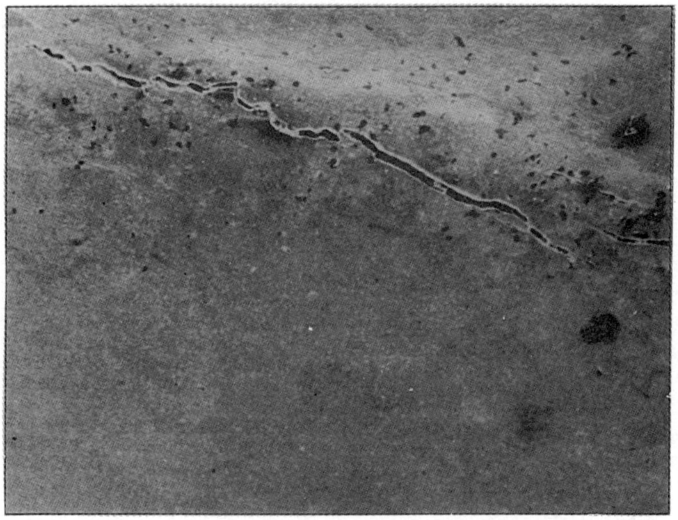

Fig. 7.40 A close-up of the zig-zag crack (× 3)

Surface Examination

Macrofractography Examination of the cracked surface under the stereoscopic microscope at x 10 magnification revealed a coarse and irregular fracture with brittle features and some faint discolouration.

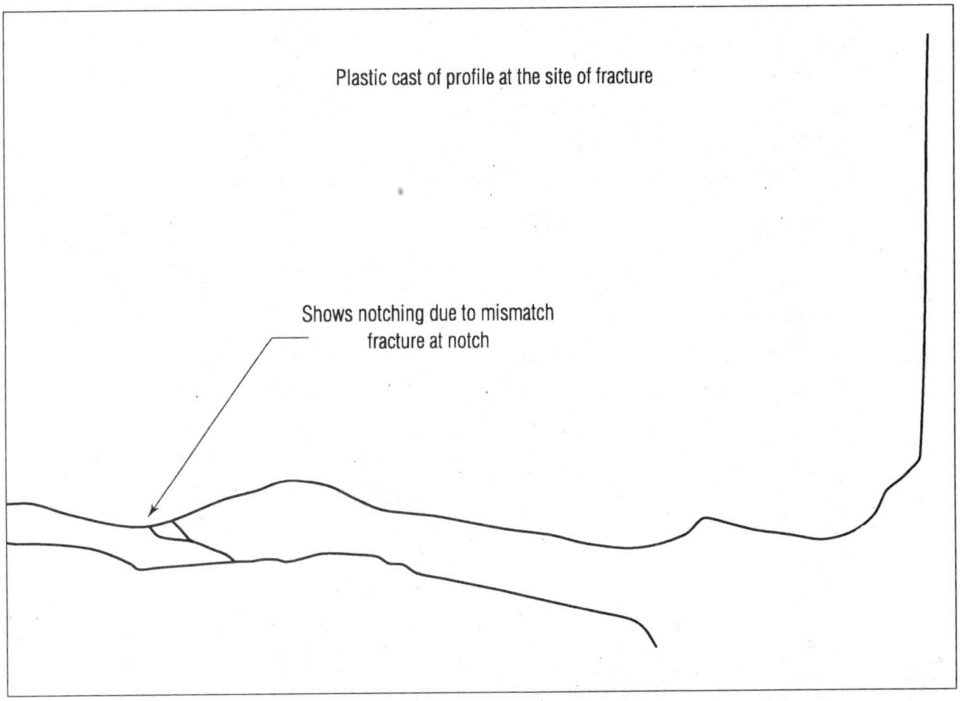

Fig. 7.41 Severe mismatch at the site of the fracture

Scanning Electron Microscopy Fractography Scanning electron microscopy examination at × 450 magnification revealed a flat faceted cleavage with virtually no sign of the dimpled features (Fig. 7.42). However, at a higher magnification of × 1500, the fracture showed secondary cracking and interface separation (Fig. 7.43). No features of a progressive type of failure, such as fatigue, was observed.

Metallography

Microstructural Analysis A section across the defect was examined under the optical microscope (at × 50). It showd a short transverse crack to a max. dèpth of 0.32 mm (Fig. 7.44), starting at the apex of the 'V' profile in the vicinity of the fusion line (i.e. between the edge of the weld bead and heat affected zone). The defect zone also showed the clear presence of an oxide entrapment in the microstructure, along with a few gas pores in the weld fusion line. The microstructure predominantly consisted of acicular alpha phase, coarse in the heat affected zone (HAZ) and finer in weld bead (Fig. 7.45 and 7.46). However, the microstructure in HAZ, away from the defect, was found to be fairly homogeneous (Fig. 7.47).

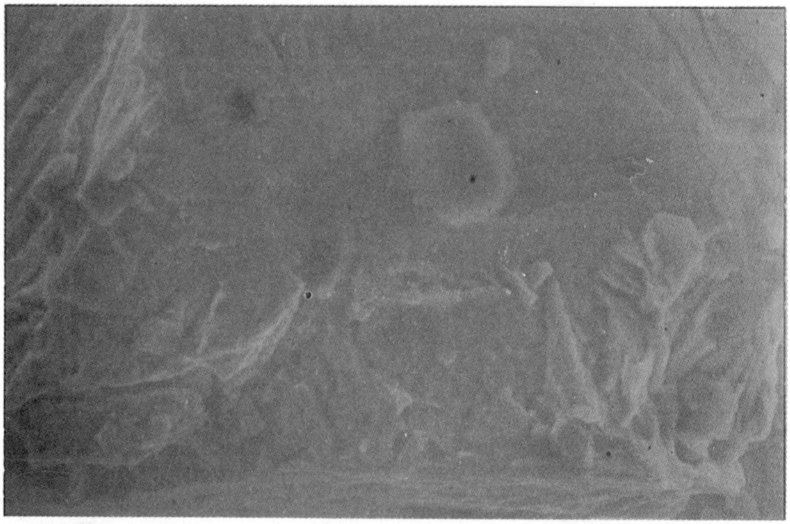

Fig. 7.42 Brittle flat faceted cleavage under the scanning electron microscope (× 450)

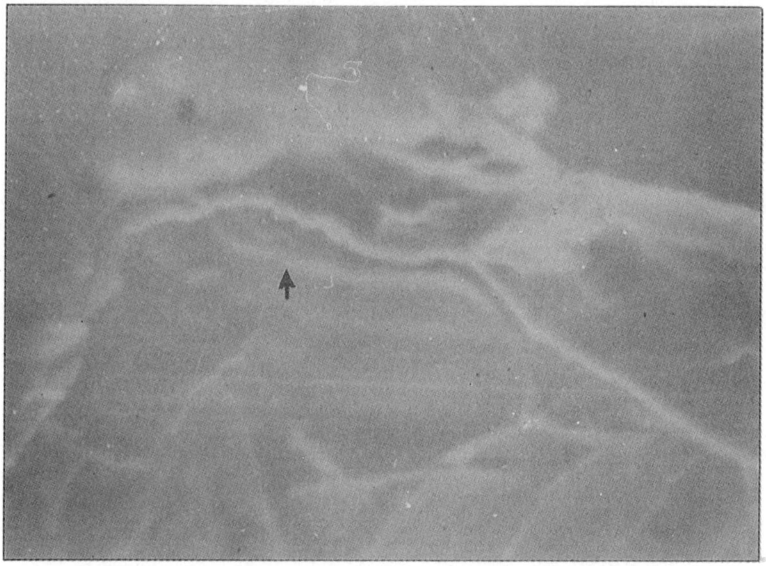

Fig. 7.43 Secondary cracking and interface separation in the same zone in Fig.7.42, at high magnification, under the scanning electron microscope (× 1500)

Fig. 7.44 Transverse crack, oxide entrapment and gas pores (× 50)

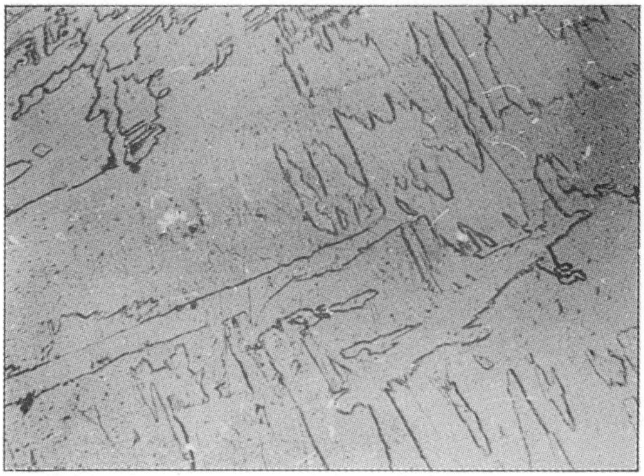

Fig. 7.45 Acicular alpha phase in HAZ (× 200)

Mechanical Properties

Hardness Microhardness survey in the regions of the weld, HAZ in the vicinity of the crack and base metal did not reveal any significant variations. The hardness was found to be in the range of 240 to 253 VPN.

Simulation Tests

Based on the fractographic analysis, a fracture was intentionally made in the sound weld region and on examination under the scanning electron microscope, it showed predominantly dimpled features, contrary to the brittle flat facets of the cracked zone.

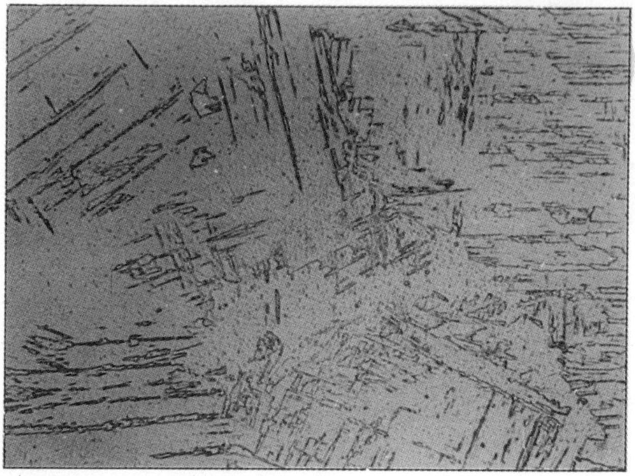

Fig. 7.46 Fine acicular alpha phase in the weldbead (× 200)

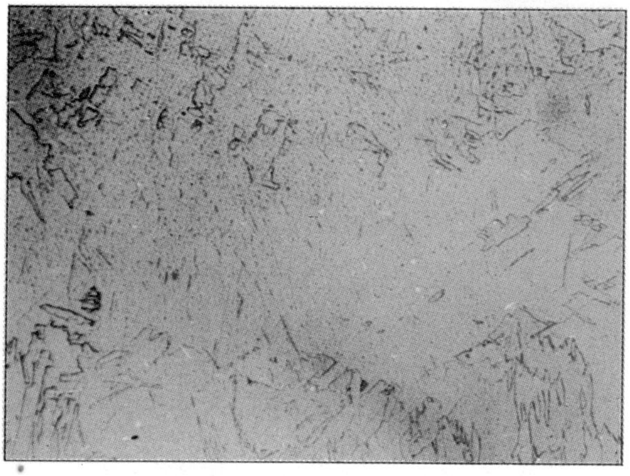

Fig. 7.47 Homogeneous microstructure in HAZ away from the defect (× 200)

Discussion

Scanning electron microscopic studies clearly established the fact that the cracking of the by pass duct was of a brittle nature.

The above observations indicated that a severe mismatch, which was present in the defective zone (as evidenced by the plastic cast impression), produced a notch-like geometry. This caused the introduction of a multi-axiality factor in the non-uniform microstructure of the weld as well as the heat affected zones, which were indicative of a differential cooling effect during welding and all these phenomena contributed to brittle cracking. Further, an intentionally fractured sample of the sound weld region (having homogeneous structure) away from the defective zone revealed only microdimples of a ductile fracture feature. This confirmed the fact that the microstructure in the defective zone was rather brittle. However, there was no evidence of any fatigue (striations) or progressive type of cracking as per the scanning electron microscopic observations.

Several undesirable geometrical features, viz. high degree of mismatch, local undulation, non-uniform weld bead and patch plate repair have all contributed to the development of a very high localised stress, resulting in surface rupture. However, the faint discolouration observed in the cracked surface indicated the possibility of an instant surface cracking during the welding.

Conclusion and Recommendations

Most Probable Cause The basic cause of the brittle cracking of the front bypass duct was attributed to a combination of non-uniform microstructure (with entrapped oxide inclusion) in HAZ and notch effect due to a mismatch.

Remedial Action It was recommended to minimise the mismatch of sheet joint and use better welding techniques to improve microstructural homogeneity.

Chapter **8**

Service Failure: Case Studies

8 Service Failure: Case Studies

8.1 FAILURE OF SHOCK STRUT ATTACHMENT BOLT OF A TRANSPORT AIRCRAFT DUE TO TENSION OVERLOAD

Summary and Recommendations

It was reported that a newly fabricated shock strut attachment bolt which was used during service, failed soon after replacement. The bolt head snapped off the shoulder radius at the end of the assembly.

Fractographic examination revealed a 'Cup and Cone' type tensile fracture due to tensional overload. Macro-examination in the head and shoulder radius zone showed that the bolt was machined out of rolled stock with grain fibres cutting across the sharp radius.

It established the fact that absence of proper grain fibre contour in the bolt-shank radius zone and presence of a sharp radius made snapping of the bolt head easy under the gradually applied tensional load due to the overtightening effect.

It was recommended to provide liberal radius and to fabricate the bolt out of forging with improved grain flow contour. At the same time, assembly operators were advised to adhere to the design stipulated torque range.

Background

Applications

Shock strut attachment bolts (Fig. 8.1) are machined out of medium carbon nickel-chromium-molybdenum steel rolled bars. These are used to connect trunion assembly

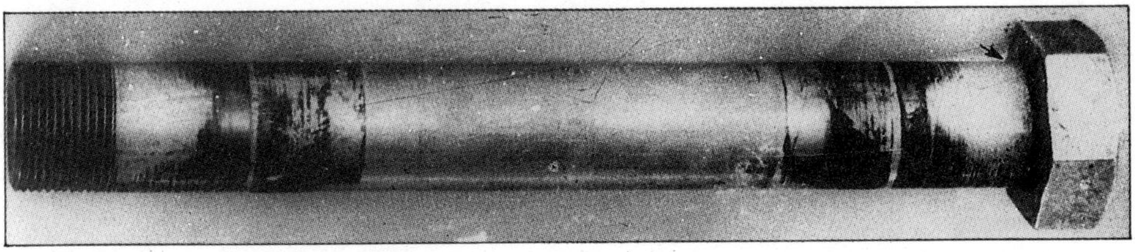

Fig. 8.1 Attachment bolt showing the location of failure and wear marks (× 3/4)

to the cylindrical main landing gear (MLG) leg assembly (Fig. 8.2). The subject bolt was newly fabricated and fitted as a replacement of an original life-expired bolt during super major servicing. After 150 hours of flight, again the MLG system was serviced by maintenance crew in a far off place.

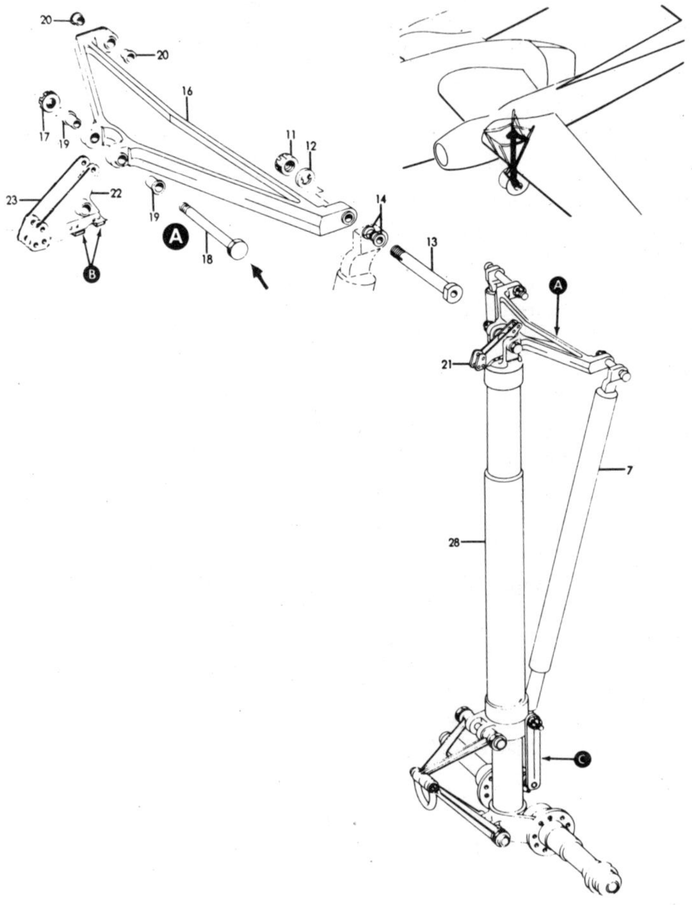

Fig. 8.2 Sketch of MLG leg assembly showing location of bolt (indicated by arrow)

Circumstances Leading to Failure

After completion of maintenance servicing and on inspection a few hours later, the hexagonal head of the bolt was found lying on the ground having snapped off the shank radius.

Specifications

The bolt material was an alloy of nickel-chromium-molybdenum steel to a British 3S99 specification.

Selection of Specimens

Broken pieces were subjected to fractographic studies by a low powered stereoscopic microscope. A section through the broken head portion was cut and polished for macro and micro examination.

Visual Examination of General Physical Features

Examination of the bolt shank and the broken off head portion revealed that the failure was circumferential and all along the shoulder radius. On placing the broken head on to the shank, it was observed that the shoulder radius was too sharp (0.25 mm) for the bolt size. The shank surface showed uneven wear marks (Fig. 8.1) all around, but confined to areas of contact with the bore surface of the component through which it was engaged. Presence of such wear marks is not an unusual phenomenon in used bolts of this type.

Testing Procedure and Results

Surface Examination

Visual A close study of the broken ends of the bolt (Fig. 8.3) indicated that the fracture was basically a' cup and cone' type.

Fig. 8.3 Side view of the shank near the fracture showing uneven wear marks in the surface and "Cup and Cone" type fracture (× 2.5)

Macrofractography Examination of the broken surface, under a low powered stereoscopic microscope at × 25 magnification, revealed coarse and irregular surfaces with no trace of any progressive type of fracture (Fig. 8.4).

Fig. 8.4 Fractured surfaces of head and shank (kept side by side) (× 2.5)

Metallography

Microstructural Analysis Micro-examination of a specimen across the failure at × 400 magnification revealed a satisfactory hardened and tempered structure with no abnormalities. However, macro-examination of a section through the hexagonal head and shank portion indicated only a straight grain flow direction without any contour grain profile in the radius zone.

Chemical Analysis

Shank material was chemically analysed and the composition was found meeting the requirements of S99 specification (Table 8.1).

Table 8.1 Chemical specification (Wt %) for the British nickel-chromium-molybdenum steel (BS 2S 99) from which bolt was made and actual composition of the failed part

Element	Min.	Max.	Failed part
Carbon	0.36	o.44	.0.38
Manganese	0.45	0.70	0.54
Silicon	0.10	0.35	0.28
Sulphur	-	0.020	0.015
Phosphorous	-	0.025	0.020

(Contd.)

Table 8.1 (Contd.)

Nickel	2.30	2.80	2.55
Chromium	0.50	0.80	0.68
Molybdenum	0.45	0.65	0.51

Mechanical Properties

Tensile Properties A tensometer specimen was fabricated out of the broken shank portion and was tested for tensile strength. The tensile strength properties obtained were 127 kg/m^2 (UTS) and 8% (%E) and were found to be well within the specification requirements.

Discussion

Fractographic examination clearly established the fact that the snapping of the head was by a 'cup and cone' type ductile fracture mode and had occurred gradually under applied tensional overloads. It was also evident, from the presence of a sharp radius in the head and shank region (which acted as a severe stress concentration point in the presence of overloads arising out of any overtightening), that a design weakness was introduced to make it amenable to fail under any overload situation. Moreover, absence of a sound grain fibre contour in the radius zone had added an easy element of failure when subjected to such tensional loads (specially when the grain fibre direction being straight and cutting across the sharp radius as was observed in the above case).

Conclusion and Recommendations

Most Probable Cause The basic cause of snapping of the hexagonal head with a cup and cone fracture, was due to the gradually applied tensile overloads arising from overtightening of the bolt, along the sharp radius.

Remedial Action In the light of the above findings, it was recommended to liberalise the radius of the head-shank region in conformity with the bolt size requirements. Further the bolt fabrication was suggested to be made out of a forging instead of rolling. Also, the importance of use of the stipulated torque in tightening of the bolt was highlighted.

8.2 FAILURE OF CROSS TUBE OF THE MAIN LANDING GEAR OF A HELICOPTER DUE TO BENDING OVERLOAD

Summary and Recommendations

An aluminium-copper alloy cross tube of the main landing gear broke during an abnormal landing.

Fracture studies revealed a characteristic bending features of flat surface with shear lips. Also a 20° angle bend was observed on matching the broken pieces of the tube. This indicated that the tube had experienced a bending load prior to breakage.

The cause of the impact bending was due to abnormal landing of the helicopter.

Background

Applications

The aluminium-copper alloy cross tube is a structural member of the helicopter. Swaged end portions of the tube are housed in the leg assembly with two ball bearings, one on either side (Fig. 8.5). The body structure is mounted on the cross tube just inboard of the leg housings.

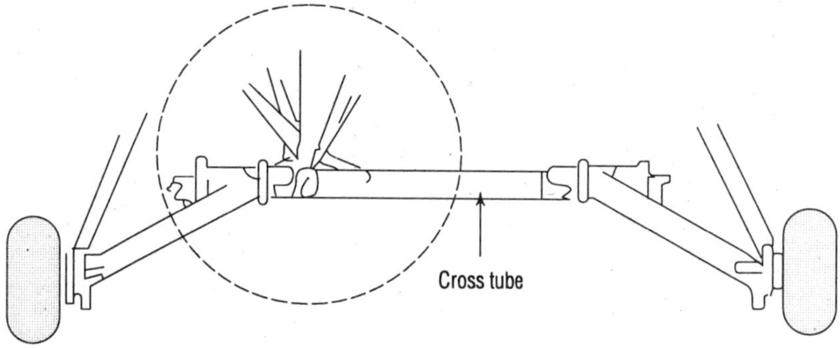

Fig. 8.5 Location of the cross tube in the helicopter leg assembly

Circumstances Leading to Failure

Bending of the cross tube reportedly occurred during landing. Bending was only on one side at a location just inboard of the in-board ball bearing and out-board of the body structure mounting (Fig. 8.6). On disassembly, the tube was found broken in two halves.

Specifications

The material specification used for the component was an American alloy to 24 ST.

Selection of Specimens

The broken halves were subjected to fracture analysis under a low power stereoscopic microscope. A section across the failure was polished and etched for micro-examination under the optical microscope. A tensile test specimen was fabricated out of the tube body for strength measurements.

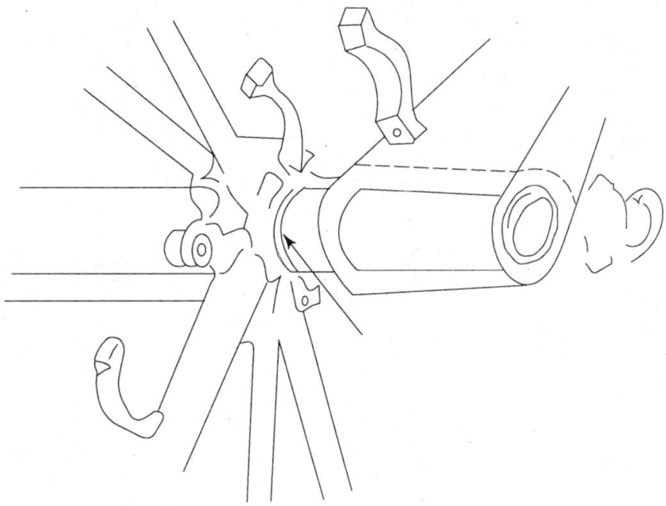

Fig. 8.6 Enlarged view of Fig. 8.5 at the circled zone showing the location of the bent portion

Visual Examination of General Physical Features

On matching the broken ends together, it was found that the swaged end portion of the cross tube had severed transversely just inboard of the in-board ball bearing set area. The tip of the cross tube was found lifted up (angle of bend about 20°), and the tube wall was buckled down (Fig. 8.7). The circular cross-section of the tube at section 'AB' (Fig. 8.7) had become oval in shape (Fig. 8.8).

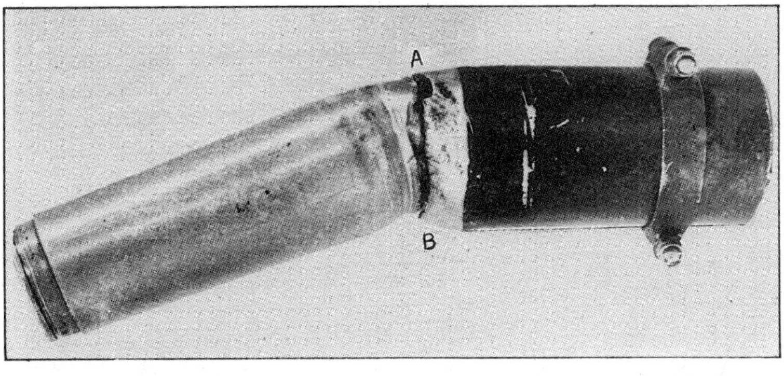

Fig. 8.7 The failed swaged end of the cross tube with the severed pieces matched together. The bright end portion was housed in the main landing gear assembly (× 1/4)

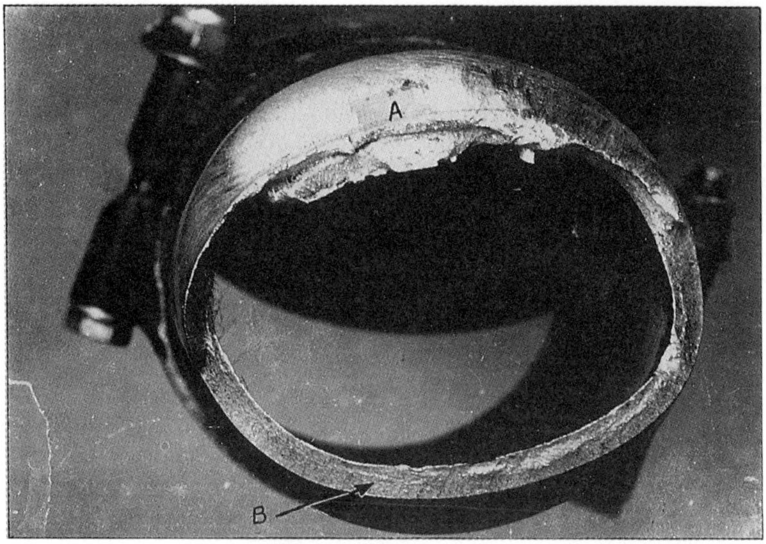

Fig. 8.8 The cross-section of the broken tube at section 'AB' (with ovality) in Fig. 8.7 (× 1/3)

Testing Procedure and Results

Surface Examination

Macrofractography On fractographic examination at × 10 magnification, the bottom half of the tube cross-section at 'AB' (Fig. 8.7) showed a flat-faced fracture confined to half the wall thickness from the outer diameter, the remainder being shear lips.

The top half showed a simple tear due to buckling/crashing, both characteristic of failure under instantly applied bending overloads. No trace of any progressive type of failure, such as fatigue or stress-corrosion, nor any evidence of initial discontinuity was noticed.

Metallography

Microstructural Analysis Micro-examination of the section across the flat faced fracture under the optical microscope at × 100 magnification, revealed a satisfactory age-herdened structure with no abnormalities.

Chemical Analysis

The tube material was found conforming to 24 ST specification requirements (Table 8.2).

Mechanical Properties

Tensile strength properties were found satisfactory as per specification requirements. (0.2% PS 345 (mpa), UTS 425 (mpa) and %E 5 minimum).

Conclusion and Recommendations

Most Probable Cause The cause of the cross tube breakage was due to accidental impact-bending overloads, presumably due to abnormal landing conditions.

Table 8.2 Chemical specification (Wt %) for the American aluminium-copper alloy (24ST) from which the cross tube was made and actual composition of the failed part

Element	Min.	Max.	Failed part
Copper	3.80	4.90	4.25
Magnesium	1.20	1.80	1.60
Manganese	0.30	0.90	0.65
Silicon	-	0.50	0.40
Iron	-	0.50	0.45
Chromium	-	0.10	0.08
Zinc	-	0.25	0.20

8.3 FAILURE OF AN ATTACHMENT BOLT OF THE MAIN LANDING GEAR OF A TRANSPORT AIRCRAFT, DUE TO SHEAR

Summary and Recommendations

A newly fabricated steel attachment bolt of the main landing gear after installation gave way along the stress-relief groove within a few days of the aircraft moving out to the field.

Fractographic studies revealed the fracture to be typical of a shear overload failure. Evidence of spanner slippage marks on the nut corners indicated an overtightening effect. The material used for the bolt fabrication was satisfactory and as per specification.

In light of this, assembly operators were advised to avoid unnecessary overtightening during installation.

Background

Applications

The attachment bolt (Fig. 8.9) is a rather large sized one (31 mm dia, 130 mm long), which was newly fabricated out of a nickel-chromium-molybdenum alloy steel and installed during overhaul. The location of the bolt in the main gear assembly is shown in Fig. 8.10.

Circumstances Leading to Failure

When the aircraft was moved out to the field, after installation of the bolt, and was inspected after a few days, the nut portion along with the threaded end of the bolt was found broken off and lying on the ground.

Fig. 8.9 Breakage along the stress-relief groove exposed under the nut (× 1)

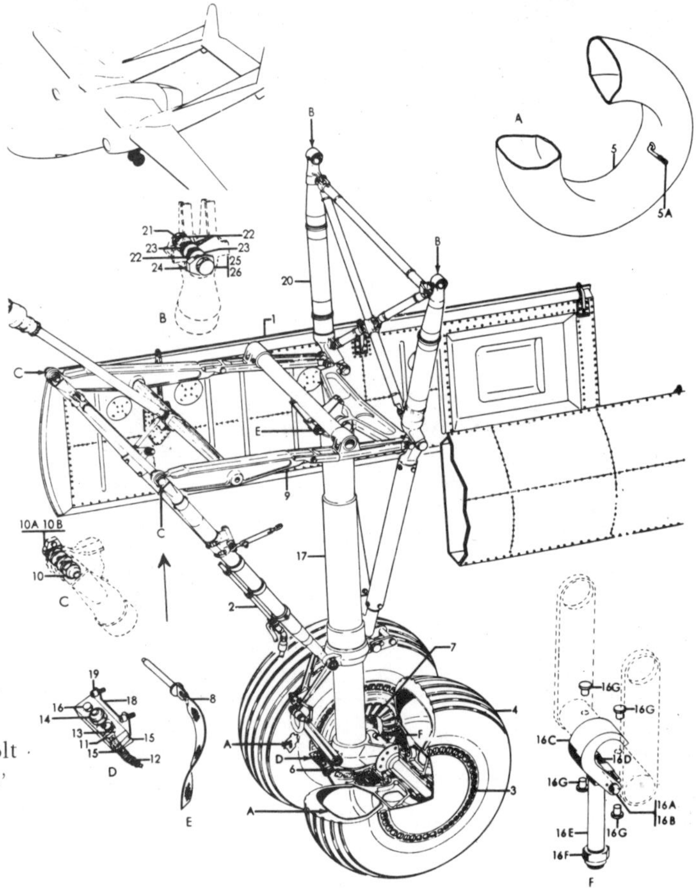

Fig. 8.10 The sketch of the main landing gear assembly with location of the bolt as indicated by 'C'

Specifications

The bolt material, as per drawing was an alloy of medium carbon nickel-chromium-molybdenum steel to a British BS S 99 specification.

Selection of Specimens

Broken pieces were subjected to fractographic examination. A section cut across the fracture was taken for micro-examination under the optical microscope. A tensometer specimen was fabricated out of the shank portion to determine tensile strength properties.

Visual Examination of General Physical Features

Visual examination of the bolt and nut revealed that the failure was circumferential. along the stress-relief groove between the threaded end and the shank, in a plane at right angles to the axis of the bolt (Fig. 8.11). No damage was observed in the bolt and nut, except for spanner-slippage marks at the nut hexagon corners.

Fig. 8.11 Fracture flow lines (broken ends kept side by side) with shear lips ($\times$ 1)

Testing Procedure and Results

Surface Examination

Macrofractography Evaluation of the broken surface under a low-powered stereoscopic microscope at $\times$ 10 magnification, revealed a clear surface fracture flow and shear lips at the end of the fracture (Fig. 8.11). This was indicative of a typical shear overload failure. The origin of the crack, from the lines of surface flow, was traced to a location in the stress-relief groove indicated by the arrow (Fig. 8.9).

Metallography

Microstructural Analysis Micro-examination of the section across the origin of fracture revealed a satisfactory hardened and tempered structure with no abnormality.

Chemical Analysis

Chemical analysis of the bolt confirmed that the material was as per BS S 99 specification (Table 8.3).

Table 8.3 Chemical specification (Wt. %) for the British nickel-chromium-molybdenum steel (BS S 99) from which attachment bolt was made and actual composition of the failed part

Element	Min.	Max.	Failed part
Carbon	0.36	0.44	0.40
Manganese	0.45	0.70	0.55
Silicon	0.10	0.35	0.22
Sulphur	-	0.020	0.015
Phosphorus	-	0.025	0.015
Nickel	2.30	2.80	2.55
Chromium	0.50	0.80	0.72
Molybdenum	0.45	0.65	0.52

Mechanical Properties

Tensile Properties Tensile strength properties measured on the tenometer specimen met the specification requirements.

Discussion

Fractographic analysis clearly established the fact that the failure of the bolt was due to shear overloads. Detailed tests such as chemical analysis, tensile strength test and micro-examination confirmed that the bolt material was satisfactory as per drawing stipulations. The breakage of the bolt head along the stress-relief groove on its own (i.e. without being subjected to service loads) implied a built-in severe overtightening condition, resulting in a shear mode of failure. The presence of the stress-relief groove only facilitated failure under overtightening effect, as evidenced from the presence of spanner slippage marks at the nut hexagon corners.

Conclusion and Recommendations

Most Probable Cause The basic cause of the bolt failure was attributed to shear overload as a result of the overtightening.

Remedial Action It was recommended to avoid unnecessary overtightening of the bolt during assembly.

8.4 BREAKAGE OF ROOT RIB AND CENTRE FITTING OF FIGHTER AIRCRAFT DUE TO HIGH VELOCITY IMPACT

Summary and Recommendations

A fighter aircraft crashed and the broken root rib and the cracked centre fitting pieces (both made of forged aluminium alloy) were collected out of the debris by experts for fracture examination to analyse the cause of the accident. Fractographic analysis on both the fractures confirmed them as brittle failure due to high velocity impact, as evidenced by the presence of clear 'chevron' pattern markings in the fractures.

Since there was no sign of a progressive type of failure and the breakage/cracking occurred under impact, the cause of the mishap could not be ascertained from the failure of the two fittings (belonging to tail plane), which were rather almost intact among the debris.

Background

Applications

A fighter aircraft crashed and out of the debris the investigators picked up a broken root rib piece and a cracked centre fitting belonging to the tail assembly for fracture analysis. As per the drawing details, both the parts were forged and heat treated out of a high strength aluminium alloy.

Circumstances Leading to Failure

Breakage of the root rib and cracking of the centre fitting were reported after the crash of the aircraft (Figs 8.12 and 8.13).

Fig. 8.12 Root rib piece showing the fracture through the large bolt hole (×2/5)

Fig. 8.13 Centre fitting showing location of crack through the small bolt hole in the mid-web (× 1/4)

Specifications

Both the root rib and the centre fitting material were aluminium-zinc-magnesium alloy to a British DTD 5044 specification (Table 8.4).

Table 8.4 Chemical specification (Wt.%) for the British aluminium-zinc-magnesium alloy (DTD 5044) from which root rib and centre fittings were made and the actual composition of the failed part

Element	Min.	Max.	Failed part (Both)
Copper	0.30	0.70	0.55
Magnesium	2.20	3.20	3.10
Manganese	0.30	0.70	0.50
Silicon	-	0.50	0.38
Iron	-	0.50	0.38
Nickel	-	0.10	0.08
Chromium	-	0.05	0.03
Zinc	5.20	6.20	5.90
Titanium	-	0.30	0.23
Lead	-	0.05	0.03
Tin	-	0.05	0.03

Selection of Specimens

The crack of the centre fitting was opened out and the fracture portions of both the centre fitting as well as the root rib were subjected to fractographic analysis with the aid of a low powered stereoscopic microscope.

Visual Examination of General Physical Features

Visual examination of the fittings revealed that the breakage had taken place through the large diameter hole in the root fitting, and cracking in the centre fitting was also through the large cut out bolt holes in the mid web. Both the fractures were found almost intact, i.e. without abrasion, etc.

Testing Procedure and Results

Surface Examination

Macrofractography Fractographic examination of both the fractures under a low-powered stereoscopic microscope at × 10 magnification, showed basically brittle features with shear lips. The well defined 'Chevron' pattern in the fracture pointed towards the clear origin of the crack (Figs 8.14 and 8.15). No trace of any progressive type of failure, such as fatigue or stress corrosion was noticed.

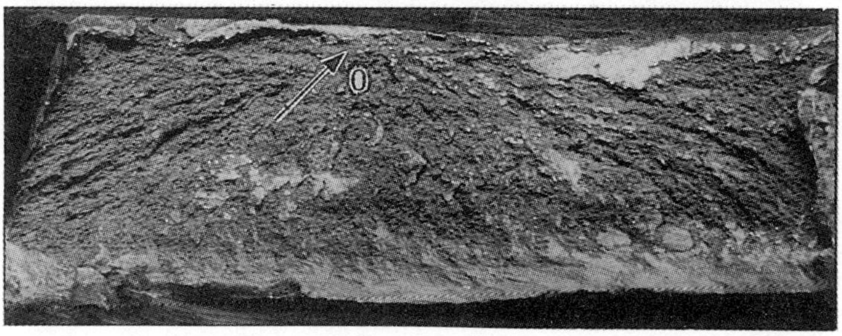

Fig. 8.14 Enlarged view of fracture 'A' in Fig. 8.12 showing 'Chevron' pattern of high velocity brittle fracture with its origin at '0' (× 2)

Discussion

Fractographic examination established the fact that the breakage of the root rib and the cracking of the centre fitting were a clear case of brittle failure under high velocity impact. Any high velocity impact failure is normally associated with 'Chevron' markings in the ensuing fracture, with the origin clearly defined as the convergence point of the emanating fracture lines (Figs 8.14 and 8.15).

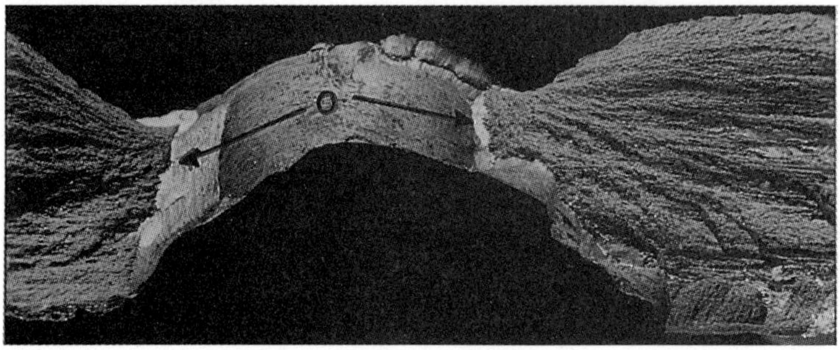

Fig. 8.15 Enlarged view of fracture in Fig. 8.13 showing typical 'Chevron' pattern of high velocity brittle fracture with its origin at '0' (×2)

Conclusion and Recommendations

The basic cause of the breakage/cracking of the fittings was due to high velocity impact as a result of the crash.

8.5 BREAKAGE OF PINION OF THE TAILPLANE ACTUATOR OF A FIGHTER AIRCRAFT DUE TO TORSIONAL OVERLOAD

Summary and Recommendations

A steel pinion of the tailplane was reported broken across the small splined portion during service. Since relevant documents regarding the material were not available, and the parts were not to be destroyed, the fracture characteristics were only analysed using a low-powered stereoscopic microscope.

The study revealed the fracture purely of a torsional shear mode and the contributory factor was found to be the use of low strength material with splines without case carburising. The intended application involved friction between the mating surfaces under torsional loads.

It was recommended to check the used (but not affected) and unused pinions for hardness. If not case-hardened in the splined surface, the subject parts could be case carburised to improve the strength as well as wear resistance properties, and used if within the dimensional limits.

Background

Applications

The pinion is a bought-out component (Fig. 8.16) for the tailplane actuator of a fighter aircraft and is made of steel. History of service and other detailed information such as

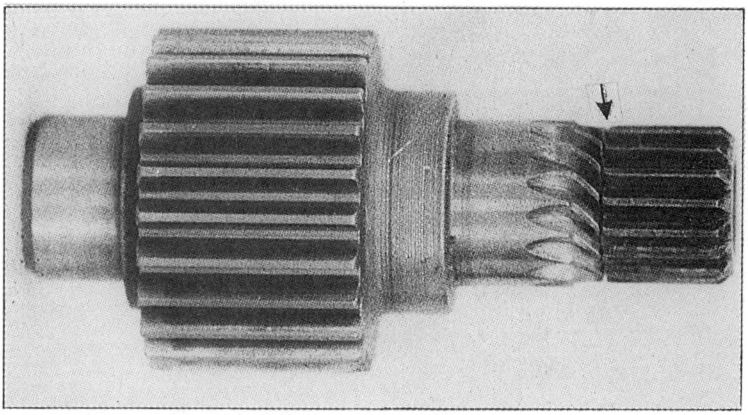

Fig. 8.16 The broken pieces of the pinion matched together showing helical deformation of the splines towards failure ($\times 3$)

material specifications and component drawings were not available for investigations. As the failed component had to be sent to the manufacturer, detailed destructive tests such as chemical, macro and micro-examination were not carried out.

Circumstances Leading to Failure

The pinion was reported deformed and broken off across the small splined end portion during service (Fig 8.16).

Visual Examination of General Physical Features

Visual examination of the broken pinion pieces on matching together revealed that the smaller pinion end had fractured transversely in a plane perpendicular to the axis of the pinion, about 1 cm off the end-face. Also the straight splines from the shank had deformed uniformly in a helical pattern (Fig. 8.16).

Testing Procedure and Results

Macrofractography Fractographic examination of the fracture surface of the pinion under a low powered stereoscopic microscope at $\times 10$ magnification, revealed that the entire surface was having a smooth texture of rotary metal smearing (Fig. 8.17).

Mechanical Properties

Hardness Hardness measurement was made on the pinion using Vickers hardness testing machine at 30-kg load, and the value obtained was 267 HV.

Discussion

Fractographic analysis of the broken surface of the small splined portion established the type and nature of the failure as torsional overload, by the characteristic deformatory condition of the straight splines and rotary metal smearing on the fracture.

Fig. 8.17 Fractured surface of the pinion showing smooth texture and rotary metal smearing (× 8)

In the absence of confirmatory tests regarding material composition and microstructure (which could have been a valuable to determine the use of right materials and case-characteristics for the intended application), only hardness measurement on the splined surface was carried out. A hardness value of 267 HV obtained was considered very low for a component of this type, which is subjected to a torsional load during service. Moreover, low hardness value reflected that the splined surface was not case-carburised, normally a prerequisite for high duty service. It may be mentioned that even the normal peak loads in service tend to become overloads for such a soft ductile material, with low torsional strength.

Conclusion and Recommendations

Most Probable Cause The basic cause of the torsional overload failure of the pinion was attributed to the use of low strength material.

Remedial Action Based on the above, it was suggested to impart case-carburising treatment to spare parts as well as those in use, after a thorough inspection by hardness measurement on the splined surface as well as core of the pinion. This would improve the strength and wear resistance properties.

How Failure could have been Avoided The failure could have been avoided if the inspection had tested the hardness measurement as routine check before use on the aircraft, for correctness of the material characteristics as per drawing specifications.

8.6 CRACKING OF FAIRING SKIN OF FIGHTER AIRCRAFT DUE TO FLEXURAL FATIGUE

Summary and Recommendations

A contoured fair skin assembled on the aircraft structure was subjected to simulated service cyclic loading for endurance test. This failed by cracking in the rivet hole area adjacent to the cut-out radius after 6500 flying hours.

Fractographic examination of the cracked surfaces revealed fatigue 'beach marks' originating from either side of the rivet hole. A close examination of the cracked rivet hole area showed a major reduction in the skin width. In other words, this seemed to have adversely affected the strength properties in the area and rendered the skin susceptible to flexing action during wing deflection.

The cracking of the skin was clearly a case of flexural fatigue failure, and was attributed to exposure to cyclic flexing actions. To prevent the flexing it was recommended to stiffen the cut-out edges in the region of cracking of the component.

Background

Applications

The fairing skin component is fabricated out of clad aliminium-zinc magnesium copper alloy sheet material and assembled over the air intake of a fighter aircraft. The component with the aircraft structure was subjected to simulated service cyclic loading tests for endurance qualification.

Circumstances Leading to Failure

The fairing skin was found cracked (Fig. 8.18) after it had served a simulated service cyclic loading for a period equivalent to 6500 flying hours.

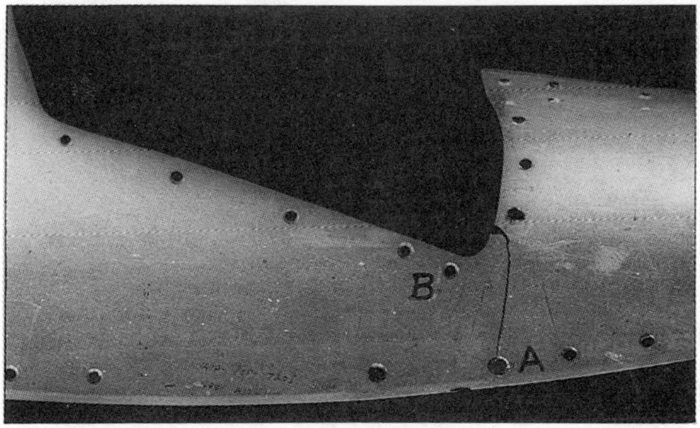

Fig. 8.18 Cracked firing skin (× 1/5)

Specifications

The sheet material used for the fairing skin was a British alloy to BS L 73 specification.

Selection of Specimens

The cracked portion was opened out and the fractured pieces were examined under a low powered stereoscopic microscope. A tensile test specimen was made out of the skin for determining the tensile strength properties. For micro-examination a sample was cut across the cracked zone.

Visual Examination of General Physical Features

Visual examination revealed one crack emanating from the rivet hole 'A' (Fig. 8.18) and extending transversely to the cut-out corner, and another 1 cm long crack starting from the cut-out corner radius adjacent to the rivet 'B'. No evidence of any fabrication defect was noticed, except that the skin width was reduced to one–third at the location of the innermost cut-out corner,

Testing Procedure and Results

Surface Examination

Macrofractography Examination of the fractured surfaces revealed that the plane of cracking was at right angles to the skin surface and 'shell' markings in the smooth fracture, typical of fatigue failure. The origins of fatigue were traced to the bottom corner surface on either side of the rivet hole (Figs 8.19 and 8.20).

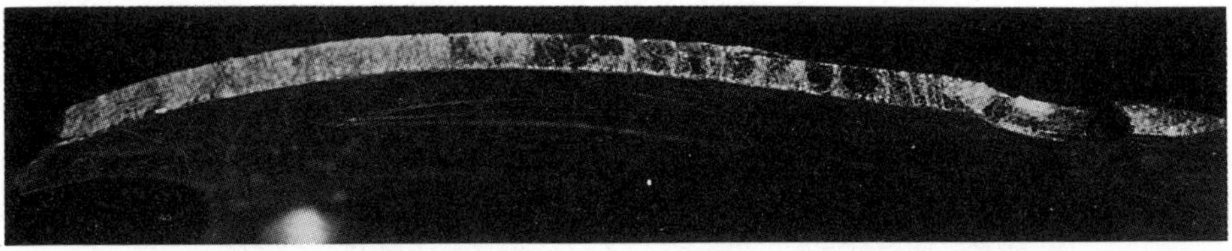

Fig. 8.19 Fractured surface at 'A' marked in Fig. 8.18 (× 2)

Fig. 8.20 Fractured surface at 'B' marked in Fig. 8.18 (× 7)

Metallography

Microstructural Analysis Micro-examination of a section across the fatigue origin at × 100 magnification, revealed transgranular cracking, which also confirmed that the failure was by fatigue. Microstructure was found satisfactory in the heat treated condition with no abnormalities.

Chemical Analysis

Chemical analysis confirmed the skin material to BS L 73 specification (Table 8.5).

Table 8.5 Chemical specification (Wt%) for the British aluminium-copper sheet (BSL 73) from which fairing skin was made and actual composition of the failed part.

Element	Min.	Max.	Failed part
Copper	3.90	5.00	4.20
Magnesium	0.20	0.80	0.68
Manganese	0.40	1.80	1.10
Silicon	0.50	0.90	0.56
Iron	-	0.50	0.15
Nickel	-	0.20	0.15
Zinc	-	0.20	0.16
Lead	-	0.05	0.03
Tin	-	0.05	0.04
Titanium + zirconium	-	0.20	0.11
Chromium	-	0.10	0.08

Mechanical Properties

Tensile Properties Tensile strength properties measured on the tensile test specimen (made from the skin away from the cracked zone) were found conforming to the specification requirements (Table 8.6).

Table 8.6 Mechanical properties of the fairing skin along with specification values (MPa)

Tensile strength properties	Specification(min)	Failed part
0.2% proof stress	360	375
Tensile strength	420	430
% Elongation	8	8
(on gauge length of 50 mm)		

Discussions

Fracture examination clearly indicated that the crack had developed as a result of fatigue failure with origins at either side of the rivet hole. These origins act as points of stress concentration during the skin deflection under flexural action.

The presence of beach marks in the wholly flat-faced fracture on either side of the rivet hole indicated fatigue cracking under (flexural) plain strain conditions, originating from the top or bottom surfaces at nominal stress intensities. Also, the reduction observed in the skin width at the location of the innermost cut-out corner had adversely affected the strength as well as fatigue resistance in the above area and rendered it easily susceptible to flexing actions during wing deflection.

Conclusions and Recommendations

Most Probable Cause The basic cause of the flexural fatigue failure was attributed to exposure of the contoured skin to cyclic flexing actions perceptibly caused by a reduction of the skin width at the cracking region of the innermost cut-out corner.

Remedial Action Stiffening of the cut-out edges in the region of cracking was recommended for consideration by design personnel for future components.

8.7 FATIGUE FAILURE OF THE REAR CAM REDUCTION DRIVE GEAR AND MAIN BEARING ASSEMBLY OF AN ENGINE

Summary and Recommendations

Rear cam reduction drive gear and main bearing assembly of an engine failed prematurely during service.

On detailed fractographic examination in a broken tooth was found to have failed by reversal bending fatigue with origin at the deep machining tool marks along the tooth root radius. Micro-examination revealed the case-carburised structure as satisfactory with a hardness of 685 HV. In the case of inner race failure of the bearing assembly, the race was found heavily spalled all over and microstructure in the zone of spalling revealed several carbide bands oriented towards the surface. This structural inhomogeneity facilitated the process of spalling by cyclic fatigue during service.

From a close study of both gear and bearing assemblies, it was suspected that fatigue failure of the gear tooth was consequential to the failure of other related components, presumably the heavily worn out bearing assembly.

Background

Applications

Both rear cam reduction drive gear and main bearing assembly are critical components in the power transmission system to the engine.

Circumstances Leading to Failure

The gear was found broken when the engine had served 645 hours of service since previous overhaul and the bearing showed severe spalling (Fig. 8.21) before the completion of first life of 700 hours.

Specifications

The bearing material used was of an American Specification to SAE 52100, whereas that of the gear was not known.

Selection of Specimens

The broken gear tooth piece was selected for fractographic examination under low powered stereoscopic microscope and micro-examination after a section was cut across the zone of fatigue. The inner race of the bearing assembly was subjected to low powered stereo microscopic examination and a section cut across the defect was taken for examination under optical microscope.

Visual Examination of General Physical Features

Examination showed that the cam reduction drive gear was broken angularly with two teeth off in one zone (Fig. 8.21) along the root radius. It also revealed that the adjacent tooth root radius from one side of the broken portion was cracked. Examination of the race way of the inner race of the bearing showed severe spalling all over (8.22).

Fig. 8.21 Breakage of the gear with spalled bearing (× 1/2)

Fig. 8.22 Close up view of the raceway with severe spalling (× 1)

Testing Procedure and Results

Surface Examination

Macrofractography On close examination under the stereoscopic microscope at × 25 magnification, both driving as well as non-driving faces of all the teeth showed wear marks. Tooth root radius portions also showed original deep machining tool marks (Fig. 8.23). Examination of the fracture of both the broken teeth revealed closely spaced

Fig. 8.23 Fatigue zones originating from the deep tool marks at the tooth radius (× 6)

and distinct 'beach' marks in the smooth surface which covered the entire thickness (Figs 8.23 and 8.24) of the tooth radius portion. Origin of the fatigue fracture was located at the deep machining tool marks along the root radius.

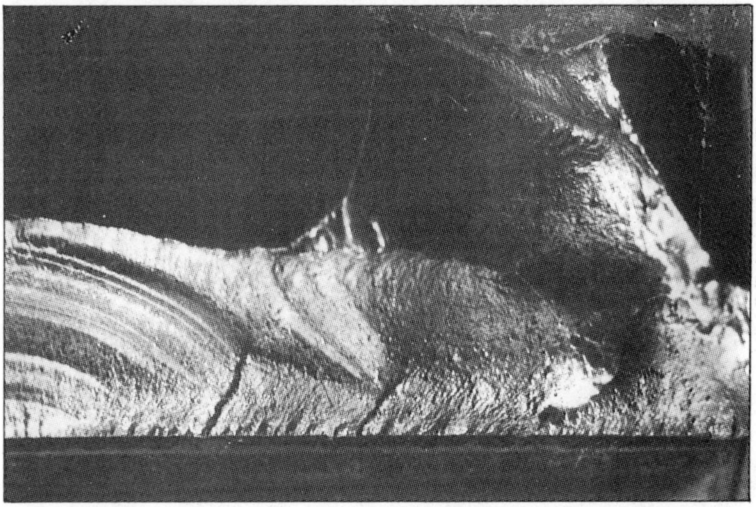

Fig. 8.24 Fatigue zones also originating from the tooth radius (× 8)

Metallography

Microstructural Analysis Micro-examination of the section of the steel gear material across the zone of fatigue revealed satisfactory case-carburised (hardened and tempered) structure in the tooth portion. In the case of inner race material the microstructure was found in satisfactory hardened and tempered condition with numerous tiny carbides interspersed in the matrix. But, however on low magnification examination at × 50, severe banding of carbides oriented towards the inner race surface was clearly discernible (Fig. 8.25). These bands, on higher magnification at × 800, were found to be a congregation of tiny carbides (Fig. 8.26).

Further examination revealed some cracks emanating from the surface damages with some of them along the carbide bands and curving back to the surface. These features indicated the phenomenon of spalling by cyclic fatigue.

Chemical Analysis

The inner race material on chemical analysis was found to conform to SAE 52100 ball bearing steel (Table. 8.7).

Mechanical Properties

Hardness A hardness survey was carried out on the microspecimen of the gear tooth to determine hardness at the case and the core, at 500 gms load.

Fig. 8.25 Severe carbide bandings in the bearing material (× 50)

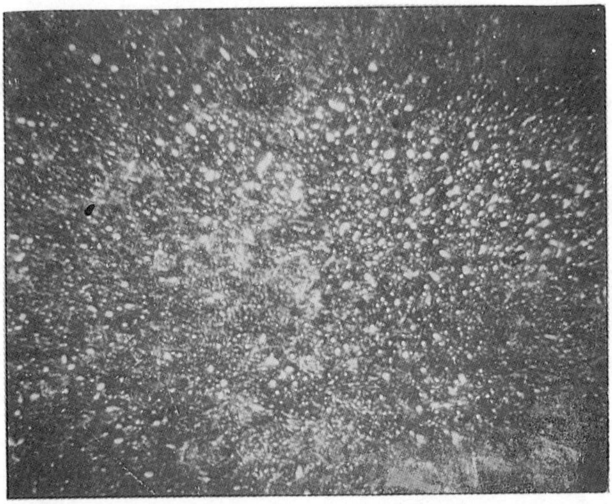

Fig. 8.26 Aggloration of numerous tiny carbides in the banded zone at higher magnification (× 800)

Table 8.7 Chemical specification (Wt. %) for the American carbon-chromium ball bearing steel (SAE 52100) from which the bearing component was made and actual composition of the failed part.

Element	Min.	Max.	Failed part
Carbon	0.95	1.10	1.10
Manganese	0.25	0.45	0.36
Silicon	0.20	0.35	0.24
Sulphur	-	0.025	0.015
Phosphorus	-	0.025	0.05
Chromium	1.30	1.60	1.48

- Case 685 HV (satisfactory)
- Core 410 HV (satisfactory)

Hardness value of the inner race was found varying from 567 to 820 HV (unsatisfactory due to inhomogeneity in the structure).

Discussion

Fractographic analysis and microstructural studies established the fact that breakage of the gear tooth was under reversal bending fatigue. From the general physical features of the broken gear it was noted that both driving as well as non-driving faces of all the teeth bore wear marks. This indicated a condition of uneven reverse loading under bending (as evidenced by the presence of fatigue "beach" marks on either face of the tooth) during service. These uneven conditions of loading, contributed excessive reverse bending stresses concentrating along the tooth radius where deep machining tool marks acted as a severe stress raiser for development of a premature fatigue crack.

In the case of failure of the inner race of the bearing assembly, it was established that the spalling was due to the development of fatigue cracks in service. The reported shorter life of the bearing was attributed to the presence of carbide bands, rendering the high hardness ball bearing material inhomogeneous. This facilitated the process of spalling by fatigue faster due to reduced fatigue resistance.

Conclusion and Recommendations

Most Probable Cause The basic cause of the fatigue failure of the gear tooth was suspected to be consequential (on a close study of the gear assembly and the bearing) to the failure of the heavily worn out bearing assembly.

8.8 FAILURE OF A RELAY VALVE GUIDE OF A TRANSPORT AIRCRAFT

Summary and Recommendations

One ball bearing steel relay valve guide component operating in the fuel supply system of a transport aircraft broke into two, through one set of elongated holes, after 26 hours of flight. Elongated holes were reported drilled by electro-discharge machining.

Examination of the hole regions indicated both the bore surfaces severely corroded and burnt, with features of erosion and craters. Fractographic studies revealed that the fracture was brittle with a few widely spaced lines of arrest marks in the crack propagation path, suggestive of a low cycle fatigue failure. Microstructural analysis revealed presence of a conspicuous white layer in the outermost surface with several transverse cracks, and also confirmed a seepage of molten white layer into the crack that had developed during machining, enveloping the fracture to a considerable depth.

Obviously, these features pointed out that the brittle separation of the valve guide across the elongated hole had occurred due to pre-existence of a crack in the outer corner of the bore, out of an unduly overheated (hard white layer) structure formed during the machining stage.

It was recommended to inspect the remaining parts using the stereoscopic microscope and/or the boroscope, for detection of surface defects of a similar nature in the bore region and if present, to reject them.

Background

Applications

A batch of relay valve guide components was imported for use in the valve assemblies of the fuel supply system of a transport aircraft for tail plane pitch control operations. It is important to note that in the event of a valve failure, there is a great risk of fire. This may endanger the safety of the aircraft due to escape of fuel to any hot zone around the engine bay. The part was reported machined to its entire length by conventional machining except for the two sets of elongated holes which were precision machined by electro-discharge process (Fig. 8.27).

Circumstances Leading to Failure

The valve broke into two after it had served only 26 hours of service (Fig. 8.27) during flight. The valve was operating at a working load of 3000 psi at the time of failure.

Specifications

Material used for the valve was an American alloy to SAE 52000 specification (a high carbon-chromium ball bearing steel) (Table 8.8).

Fig. 8.27 The relay valve guide separated across an elongated hole (× 1)

Table 8.8 Chemical specification (Wt. %) for the American carbon-chromium ball bearing steel (SAE 52100) from which relay valve guide component was made and actual composition of the failed part

Element	Min.	Max.	Failed part
Carbon	0.95	1.10	1.05
Manganese	0.25	0.45	0.32
Silicon	0.20	0.35	0.25
Sulphur	-	0.025	0.015
Phosphorus	-	0.025	0.015
Chromium	1.30	1.60	1.45

Selection of Specimens

One out of the two broken pieces was selected for fractographic analysis under the stereoscopic microscope. A section across the origin of failure was metallographically prepared for microstructural studies under the optical microscope.

Visual Examination of General Physical Features

Examination revealed that the valve had separated transversely into two through one of the two sets of elongated holes, and the hole bore surface was found irregular and discoloured.

Testing Procedures and Results

Surface Examination

Macrofractography Examination of the failed elongated hole at $\times 10$ magnification under a low powered stereoscopic microscope, revealed a rough and corroded bore surface with clear signs of melting and craters at places (Fig. 8.28). However, the

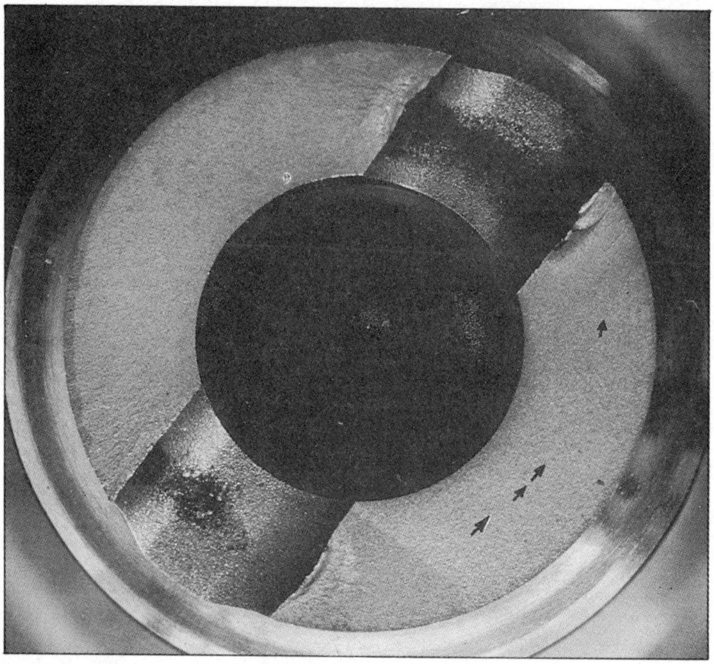

Fig. 8.28 Smooth brittle fracture surface with burning effects and crators in the bore surface of the elongated hole. Fracture surface also shows faint arrest lines of crack propogation (marked by arrows) ($\times 4$)

fracture was found with a smooth but rather coarse crystallined surface and in a plane at right angles to the valve axis. Further, both the elongated holes were found sunk in, almost identically at the bore edge, due to severe burning cum errosion effects (Fig. 8.28). Also, the fractured surface showed a conspicuous dark shaded (triangular) zone adjoining the deeply sunk hole bore edge. On close observations, several faint 'arc' like markings of rather wide spacing were observed running parallel to the profile of the dark shaded zone. This indicated that the crack had propagated in stages through several lines of arrest in the crack path, having originated at the crater region in the outer corner of the bore surface. All these features were suggestive of a typical brittle cracking under low cycle fatigue.

Metallography

Microstructural Analysis Micro-examination of a section across the zone of origin, revealed a conspicuous white layer (remelted) in the outer surface of the bore, to a maximum depth of 0.035 mm. Microstructure also revelead a 0.35 mm deep overtempered zone just beneath the white layer (Fig. 8.29). Otherwise, the core structure

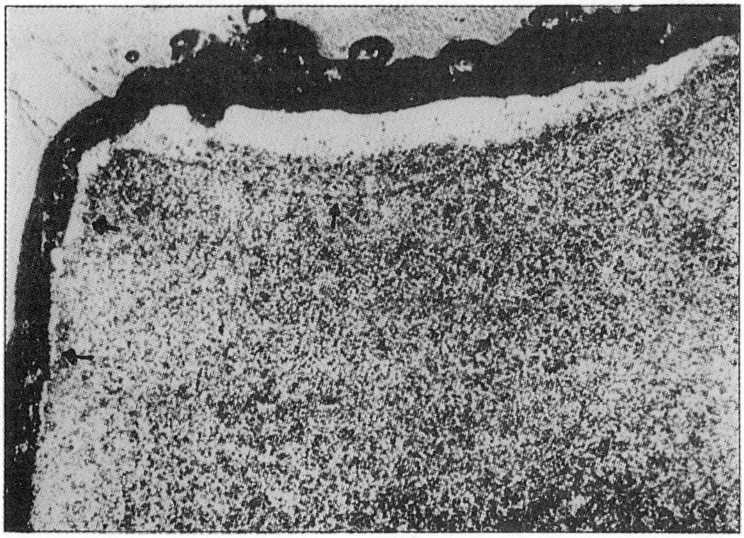

Fig. 8.29 White layer and underlying overtempered layer with a trace of white layer penetrating to a considerable depth (× 100)

showed satisfactory tempered martensite, with uniform distribution of numerous tiny globular carbides in the matrix, a typical structure of a high carbon-chromium ball bearing steel material.

Micro-examination further revealed that the fracture was enveloped with a trace amount of white layer to a considerable depth (Fig. 8.29). Moreover, the white layer showed several transverse cracks emanating from the outer bore surface and tending to penetrate into the base of the overtempered zone (Fig. 8.30).

Mechanical Properties

Hardness A hardness survey was carried out on the micro-specimen using the miniload hardness tester at 500 gms load. A maximum hardness of 1100 HV in the outermost white layer and a minimum value of 550 HV in the overtempered zone were obtained. However, the core hardness gave an average value of 800 HV.

Discussion

Fractographic studies established the fact that the failure was due to low cycle fatigue,

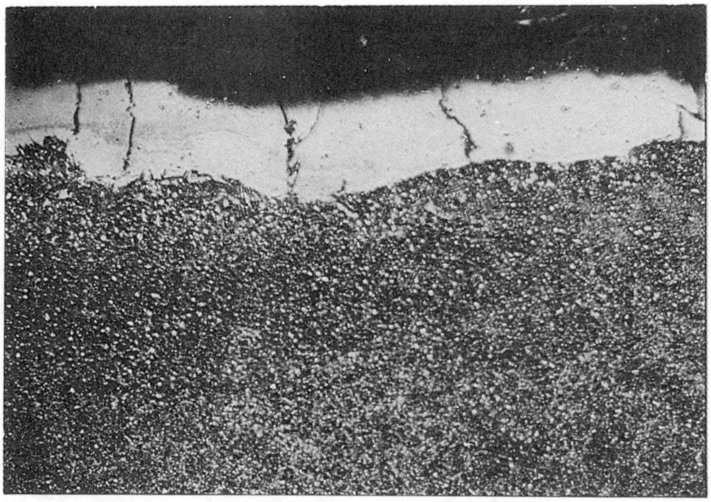

Fig. 8.30 Numerous transverse cracks in the white layer (× 400)

as evidenced by the presence of several lines of arrest marks in the crack propagation path. Further, a close analysis of the fracture path pointed out that the brittle crack had its origin at the severely erroded zone of craters in the outer corner of the bore surface. Presence of surface defects like discolouration, erosion and craters, etc., indicated a high temperature melting phenomenon that occurred during electro-discharge machining, forming a very hard and deep white layer. Further, it was evident from the microstructure that since the white layer had seeped into the fracture surface, the initiation of the crack did take place during the high temperature machining stages. On the other hand, development of transverse cracks in the outer most bore surface was a direct consequence of drastic quenching of a hard and brittle white layer. Therefore, under these conditions any minor stress during service would lead to premature brittle failure. However, the depth of the white layer was found in excess of the limit normally permissible for the finished hardenable steel products, i.e. 0.035 against 0.025 mm max. As such, the excess limit was believed to have contributed towards extra brittleness of the white layer. All these findings pointed out that the process parameters adopted for electro-discharge machining could be faulty, specially in respect of maintaining proper current density, or there could be some short circuiting, etc. leading to high temperature melting.

Conclusion and Recommendations

Most Probable Cause The basic cause of brittle cracking (low cycle type fatigue) of the valve guide was attributed to the presence of pre-existence of a crack in the outer corner of the bore, out of an unduly overheated (white layer) structure formed during machining.

How the Failure could have been Avoided Stereoscopic macro-examination as well as boroscopic inspection of the electro-discharge machined bore surfaces could have been carried out before acceptance of the imported valves to avoid such hazards.

8.9 FATIGUE FRACTURE OF THE CRANKSHAFT OF A TRANSPORT AIRCRAFT DURING FLIGHT

Summary and Recommendations

A catastrophic failure of a piston engine crankshaft was reported during flight of a transport aircraft. It was ascertained that the aircraft had undergone three major overhauls, involving 3000 hours of service prior to breakage along the pin radius zone. It was also learnt that the magnetic particle inspection by AC method was resorted to during over haul for detection of cracks.

Examination of the fracture revealed several 'conch' shell marks forming a distinct eye zone around a tiny sub-surface defect as the origin of fatigue, under complex bending cum torsional stresses. Studies with the help of the scanning electron microscope confirmed the existence of a sub-surface defect, as an isolated hard refractory type of inclusion in the zone close to the pin radius. This served as a stress raiser for initiation and development of the fatigue crack. From the cleanliness point of view, the crankshaft material was found to be of inferior quality.

The basic cause of the crankshaft failure by fatigue was due to the presence of a subsurface defect, in the form of a hard refractory inclusion, in the critically loaded pin radius. It was recommended to check for cracks by magnetic particle inspection using DC method during overhaul.

Background

Applications

The massive complex shaped crankshaft (Fig. 8.31) used in the piston engine of a transport aircraft is machined out of a triple alloy steel forging. During the flight the crankshaft experiences a combination of highly complex stresses which include bending and torsion, particularly at the critical pin radius zone. Periodically the crankshaft is checked by magnetic particle inspection, during major overhaul, for presence of any defect or crack indication in the pin radius zone.

Circumstances Leading to Failure

One of the two crankshafts of a twin piston engine catastrophically broke into two in the area close to the pin radius zone (Fig. 8.31) while in flight. It had undergone 3 major overhauls involving 3000 hours of service. The aircraft however, landed safely with the help of the other engine. The aircraft was maintaining an average flight speed of 2050 rpm at the time of the failure.

Fig. 8.31 A close-up photograph of the broken crankshaft portion on assembling them together, showing breakage through the pin shoulder radius zone (× 1/3)

Specifications

Material used for the crankshaft was an American alloy to SAE 4340 specification.

Performance of Other Parts in Same Service

Similar crankshafts used previously were reported to have completed the stipulated life of their total service satisfactorily.

Selection of Specimens

Out of the two broken pieces, the undamaged one was selected for detailed fracture analysis under a low powered stereoscopic microscope. Further, a small portion containing the origin was sectioned for an in-depth examination by the scanning electron microscope. Micro-examination was carried out on the same section used for SEM study.

Visual Examination of General Physical Features

Examination revealed that the failure had occurred transversely and in a plane at right angles to the crankshaft axis, with no evidence of plastic deformation around the zone of fracture. Further, the fracture features of the broken half on the LH side of the crankshaft (Fig. 8.32) were found more or less intact, while those of the mating surface was completely obliterated due to the rubbing action subsequent to failure.

Fig. 8.32 Smooth fracture zone of the undamaged portion of the broken crankshaft (× 2)

Testing Procedures and Results

Surface Examination

Macrofractography Examination of the fracture at × 10 magnification under the stereoscopic microscope revealed a smooth surface with a distinct 'eye zone' formed by several conspicuous 'beach' marks around a tiny sub-surface defect as the origin (black spot in Fig. 8.33). All these fracture features were indicative of a failure under bending fatigue. The smooth zone of the fatigue was found to have covered a little more than half of the circumference. The remainder exhibited a coarse and irregular surface, typical of an over-stress failure.

Scanning Electron Microscopy Fractography The fracture origin area was examined at higher magnification by SEM to identify and establish the nature of the sub-surface defect. Examination revealed that the defect was a lone refractory type of inclusion enclosed by a non-metallic film (Fig. 8.34). The inclusion appeared to be of $A1_2\ O_3$ type,

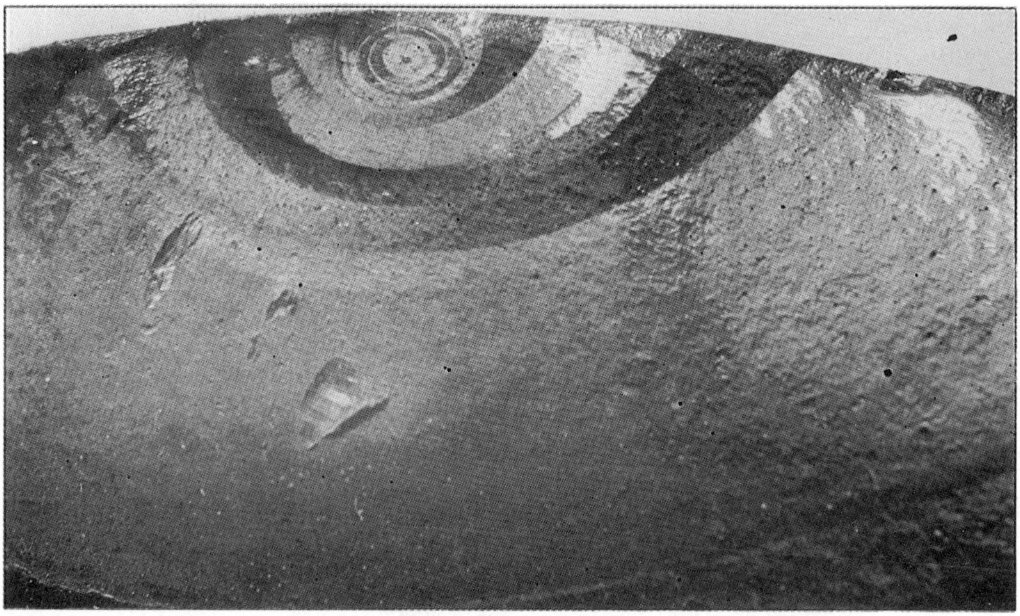

Fig. 8.33 A 'distinct eye' zone formation by fatigue cracking with origin at a tiny sub-surface defect (black dot) (× 8)

as this type of hard refractory inclusion does not undergo any kind of deformation due to heavy mechanical working, such as forging, etc. However, actual characterisation of the inclusion could not be confirmed owing to a lack of instruments at the time of investigation.

Metallography

Microstructural Analysis Micro-examination of a section across the fracture, in the region very close to the sub-surface defect, revealed the material rather unclean with numerous globular non-metallic oxide inclusions (Fig. 8.35) conforming to D5 thin series at × 100 magnification. Otherwise, the microstructure consisted of satisfactory fine grained tempered martensite.

Chemical Analysis

Chemical composition of the crankshaft material was found conforming to SAE 4340 specification (Table. 8.9).

Mechanical Properties

Hardness Hardness measurement was carried out on the micro-specimen using Vickers pyramid hardness testing machine at 30 kg load and showed an average value of 383 VPN, against a minimum requirement of 360 VPN as per specification.

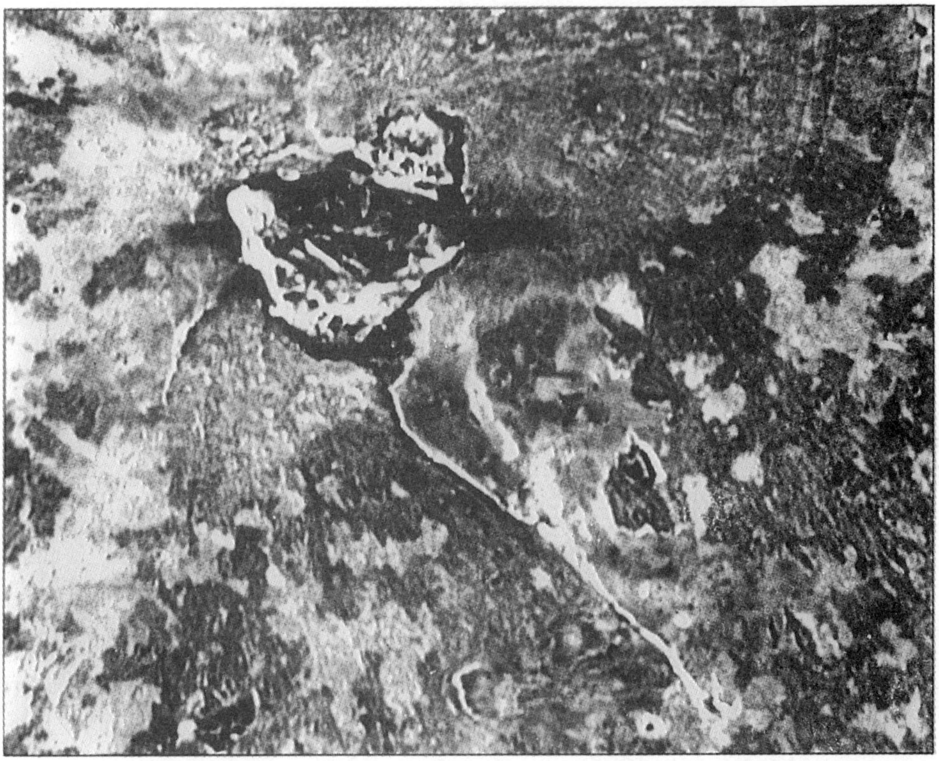

Fig. 8.34 A hard refractory type inclusion enclosed by a non-metallic film inclusion, around which fatigue beach marks are clearly observed under scanning electron microscope (×450)

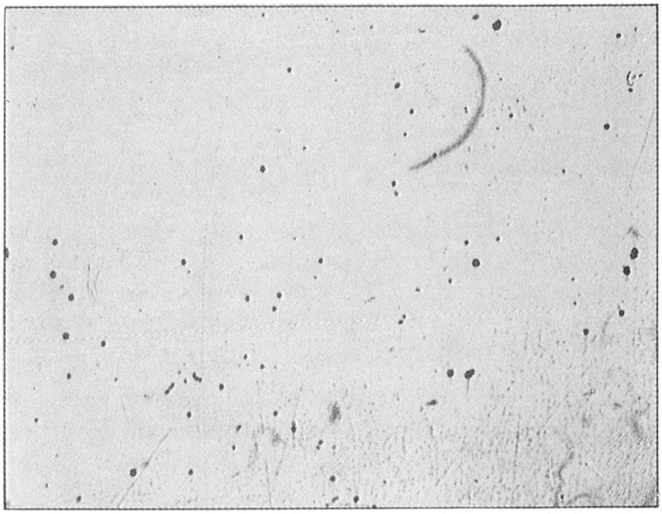

Fig. 8.35 Numerous globular non-metallic oxide inclusion in the material (×100)

Table 8.9 Chemical specification (Wt. %) for the American nickel-chromiun molybdenum steel SAE 4340 from which the crankshaft was made and actual chemical composition of the failed part

Element	Min.	Max.	Failed part
Carbon	0.38	0.43	0.41
Manganese	0.65	0.80	0.75
Phosphorus	-	0.025	0.015
Sulphur	-	0.040	0.025
Nickel	1.65	2.00	1.75
Chromium	0.70	0.90	0.82
Molybdenum	0.20	0.30	0.25

Discussion

The fractographic studies clearly established the mode of failure as fatigue, under a complex combination of bending and torsional stresses. Its origin was traced to a sub-surface defect of a hard refractory (Al_2O_3) type inclusion. It is evident that the sub-surface defect had acted as a severe notch for initiation of the fatigue crack, under the dynamically loaded condition. Further, overall cleanliness of the material was found to be of a inferior quality, due to a high content of oxide inclusions.

However, it is important to note that during three successive overhauls the failed part was checked by magnaflux using AC method, but no crack indication was reported even though the crack might have existed in the sub-surface in the dormant state. The reason for failure to detect the presence of the defect or crack, was the miniature size of the inclusion (measuring about 0.01 mm) and the depth of its location (at 1 mm below) from the surface to which AC method magnaflux is insensitive. In such cases, the DC method is employed for detection of sub-surface cracks.

Conclusion and Recommendations

Most Probable Cause The basic cause of the catastrophic fracture was attributed to the presence of a sub-surface defect in the form of a lone hard refractory inclusion in the critically loaded pin radius zone. This was the ultimate source for the initiation of a hidden fatigue crack. In addition, the dangerously unclean material provided the underlying cause for further reduction of its fatigue resistance capability.

Remedial Action Utmost care on the part of the supplier to maintain the quality of products for use in critical applications was recommended in order to prevent recurrence of failure of such catastrophic nature.

How Failure could have been Prevented Had it been known to the inspector while magnafluxing the part by AC method (which is commonly used to detect surface crack), that fatigue cracks may sometimes develop from defects in the sub-surface zone, he could have resorted to DC method magnaflux check (which is a better method for detecting sub-surface crack). Thus the failure could have been prevented.

8.10 FATIGUE FAILURE OF THE THRUST BEARING RACE OF THE BLADE SPACING SYSTEM OF A HELICOPTER

Summary and Recommendations

A repetitive number of thrust bearing races of the blade spacing system of helicopter were reported diametrically broken. To find out the cause of the failure, one of these broken races was investigated.

Fractographic examination revealed a small fatigue zone originating from the flat inner surface, between the race groove and the ID. Visual examination of the affected area, on the other hand established the presence of annular uneven wear marks, indicating the existence of a play in the assembly. The bearing race material was found satisfactory with respect to composition, structure and hardness. Breakage of the race was attributed to fatigue, which developed due to concentration of repeated uneven compression loads caused by some play in the assembly.

It was recommended to check the cause of play in the assembly and ensure measures against it.

Background

Applications

Thrust bearing race made of ball bearing steel is a critically loaded component in the trunnion assembly of the blade spacing system of the main rotor head of the helicopter. The assembly of the trunnion and its location in the system along with the race is shown in Fig. 8.36. Service life of the component was not known.

Circumstances Leading to Failure

The race which was in contact with the larger dia base (marked 'X' in Fig. 8.36) was found broken during inspection of the blade spacing system.

Specifications

The race material was an American alloy to SAE 52100 specification as per the drawing requirements.

Performance of Other Parts in Same Service

Similar failure of the thrust bearing races was reported repeatedly.

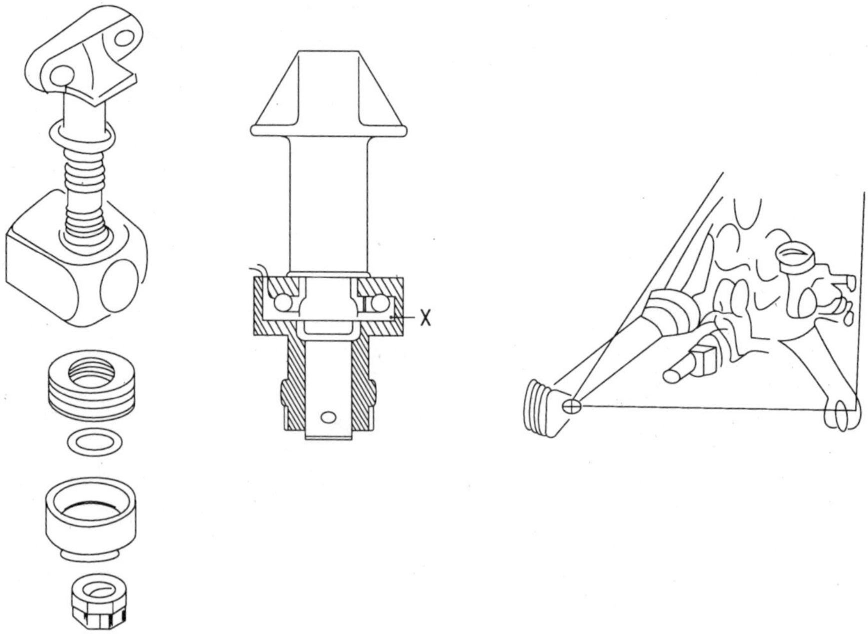

Fig. 8.36 Break up view of the trunion and bearing assembly of the blade spacing system with the type of bearing under investigation marked 'X'

Selection of Specimens

Broken halves of the race were taken for fractographic analysis by a low-powered stereoscopic microscope. A section was cut across the origin of the fracture for micro-examination and hardness property measurement purposes.

Visual Examination of General Physical Features

Examination revealed that the breakage of the bearing race had taken place almost diametrically (Figs 8.37 and 8.38). Deep indentation marks observed on the flat surface were found incidentally outside the line of fracture (normally which act as weak/stress concentration points for sudden breakage or progressive failures). On close observation it showed annual uneven wear marks on the flat surface adjoining both the ID as well as OD but there were no brinelling marks in the race way groove (Fig. 8.38). The failed thrust bearing was found consisting of two identical races with a groove and a ring of caged balls in the groove. However, slight corrosion patches seen on the flat outer surface had developed subsequent to breakage.

Fig. 8.37 Flat outerside of the broken bearing race (× 1 1/2)

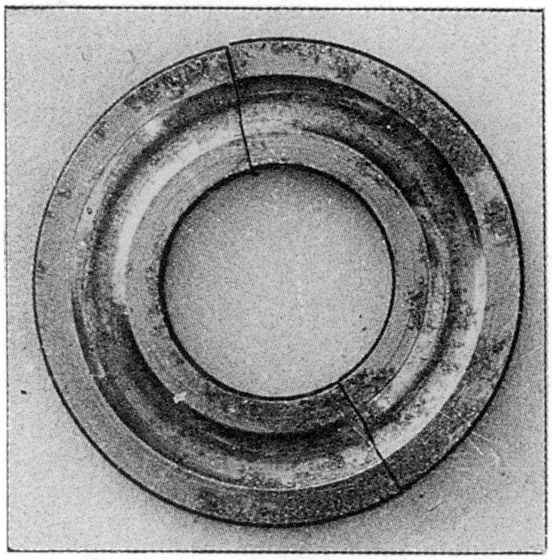

Fig. 8.38 Inner grooved side of the broken bearing race (× 1 1/2)

Testing Procedure and Results

Surface Examination

Macrofractography Fractographic examination of the broken race, in both the identical regions revealed distinct radial markings. They were found originating from the flat inner surface between the groove and the ID (Fig. 8.39). At the origin of the converging fracture lines, a small fatigue zone with 'beach' marks was detected during examination under the stereoscopic microscope at × 25 magnification.

Metallography

Microstructural Analysis Micro-examination of the section across the zone of fatigue revealed that the cracking was transgranular (which confirmed the initiation of breakage by fatigue process) and satisfactory hardened and tempered structure interspersed with numerous tiny globular carbides, typical of a high carbon chromium ball bearing steel.

Chemical Analysis

The race material on spectro-graphic analysis was found to be of 1.1% carbon and 1.2% chromium ball bearing steel, meeting the specification requirements (Table 8.10).

Mechanical Properties

Hardness Hardness survey was carried out on the microspecimen using mini-load hardness tester at 500 gm, and an average value of 780 HV was obtained.

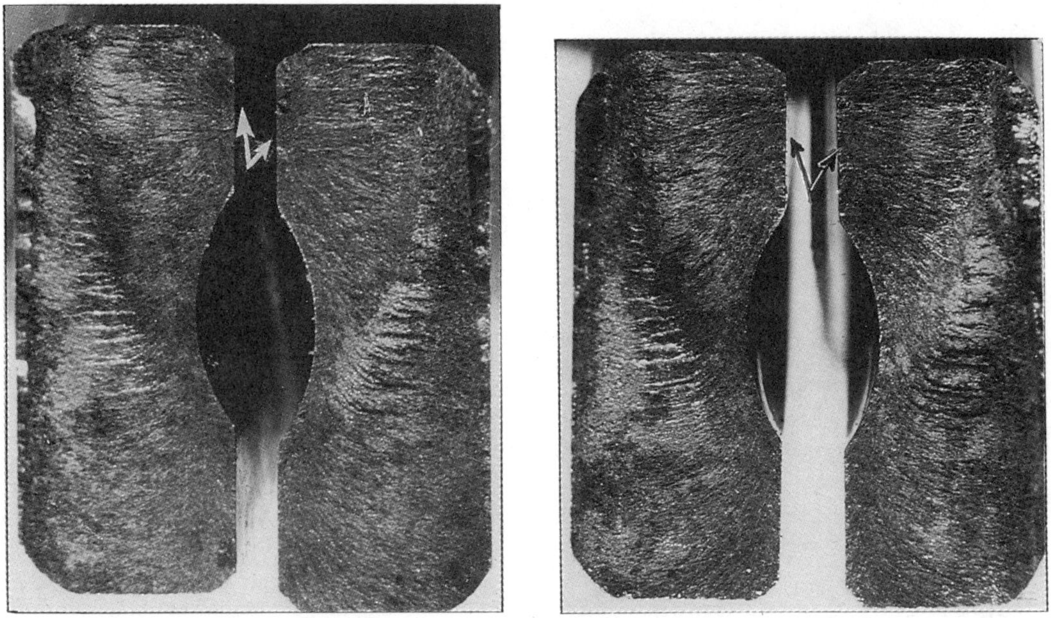

(a) (b)

Fig. 8.39 Small fatigue zone as fractured lines indicating the origin of cracking from the flat region between ID and groove (fractured surfaces kept side by side) (× 1 0)

Table. 8.10 Chemical specification (Wt. %) for the American carbon-chromium ball bearing steel (SAE 52100) from which thrust bearing race was made and actual compo sition of the failed part

Element	Min.	Max.	Failed part
Carbon	0.95	1.10	1.10
Manganese	0.25	0.45	0.32
Silicon	0.20	0.35	0.22
Sulphur	-	0.025	0.018
Phosphorus	-	0.025	0.020
Chromium	1.30	1.60	1.20

Discussion

Fracture studies clearly established the fact that the breakage of the bearing race was due to brittle failure, initiated by a small fatigue zone.

It was pertinent to note that the presence of annular uneven marks on the flat surface adjoining both the ID as well as the OD, indicated a play in the assembly prior to failure. This play in the form of repeated concentric compression loads, caused stress concentration for development of a fatigue crack from the flat inner surface of the race. Metallurgical tests such as chemistry, microstructure and hardness properties indicated that the bearing race material was satisfactory for the type of service application.

Conclusion and Recommendations

Most Probable Cause The basic cause of the failure was due to development of fatigue by the repeated concentric uneven compression loads, as a result of some play in the assembly during service.

Remedial Action In view of the fact that the trunnion assembly had experienced some play, it was suggested to examine the cause and ensure measures against its recurrence.

8.11 FATIGUE FAILURE OF SELF SEALING COUPLING HALF FLANGE OF A FIGHTER AIRCRAFT

Summary and Recommendations

A number of self sealing coupling half flanges were reported broken in an identical manner along the inner radius of the inner tube in the coupling assembly. One such failed part was analysed to find out the cause of breakage. Fracture studies revealed that the circumferential fracture had taken place due to fatigue cracking along the sharp inner radius. Macro and micro-examination indicated that the fracture lines were along the grain fibres, rendering the fatigue cracking more vulnerable, with turning marks as notches in the initiation zone.

On close observation of the actual ground operation of the landing gear, violent whipping actions were noticed and this invariably contributed to the concentration of excessive fluctuating bending stresses along the inner radius.

Based on these observations and conclusions, it was suggested to liberalise the inner radius with a good finish, and to modify the grain fibre orientation in the longitudinal direction of the tube component.

Background

Applications

Self sealing fixed half flange coupling (Fig. 8.40) is fitted on a fighter aircraft and serves as a self sealing valve for connection to external hydraulic line for ground servicing of the landing gear system, etc.

Circumstances Leading to Failure

Failure of the coupling during flight, resulting in total failure of the hydraulic system, was reported (Fig. 8.40).

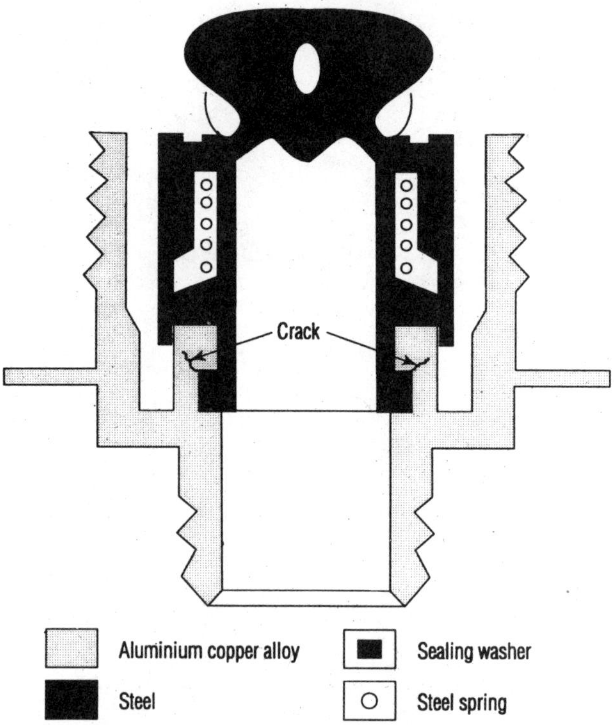

Fig. 8.40 A schematic construction of the self sealing coupling half flange and the location of crack

Specification

The material used for fabrication of the failed flange coupling was a British alloy to BS L 77 specification.

Performance of Other Parts in Same Service

A number of couplings were reported broken repeatedly in an identical manner.

Selection of Specimens

The broken half of the coupling was examined under a low powered stereoscopic microscope for fracture studies. Sections were taken along and across the origin of the fracture for macro and micro-examination under the optical microscope and also for hardness measurement.

Visual Examination of General Physical Features

Visual examination of the failed coupling after disassembly revealed that the breakage was circumferencial and along the fillet radius between the thin wall of the inner tube and its thick swaged end of the aluminium alloy housing (Fig. 8.41). The fillet radius was found rather sharp and with turning marks.

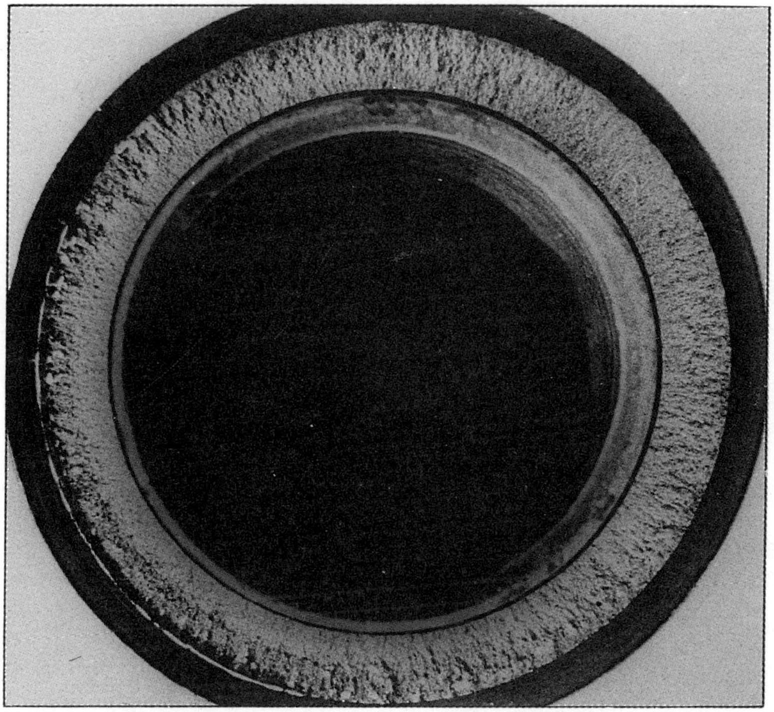

Fig. 8.41 The fractured surface of the inner tube with fatigue originating from the inner radius (× 5)

Testing Procedure and Results

Surface Examination

Macrofractography Examination of the fractured surface of the inner tube showed smooth bright segments (at × 20 magnification) covering more than three fourths of the area of cross-section, although radial fracture lines were conspicuous. Traces of 'beach' marks of fatigue zones were observed in bright segments, with origins at the inner radius. The remainder portion was rough and coarse with shear lips in places along the outer periphery and was typical of an overstressed fracture (Fig. 8.41). All these indicated that the origins of fatigue had emanated from the various segments of the inner radius and progressed as multi-fatigue cracking.

Metallography

Microstructural Analysis Macro and micro-examinations of both longitudinal and transverse sections of the coupling half flange revealed that the line of fracture was transgranular and followed the grain fibre. No structural abnormalities were observed in the heat-treated aluminium copper alloy material.

Chemical Analysis

The chemistry of the material was found meeting the requirements of BS L 77 specification (Table 8.11).

Table 8.11 Chemical specification (Wt%) for the British aluminium-copper alloy (BSL 77) from which half flange was made and actual composition of the failed part

Element	Min.	Max.	Failed part
Copper	3.90	5.00	4.15
Magnesium	0.20	0.80	0.62
Manganese	0.40	1.20	0.75
Silicon	0.50	0.90	0.48
Iron	-	0.50	0.38
Nickel	-	0.20	0.15
Chromium	-	0.10	0.08
Zinc	-	0.20	0.13
Titanium + zirconium	-	0.20	0.16

Mechanical Properties

Hardness Hardness measurement made on the microspecimen was found to be 142 BHN, which met the specified requirements.

Discussion

Fractographic examination clearly indicated that the breakage of the aluminium-copper alloy inner tube of the coupling had occurred due to fatigue. Since repeated failures of this nature were reported, the investigator closely examined the actual ground operation of the landing gear and noticed violent whipping actions during cut-in and cut-out of the hydraulic pressure. Initiation of the fatigue crack along the inner radius was accelerated by the presence of turning marks in the sharp radius zones, which acted as multi point notches under repeated whipping actions and fluctuating hydraulic pressures during ground servicing. It gradually progressed a fairly long period of ground service. Moreover, unfavourable grain fibres (cutting across the wall thickness of the tube, instead of running along the tube length) had definitely weakened its fatigue resistance to repeated imposed loads. They were concentrating under constant whipping action at the sharp radius zone, and resulting in early fatigue initiation.

Conclusion and Recommendations

Most Probable Cause The basic cause of the failure of the fixed half coupling was attributed to occurrence of fatigue crack, rendered easy by the presence of undesirable turning marks in the sharp inner radius and unfavourable grain flow pattern, under vibratory stresses.

Remedial Actions It was recommended to liberalise the fillet radius with a good finish, i.e. without any turning marks, together with a measure to change the grain fibre pattern of the forging in such a way that they run longitudinally in the failed tubular portion.

8.12 FAILURE OF THE FIRST STAGE SUN GEAR OF A HELICOPTER DURING THE ENDURANCE TEST DUE TO SPALLING

Summary and Recommendation

A high duty gear designed and fabricated out of a case hardening steel for service in the magnesium gear box assembly of a helicopter, was subjected to 50 hours of design qualification test along with an imported one. The imported gear withstood the test whereas the indigenous one failed.

Visual examination showed severe wear/scuffing in the pitch line area of the driving faces of all the teeth. Micro-examination of the zone of failure revealed several subsurface as well as angular cracks which were found lifting up several chunks of metal from the carburised layer. This indicated that the cause of surface deterioration was due to spalling. Further, hardness measurement on the section, across the carburising depth, indicated a lower hardness in the failed gear, compared to the imported one.

The basic cause of spalling was attributed to the lower hardness (wear resistance) characteristics of the carburised layer.

Background

Applications

The first stage sun gear (Fig. 8.42) is machined out of a low carbon nickel-chromium-molybdenum alloy forged steel blank and case hardened in the tooth portion. The gear is assembled in series with second and third stage gears in the main gear box for transmission of power to the engine. For qualification purpose the part was subjected to an endurance test under an accelerated simulating condition of service.

Circumstances Leading to Failure

After 50 hours of the qualification test, the sun gear showed excessive wear and scuffing along the entire mating tooth contact surfaces of the splines.

Specifications

Material used for the gear was a French alloy to 16 NCD 13 specification.

Fig. 8.42 Sun gear part which was subjected to the qualification test (× 3/4)

Performance of Other Parts in Same Service

An imported first stage sun gear was also subjected to the same qualification test for comparison purposes and found to have withstood the accelerated test successfully.

Selection of Specimens

Two severely affected tooth portions were cut off from the gear body for clear observation of the surface defects. A portion was sectioned across the defective zone for metallographic examination. A similar section was made on the imported gear for comparison of microstructural and hardness characteristics.

Visual Examination of General Physical Features

Examination showed that the driving faces of all the teeth were severely worn out and scuffed, mostly confined to the region of contact with the pinion (mating part) teeth at the pitch line (Fig. 8.43). Further, broken metal pieces, in small chunks, were found fallen off the teeth surface at places.

Testing Procedure and Results

Metallography

Microscopic Analysis Micro-examination of the section across the spalled zone at × 100 magnification revealed several sub-surface as well as angular cracks tending to lift

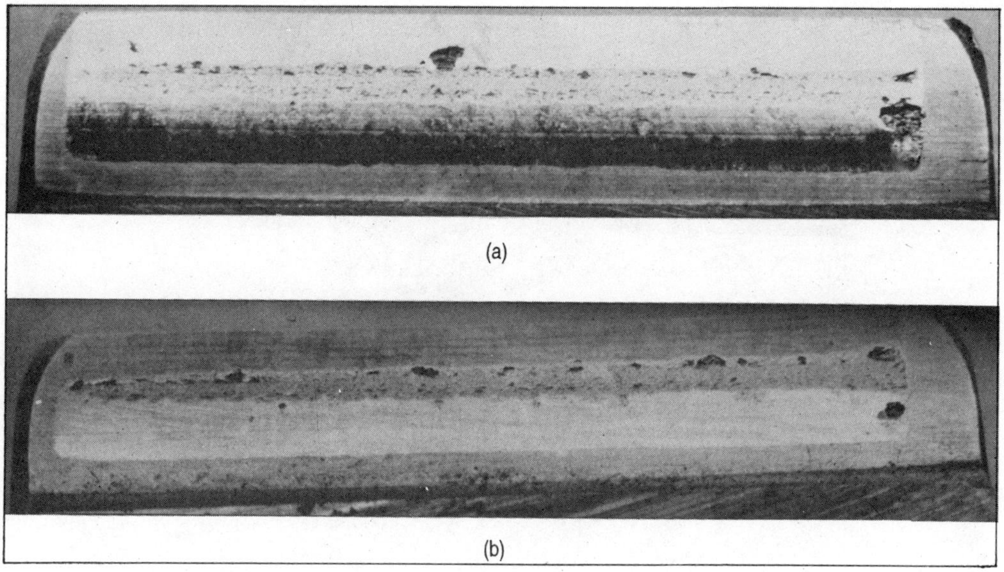

Fig. 8.43 Severe spalling on the driving face of the sun gear teeth (×8)

up the carburised layer (Figs 8.44 and 8.45). Further, tempered martensitic structure of the case was found free from any carbides, either in the form of network or isolated particles, whereas structure of the case in the imported gear revealed carbides in network form at the grain boundaries as well as in isolation.

Fig. 8.44 Sub-surface crack (×50)

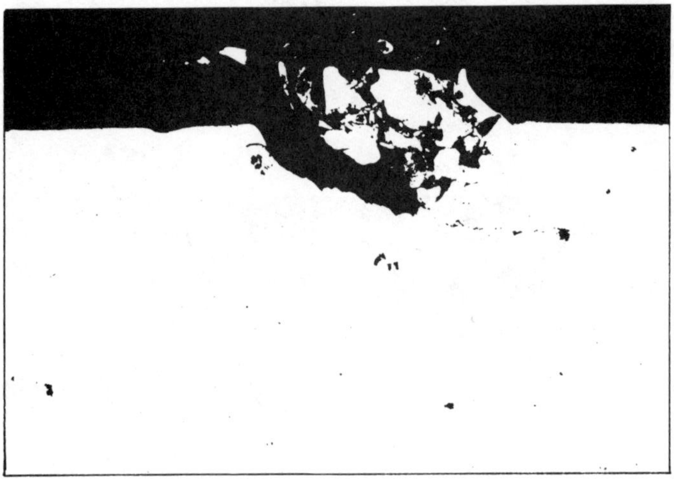

Fig. 8.45 Angular cracks and lifting up of the spalled piece (× 100)

Chemical analysis

Chemical composition of the gear material was found conforming to specification (Table 8.12).

Table 8.12 Chemical specification (Wt.%) for the French nickel-chromium-molybdenum steel (16 NCD 13) from which the 1st stage sun gear component was made and actual chemical composition of the failed part.

Element	Min.	Max.	Failed part
Carbon	0.12	0.17	0.15
Manganese	0.30	0.60	0.45
Silicon	0.15	0.40	0.33
Sulphur	-	0.020	0.014
Phosphorus	-	0.025	0.018
Nickel	3.0	3.50	3.23
Chromium	0.80	1.10	0.95
Molybdenum	0.25	0.30	0.27

Mechanical Properties

Hardness The case hardening characteristics of the failed gear as well as the imported one were determined by hardness measurements (Fig. 8.46), using mini load hardness

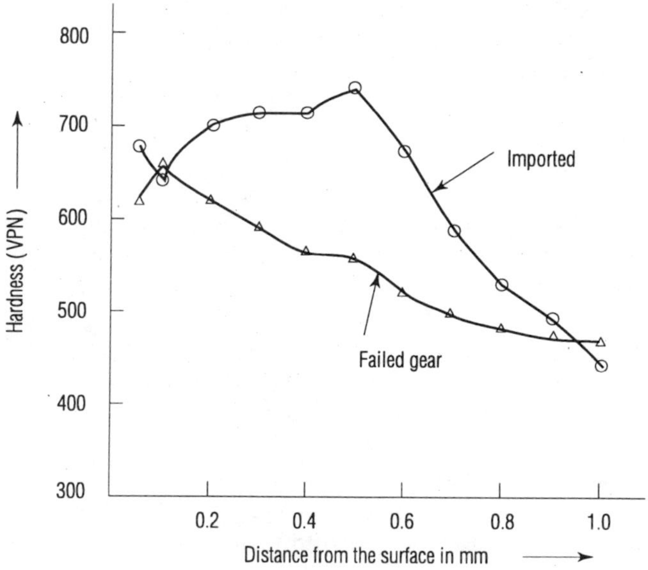

Fig. 8.46 Comparative hardness survey of failed gear with imported gear

tester at 500 gms. The results indicated a lower hardness in the case hardened layer compared to that of the imported one.

Discussion

Macroscopic as well as micro-examination clearly established the nature of surface deterioration in the pitch line area as spalling. The presence of sub-surface as well as angular cracks which were seen lifting up chunks of metal zone from the carburised layer, indicated that the cyclic loading during testing had induced several fatigue crack nucleations at the sub-surface and propagated upwards in an angular manner. Further, from the study of comparative hardness characteristics of the failed as well as imported gears, it was found that the requisite hardness (which is necessary for achieving the wear resistance property of a dynamically loaded heavy duty gear) was comparatively low in the case hardened portion of the failed gear. The reason for low hardness was confirmed by the microstructure which was found free from any carbide formation, unlike the imported sample. It is known that carbides provide high wear resistance and hardness at the working surface. Hence, low hardness of the case was found responsible for easy development of the spalling phenomenon.

Conclusion and Recommendations

Most Probable Cause The basic cause of surface deterioration of the gear teeth was attributed to low hardness of the carburised layer, which resulted in spalling failure during testing.

8.13 CRACKING AND SPALLING OF THE CONNECTING ROD BEARING OF A TRAINER AIRCRAFT ENGINE DUE TO THERMAL FATIGUE

Summary and Recommendations

A number of bearing bottom halves used in the pre-mod connecting rod with a dowel hole were reported deteriorated identically in the mid zone of the bearing surface. One of these failed bearing halves was investigated to ascertain the reason for failure.

Examination of the bearing bottom and the corresponding connecting rod revealed the location of deterioration (i.e. spalling and flaking off) exactly in line with that of the dowel pin hole and the copper back plate of the bottom half severely heat affected in the mid-zone. Micro-examination of the affected zone showed numerous parallel as well as transverse cracks in the interface zone of the otherwise satisfactory structure.

These characteristic surface deteriorations indicated a thermal fatigue phenomenon, the basic cause of which was attributed to undue heat transfer caused by the presence of a dowel pin hole in the mid zone, resulting in the development of a high thermal gradient.

Under the circumstances, it was recommended to use the modified connecting rod without the dowel pin hole.

Background

Applications

Semi-circular bearing halves are made out of copper backplate with an overlay of bearing metal on the convex surface and are assembled onto the connecting rod housing, which engages the driving mechanism to the engine. As per original design, there was a dowel pin hole in the mid-zone of the bearing housing of the heat-resistant aluminium alloy connecting rod (Fig. 8.47). Although, modified design had eliminated

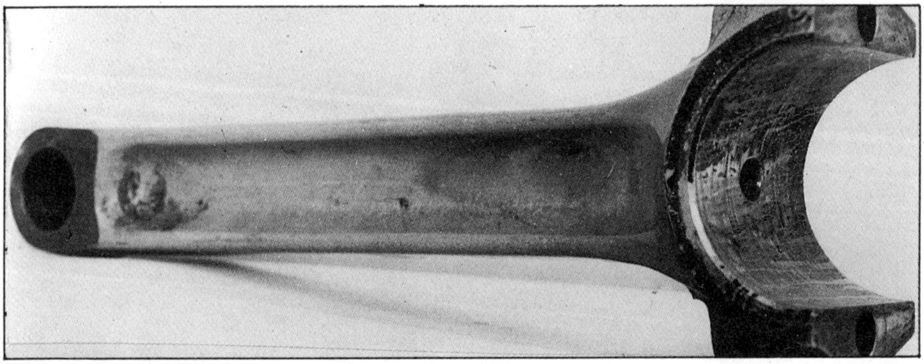

Fig. 8.47 Pre-mod connecting rod showing the dowel pin hole (× 1/2)

the dowel pin hole due to change of bearing locking system, it was reported that pre-mod connecting rods were permitted for use.

Circumstances Leading to Failure

One bottom bearing (Fig. 8.48) was found deteriorated after 400 hours of service since installation.

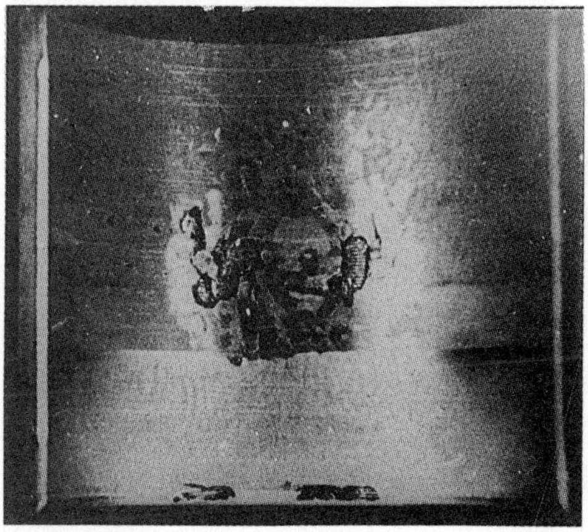

Fig. 8.48 Bottom half bearing showing cracking and spalling (× 1)

Specifications

Since the bearings were imported, the specification of the bearing metal was not available.

Performance of Other Parts in Same Service

It was reported that identical failures were occurring rather repeatedly.

Selection of Specimen

Sections across the defective zone were taken for micro-examination under the optical microscope.

Visual Examination of General Physical Features

Visual examination of the bearing half revealed cracking and spalling of the bearing overlay (chunks of bearing metal flaked off in a few zones, exposing the original deep turning marks in the copper back plate) confined to the central portion only (Fig. 8.48). Also a circular discolouration (heat affected) was observed on the outside of the back plate in the same central portion, which was found coinciding with the dowel pin hole location of the pre-mod connecting rod housing.

Testing Procedure and Results

Metallography

Micro-examination of sections of the deteriorated bearing metal together with copper back plate revealed satisfactory adhesion of the bearing metal to the back plate. Numerous cracks, some parallel to the back plate and others at right angles to it, were observed in the bearing layer-microstructure (Fig. 8.49). Such features were typical of thermal fatigue cracks in bonded bearings.

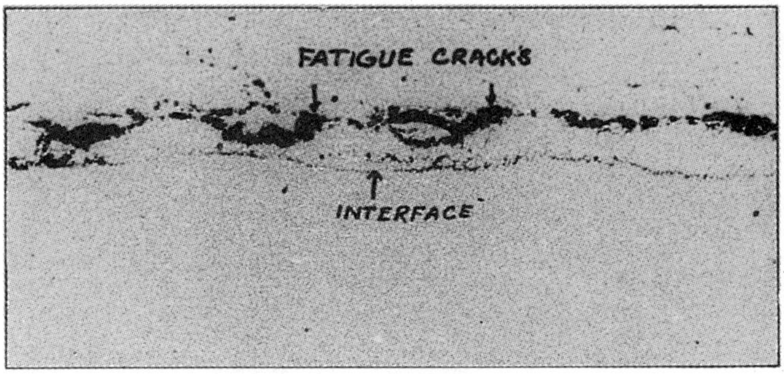

Fig. 8.49 Fatigue cracks in the interface zone (× 100)

Chemical Analysis

The bearings metal on chemical analysis was found to be a lead base alloy with 5% copper, 3.5% antimony and 0.25% aluminium.

Discussions

The very nature of spalling and flaking off of the bearing metal surface as well as crackings observed in the microstructure in the defective zone, indicated that the surface deterioration was due to thermal fatigue phenomenon. It was also evident from the fact that uneven heat transfer, by virtue of the presence of dowel pin hole, seemed to have resulted in undue thermal gradient in the central zone (as observed by the presence of circular discolouration, i.e. heat effects, on the outside of the back plate in the zone coinciding with the dowel pin hole location), rendering development of thermal cracks easy.

Conclusion and Recommendations

Most Probable Cause The basic cause of deterioration of the bearings was suspected to be due to development of thermal fatigue cracks.

Remedial Action In light of the above circumstances, it was recommended to use the modified connecting rod without the dowel pin hole, in consultation with design personnel.

8.14 FAILURE OF INTAKE VALVE OF A PISTON ENGINE DUE TO HOT CORROSION AND THERMAL FATIGUE

Summary and Recommendations

One intake valve of a cylinder head belonging to the piston engine was reported cracked in the valve seat area after a long period of service.

Fractographic studies of the fracture revealed a fatigue crack originating from the deep corrosion pits at the outer periphery of the rim. The valve seat was found heavily oxidised/scaled due to hot combustion gases. Micro-examination of the zones in the rim portion as well as on the surface of the valve seat adjoining the neck and the fatigue crack revealed deep corrosion pits (in the rim portion) and secondary cracks running parallel to the main crack. These microstructural features suggested a thermal fatigue phenomenon. Further, chemistry of the valve material indicated that the presence of molybdenum in the austenitic stainless steel was very low. Molybdenum had a marked influence on the improved corrosion resistance property. Hence the very low content (0.15%) of molybdenum had rendered the material susceptible to corrosion in the corrosion prone environment. Based on the above facts, the cracking of the intake valve was attributed to a combined effect of hot corrosion and thermal fatigue. Since stainless steel is more susceptible to hot corrosion attack, it was recommended to use a nimonic alloy in place of stainless steel.

Background

Applications

Intake valve of the stainless steel variety is used in the cylinder head of the piston engine. It operates to and fro inside the cylinder head (Fig. 8.50) usually in the temperature range of 500–1200 °C.

Circumstances Leading to Failure

During a routine inspection of the engine, one intake valve was found cracked (Fig. 8.51), after 107 hours of service since the previous overhaul (total period of service was not known).

Specification

Material specification of the intake valve was not known.

Selection of Specimen

The crack was opened and one segment of the fracture was cleaned and examined under the stereoscopic microscope. Another segment was sectioned for microstructral observation.

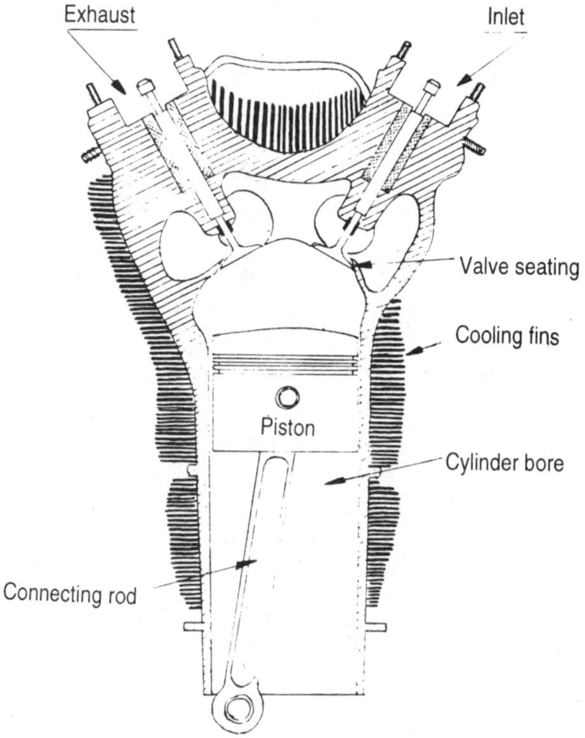

Fig. 8.50 The profile of the cylinder with valves and piston

Visual Examination of General Physical Features

The intake valve was found cracked radially in the seat region, starting from the rim and extending deep into the neck portion (Fig. 8.52). Heavy deposits of oxide/corrosion products were noticed, particularly towards the rim surface.

Testing Procedure and Results

Surface Examination

Macrofractography The crack was carefully opened out and fracture was found in a plane at right angles to the valve seat. As the fracture was found corroded, scaled/oxidised, more severely towards the rim, the surface was cleaned in 10% nitric acid solution. Examination of the surface under the stereoscopic microscope, at × 25 magnification, revealed 'beach marks' (typical of a fatigue failure) in the rather smooth region towards the neck (Fig. 8.52). The fracture region adjoining the rim was found completely

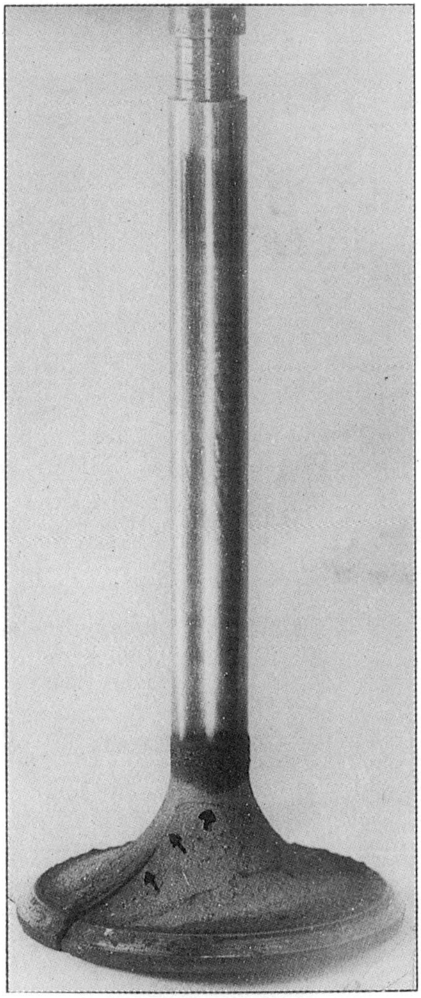

Fig. 8.51 Intake valve showing radial crack in the seat portion (× 3/4)

obliterated due to excess scaling and corrosion. From the traverse of the beach marks, the origin of fatigue was located at the tapered rim end that contained deep corrosion pits. The fatigue crack was found covering almost the entire section thickness (Fig. 8.52).

Metallography

Microstructural Analysis Micro-examination of a section of the intake valve in the seat region adjoining the fracture as well as the tapered rim end,. revealed numerous deep corrosion pits along the outer periphery of the rim (Fig. 8.53) and secondary cracks (on the surface) running parallel to the main crack (Fig. 8.54) of thermal fatigue nature. Otherwise, the austenitic structures were found satisfactory.

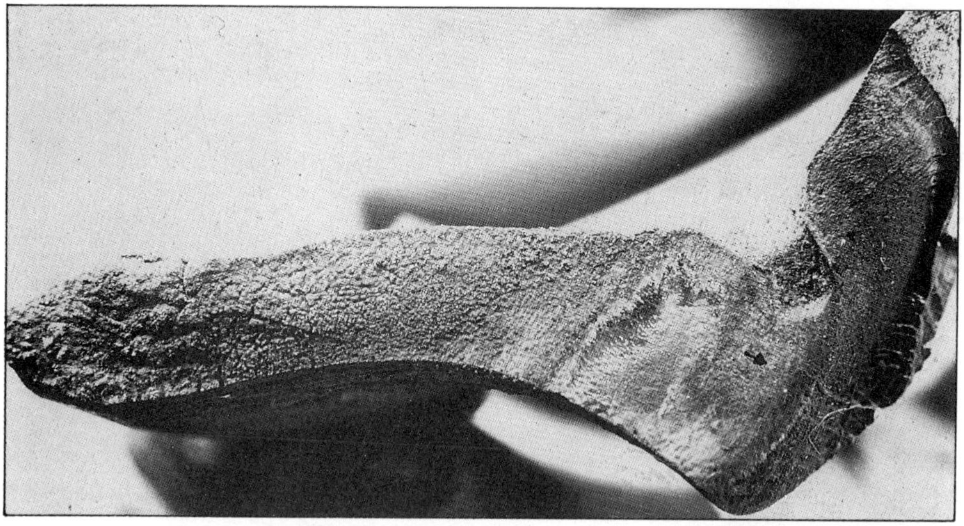

Fig. 8.52 Fractured surface of radially cracked portion of the seat region showing beach marks with origin at the corrosion pits at the outer periphery of the rim (× 5)

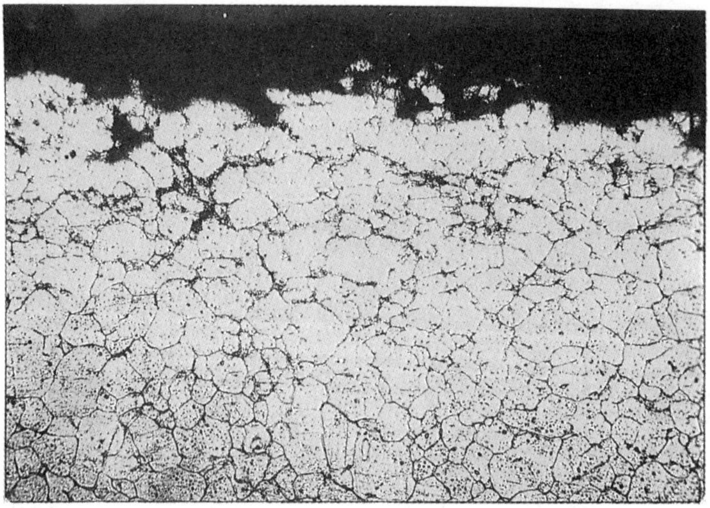

Fig. 8.53 Deep corrosion pits in the outer periphery of the rim (× 50)

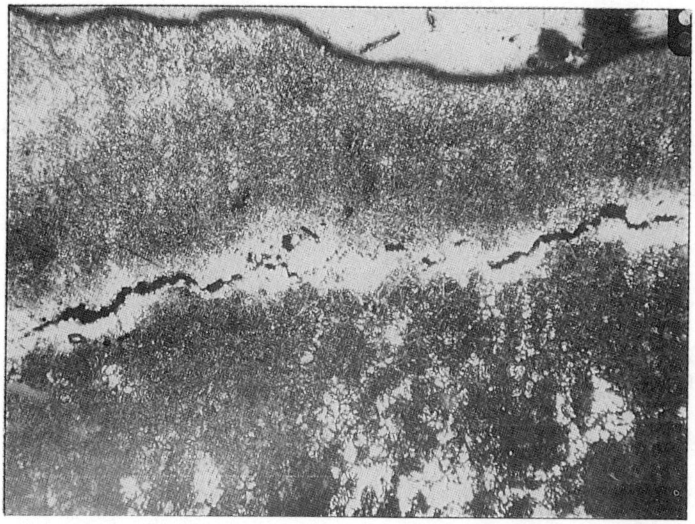

Fig. 8.54 Shows secondary cracks parallel to the main fracture (×5)

Chemical Analysis

Chemical composition of the intake valve was found conforming to an American alloy No. 20 (C : 0.29%.Cr : 13.20% Mo : 0.15%, W : 0.31% Co : 0.22%)

Discussion

Fractographic studies as well as microstructural observations clearly established the fact that the radial cracking of the seat region of the intake valve had occurred due to the combined effect of hot corrosion and thermal fatigue.

As the chemistry of the material clearly indicated that molybdenum content was very low (0.15%) in Cr.Ni. type stainless steel, the corrosion resistance property of the intake valve operating at a high temperature range was drastically reduced yielding to a hot corrosion attack, particularly under combustion gases containing sulphur, bromine, chlorine, etc. It is pertinent to note that the presence of a significant amount of molybdenum is known to have a marked influence of the resistance to corrosion. Heavy scaling, as observed in the valve seat, indicated that the valve seat was subjected to high temperatures which perhaps tended to accelerate the absorption of oxygen from the gases, particularly in the presence of lead oxide. Apart from the hot corrosion attack, the intake valve experienced thermal fatigue cracking due to intermittent heating and cooling in the operation.

Conclusion and Recommendations

Most Probable Cause

The basic cause of hot corrosion attack and thermal fatigue in the valve seat zone was attributed to lowered resistance property of the material in an environment of high service temperature and combustion gases.

Remedial Action Since the hot corrosion properties of the stainless steel, particularly of the type under investigation is very poor, a suitable nimonic alloy was recommended for use in place of the existing one.

8.15 FATIGUE CRACKING OF THE STEEL BARREL OF A LINEAR FOUR CYLINDER AERO ENGINE

Summary and Recommendations

One cylinder barrel was found circumferentially cracked in the flange radius during a major overhaul.

Visual and fractographic studies revealed that the crack was due to fatigue phenomenon. The cause of fatigue was traced to the presence of notch rendered by the severe biting effect due to the fouling action of the oversized steel washer on the radius. As a result, a localised strained zone was produced on the radius, where the service vibratory stresses were concentrated maximum, leading to progressive fatigue cracking.

It was recommended to ensure the use of the correct size of steel washer during assembly.

Background

Applications

The cylinder barrel (Fig. 8.55) is made of forged steel blank by the machining route, and is used in four-cylinder aero engine of transport aircraft.

Circumstances Leading to Failure

During a major overhaul of the aircraft, one of the four cylinders was found circumferentially cracked in the flange radius portion of the barrel (Figs 8.55 and 8.56).

Specifications

The barrel steel material was an American alloy to MIL-S 6758 specification.

Selection of Specimens

The circumferential crack was opened up and one of the fracture segments with its bottom flange portion in tact, was studied under a stereoscopic microscope for analysis of the mode of fracture. A small portion across the fatigue zone was sectioned for micro-examination using the optical microscope.

Fig. 8.55 Cylinder barrel cracked at the bottom groove radius (× 1/3)

Fig. 8.56 A close-up of the cracked portion (× 2)

Visual Examination of General Physical Features

Examination of the failed barrel revealed cracking along the bottom radius of the thick flange for about one fourth of the circumference (Figs 8.55 and 8.56).

On close examination, it was found that the machining of the barrel was poorly finished, leaving deep tool marks and an improper flange fillet radius (Fig. 8.57).

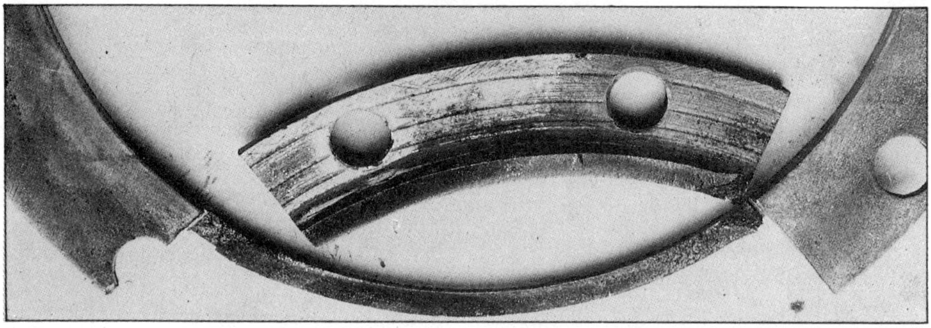

Fig. 8.57 A distinct circular impression mark around the bolt hole with a severe bite at the flange radius (× 1 ½)

Testing Procedure and Results

Surface Examination

Visual Examination Examination under a magnifying lens further revealed a distinct circular impression corresponding to that of the steel washer used for bolting the barrel. The impression was around a bolt hole, with a prominent deep bite at the flange radius (Fig. 8.57). Other holes were found free from such dents impressions.

Macrofractography Examination of the fracture surface revealed faint but clearly discernible beach marks of fatigue failure type under bending loads, with the origin at the fouled radius zone (Figs 8.57 and 8.58).

Metallography

Microstructural Analysis Micro-examination of the section across the fatigue revealed transgranular cracks, further confirming that the crack propagation was due to fatigue. The microstructure was found satisfactorily heat treated in the hardened and tempered condition.

Chemical Analysis

The steel barrel material was analysed and found conforming to MIL-6758 specification (Table 8.13).

Fig. 8.58 Clear fatigue beach marks originating from the fouled radius zone marked by arrow (× 6)

Table 8.13 Chemical specification (Wt. %) for the American chromium-molybdenum steel (MIL-S-6758) from which cylinder barrel was made and actual composition of the failed part

Element	Min.	Max.	Failed part
Carbon	0.28	0.33	0.31
Manganese	0.40	0.60	0.45
Silicon	0.20	0.35	0.24
Sulphur	-	0.025	0.18
Phosphorus	-	0.025	0.019
Chromium	0.80	1.10	1.02
Molybdenum	0.15	0.25	0.21

Mechanical Properties

Hardness A hardness check was carried out on the microspecimen and the average hardness was 310 BHN.

Discussions

Fracture analysis clearly established the fact that the circumferential cracking along the flange radius occurred due to fatigue.

Analysis further showed that the cause of the fatigue was not due to metallurgical factors but was manifested due to the ignorance of the man at the assembly level. In this particular case, on non-availability of the steel washer of proper size, the assembly man used his discretion in selecting an oversized one and fixed it over the bolt hole, without taking into consideration the effect that might arise due to interference or fouling of the washer with the radius zone.

Deep washer biting marks close to the radius indicated a built-in strained zone where vibratory stresses in service were forced to concentrate maximum, yielding a localised crack by fatigue.

Conclusion and Recommendations

Most Probable Cause The basic cause of fatigue cracking of bottom flange radius of the steel barrel was due to the localised (built-in) straining of radius zone by fouling action of the oversized steel washer.

Remedial Action It was suggested that during assembly of the cylinders, inspection should ensure the correct size of the steel washer before fixing the bolt, to avoid creation of a notch factor in the hidden assembly.

How Failure could have been Avoided A more indepth metallurgical knowledge of the consequences of poor design could have averted this failure.

8.16 FATIGUE CRACKING OF THE TRAILING EDGE STRIP OF THE MAIN ROTOR BLADE OF A HELICOPTER

Summary and Recommendations

One of the three main rotor blades of a helicopter showed cracks at the first two rivet hole joints of the trailing edge strip and the skin, from the root end of the blade during routine inspection. Fractographic studies as well as micro-examination of the cracked portion confirmed the failure due to fatigue under a bending load. The fatigue crack was found to have developed from the offset drilled hole region which was filled with araldite subsequent to pilot drilling operation and prior to assembly. The premature fatigue failure was attributed to the accumulation of a complex combination of stresses, including centrifugal forces at the root end of the blade with the offset drilled hole area as the weak point of stress concentration.

In light of repeated failures of this nature, a fool proof system during drilling operations was recommended to be evolved to avoid such offset drilling and cause of stress concentration.

Background

Applications

Rotor blades were manufactured for use in helicopters. As per the method sheet, the trailing edge strip was initially drilled to dimension in the required locations using a template. Keeping the strip in position on the inside edge of the flap of the 'U' formed skin, undersized drilled holes in the strip were transferred to the skin edge by undersized pilot holes using 'clicko' pins in the holes at the extremities and the middle. Dummy rivets were driven on removal of the clicko pins and the detail parts were glued on with araldite. After this the correct size drill holes were made through the pilot holes and dummy rivets. This procedure was followed for accurate alignment of the originally drilled holes of the trailing edge strip with those of the skin edge. A cross-section with details of assembly is shown in Figs 8.59 and 8.60.

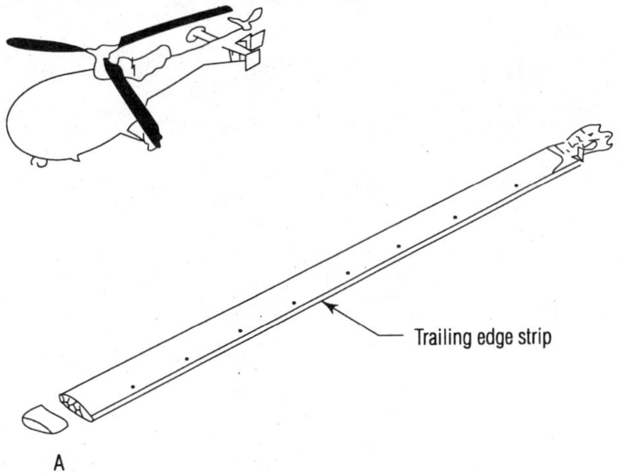

Fig. 8.59 The sketch of a main rotor blade

Circumstances Leading to Failure

During inspection, after 1000 hours of flight, the trailing edge strip of one of the three rotor blades was found transversely cracked through the rivet hole (Fig. 8.61).

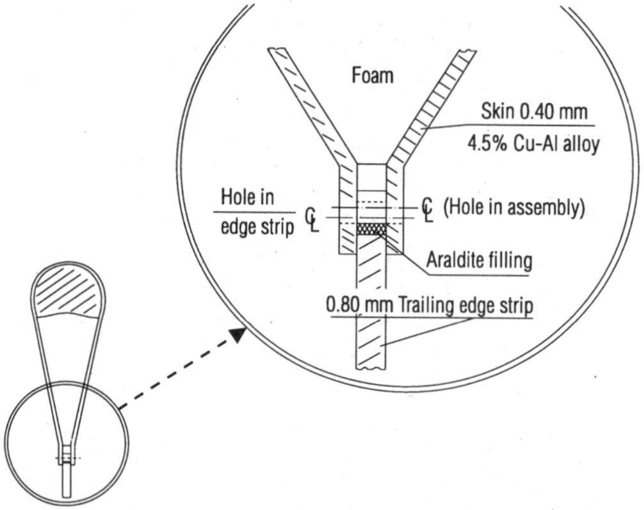

Fig. 8.60 The sketch of the assembly details of trailing edge portion

Fig. 8.61 Straight transverse crack in the trailing edge through the first rivet hole (× ½)

Specifications

The skin material used was a French alloy to Au4G and that of the strip to AG3 specifications.

Selection of Specimens

The cracked portion was opened out for fracture examination under stereoscopic microscope. A section across the fracture was made for microstructural studies under the optical microscope.

Visual Examination of General Physical Features

The skin was peeled off at the cracked trailing edge strip section and a transverse crack was observed to have emanated from the first drilled hole, from the root end of the blade (Fig. 8.61). The bucked rivet did not show any abnormalities in the bucking process. Further, the rivet hole was found elliptical due to offset drilling with the bottom basement below the rivet shank, filled with araldite (Fig. 8.62). This indicated that there existed a discontinuity just below the rivet shank (Figs 8.61 and 8.63).

Fig. 8.62 Offset drilling at first rivet hole (i.e. space outside the rivet filled with araldite) (× 10)

Testing Procedure and Results

Surface Examination

Visual On close examination, another tiny crack was observed on the opposite edge of the hole (Fig. 8.61). The strip was also found cracked in an identical manner from the second rivet hole (Fig. 8.63). However, the third rivet hole showed no such cracking, but the hole was found drilled offset horizontally (Fig. 8.64).

Macrofractography Examination of the fracture at the rivet hole No.1 under a low powered stereoscopic microscope revealed smooth surface with signs of 'beach' marks under bending fatigue mode with origins from either side of the outer surface of the hole edge, adjoining the araldite filled zone. (Fig. 8.65)

Matallograpby

Microstructural analysis Micro-examination of the section across the fatigue origin area revealed transgranular cracks emanating from the hole edge. These transgranular

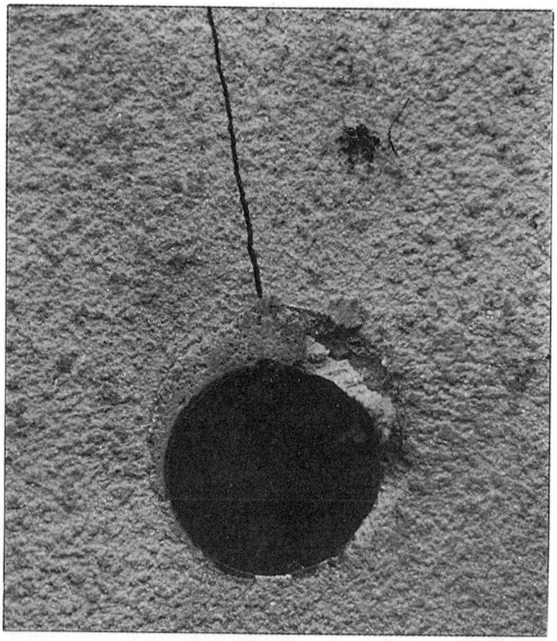

Fig. 8.63 Offset drilling in the second rivet hole (× 10)

cracks also confirmed the failure by fatigue. Microstructure confirmed the use of strip material in non-heat treated (soft) condition.

Chemical Analysis

Chemical composition of the cracked strip material was found conforming to AG3 specification (Table 8.14).

Table 8.14 Chemical specification (Wt. %) for the French aluminium magnesium alloy (AG3) from which the trailing edge strip was made and actual composition of the failed part

Element	Min.	Max.	Failed part
Copper	-	0.10	Trace
Magnesium	2.40	3.00	2.65
Manganese	0.50	1.00	0.70
Silicon + Iron	-	0.40	0.30
Chromium	0.05	0.20	0.12
Zinc	-	0.25	0.17
Titanium	-	0.20	0.15

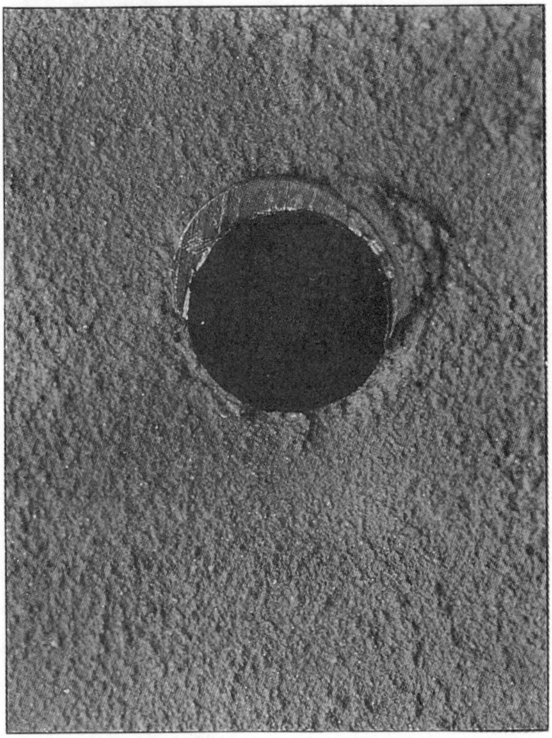

Fig. 8.64 Offset drilling in the third rivet hole (× 10)

Discussion

Fractographic as well as micro-examination of the cracked strip zone clearly established the fact that identical cracking of the first two successive rivet holes from the root end of the blade occurred due to bending fatigue phenomenon. It is well known that fatigue crack develops prematurely in a highly loaded dynamic structure, specially in presence of a notch. In this case, bending stresses by virtue of repeated flapping actions of the blade in flight had concentrated towards the root end portion. In addition, presence of discontinuities in the form of soft filled-up (araldite) zone around the trailing edge side of the first and second rivet holes (due to offset drilling) had served as severe stress raisers for the fatigue crack initiation.

Conclusion and Recommendations

Most Probable Cause The basic cause of the premature failure of the main rotor blade was due to fatigue under complex combination of bending and centrifugal stresses with offset drilled hole areas (filled with soft araldite) as stress raisers.

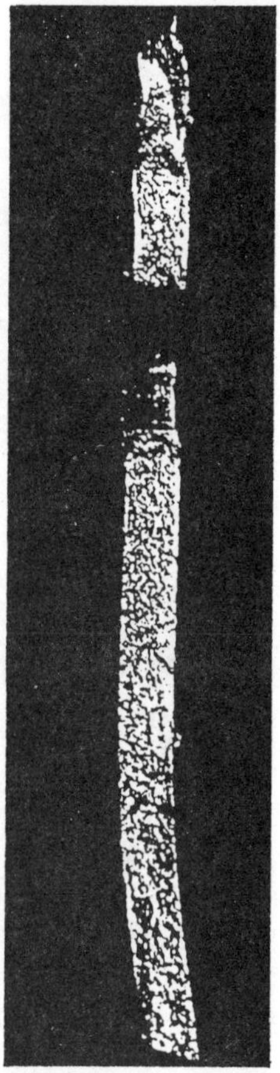

Fig 8.65 Fatigue zones on either side of the first rivet hole (× 6)

Remedial Action Since a number of cases of such cracking have been reported in the main rotor blades, the process of drilling operation of the trailing edge as well as the skin were recommended to the modified, so as to avoid offset drilling.

Also, it was recommended to ensure freedom from such offset drilling after completion of rivetting operation by suitable NDT methods such as the X-ray.

8.17 CORROSION FATIGUE FAILURE OF ALUMINIUM COPPER ALLOY PROPELLER BLADE OF PISTON ENGINE

Summary and Recommendations

One of the aluminium 5% copper alloy forged propeller blades belonging to the piston engine of a transport aircraft, catastrophically broke into two along the root radius during the engine test run.

The broken hub portion of the blade was subjected to fractographic as well as microscopic examination. These studies revealed that the failure was caused due to corrosion fatigue phenomenon. Further, it was interesting to find that almost unnoticeably minor atmospheric corrosion pits in the critically loaded blade root radius zone had caused the development of numerous fatigue cracks as points of stress concentration during long service of aircraft. Since corrosion pits were too small to be identified even under a magnifying lens during periodic inspection, the defects could not be detected until failure occurred.

Hence it was recommended to enforce periodic inspection of the critical root radius portion at suitable magnification, preferably under the low powered stereoscopic microscope for detection of corrosion pits or cracks.

Background

Applications

Propeller blades (Fig. 8.66) are machined out of an aluminium-copper alloy forging and used in the piston engines of the transport aircraft. The propeller assembly under

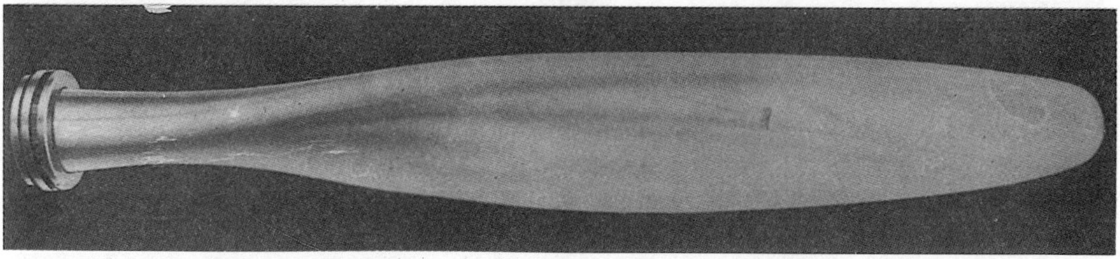

Fig. 8.66 An unused propeller blade (× 1/8)

investigation had served for several years and had been periodically overhauled. During overhaul the propeller blade was subjected to a test run in the overhaul shop. In this case, the blade portion of the propeller assembly was well protected against corrosion by anodising and painting, but the hub portion which had only anodic protection was exposed to the atmosphere. The propeller is subjected to complex centrifugal forces during flight.

Circumstances Leading to Failure

During the test run, after a period (not known) the blade portion of one propeller broke off from the hub (Fig. 8.67), resulting in considerable damage to the surroundings. However, it could not be ascertained whether the propeller blade had been thoroughly checked for any crack prior to the engine test run.

Specifications

Material specification of the alloy used for the propeller blade was 20256.

Selection of Specimens

The only available hub portion of the broken propeller blade was subjected to fractographic examination under a low powered stereoscopic microscope. A section cut across the origin of the fracture was used for microstructural analysis.

Visual Examination of General Physical Features

Visual examination revealed that the breakage had occurred circumferentially along the radius portion of the hub, in the region of minimum OD (Fig. 8.67). The surface in the region of the fracture showed no apparent defect or abnormality. However, examination under the stereoscopic microscope showed numerous corrosion pits all around the hub radius. On cleaning them, deep pittings were found in places with interconnecting fine cracks (Fig. 8.68).

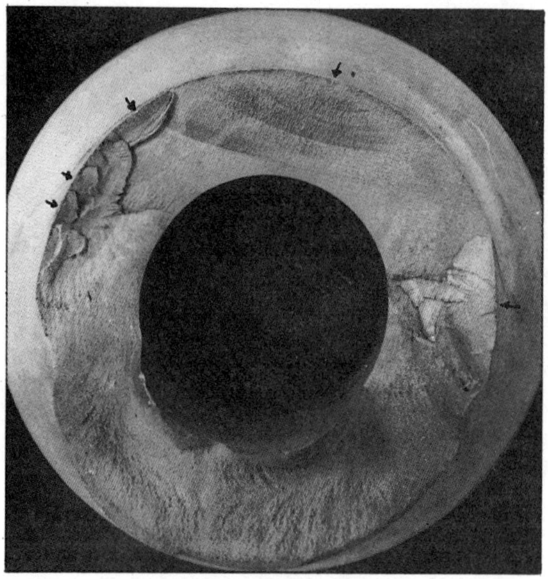

Fig. 8.67 Distinct fatigue zones at several places marked by arrows (× 3/4)

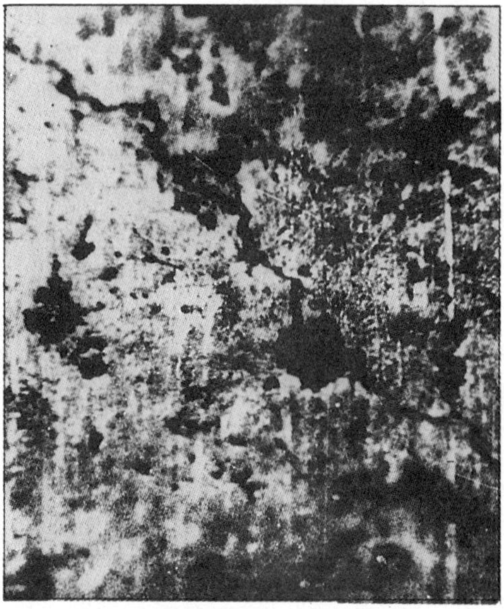

Fig. 8.68 Deep pittings and connecting cracks (× 6)

Testing Procedure and Results

Surface Examination

Visual Examination Visually the fracture surface showed two distinct zones of disintegration. One portion was showing a very smooth surface with 'conch' shell markings over almost half of the fracture and the other half with very irregular and coarse fracture lines (Fig. 8.67). The smooth area of the fractured surface also showed several minor segments of fatigue origins along the radius periphery.

Macrofractography Examination of the fractured surface under the stereoscopic microscope at × 20 magnification, revealed distinct and uniformly closely spaced 'beach' marks of a typically high cycle fatigue failure in the major smooth segment (which covered half of the section thickness, originating from the deep corrosion pits (Fig. 8.68) at the hub radius portion. In addition, at least five other minor zones of fatigue were observed to have originated along the periphery of the hub radius (Fig. 8.67). These secondary fatigue cracks after an initial propagation for a short while were seen merging with the main crack front. This propagated as a single major crack until the reduced section thickness could no longer bear the sustained load. All the fatigue origins were traced to corrosion pits in the hub radius. The rough, irregular and fibrous area with 'chevron' markings indicated the zone of final overload fracture (Fig. 8.67).

Metallography

Microstructural Analysis Micro examination of the section across the main fatigue zone revealed the crack path as transgranular. Transgranular mode of cracking also confirmed that the failure was due to fatigue. Otherwise, the aluminium copper alloy material was found satisfactorily heat treated with no structural abnormalities. However, corrosion attack on the surface in the radius portion was found intergranular, forming deep fissures up to a maximum size of 1.5 mm dia and 1 mm depth in the section examined.

Discussion

Although fractographic analysis of the hub fracture clearly established the fact that the sudden failure of one of the propeller blades was due to a high cycle fatigue phenomenon, the cause of fatigue was traced to the presence of unnoticed corrosion pits which acted as severe stress raisers in the highly rotating blade root radius zone. Fractograph (Fig. 8.67) revealed a multipronged attack from the several sites of the corrosion intrusion in the outer periphery of the root radius. The fatigue crack growth continued till the section thickness got reduced to half its diameter and failed due to overload. This caused an instant high velocity fracture in a random manner. This fact indicated that periodic inspection of the hub was not maintained to detect corrosion in the critical areas of the root radius portion during the long service of the aircraft, which is constantly exposed to extreme atmospheric corrosion factors like rain water/moisture and heat. Under these circumstances, the critically loaded hub radius portion was more susceptible to development of corrosion pits since its anodic coating had lost its corrosion resistance due to prolonged usage. On the other hand if the propeller blade surface had been protected by both anodic treatment and painting the problem of atmospheric corrosion would not have arisen.

Conclusion and Recommendations

Most Probable Cause The above was clearly a case of corrosion fatigue failure with deep corrosion pits serving as stress raisers in the presence of a complex combination of torsional and cyclic stresses developed in service.

Remedial Action It was recommended to enforce periodic inspection of the hub root radius portion of the propeller blade for presence of any corrosion pits or cracks, preferably at high magnifications under the low powered stereoscopic microscope.

How Failure could have been Avoided In view of the fact that corrosion pits were too small to be detected either by the naked eye or the low power magnifying lens, it appeared to be a case of ignorance on the part of the maintenance as to how tiny corrosion pits could be critical particularly in a dynamically loaded part. Had they known the magnitude of the effect of corrosion on the fatigue life, probably the catastrophic failure could have been averted by paying more attention to the corrosion pits.

8.18 CRACKING OF MAIN WHEEL HUB ASSEMBLY OF A TRANSPORT AIRCRAFT DUE TO CORROSION FATIGUE

Summary and Recommendations

A cast magnesium alloy main wheel hub assembly was reported cracked in the rim portion after a long period of service. Fractographic examination revealed several ratchet marks along the 25 cm long crack and clear shell markings on the fractured surface covering the entire cross sectional thickness. Numerous corrosion pits in and around the cracked zone were observed to have acted as stress raisers/origins for development of fatigue ratchet as well as beach marks. Micro-examination revealed the corrosion attack of intergranular type, forming deep fissures and the fatigue crack typically transgranular.

The cracking of the wheel hub was attributed to corrosion fatigue. As the protective coating of the groove portion was abraded during assembly and disassembly, the magnesium alloy part became susceptible to corrosion attack.

It was recommended to provide immediate protection of the abraded regions and periodic check for detection of corrosion pits/cracking.

Background

Applications

The main wheel hub assembly of the transport aircraft is machined out of a large complicated magnesium aluminium alloy casting. The alloy material is known for its good corrosion resistance in marine as well as other environments, with chromate treatment. Assembly break up is shown in Fig. 8.69. The part was reported to have served a long period before failure, but the exact time is not known.

Circumstances Leading to Failure

The main wheel assembly was reported cracked in one segment of the machined groove portion (Fig. 8.70) near the rim.

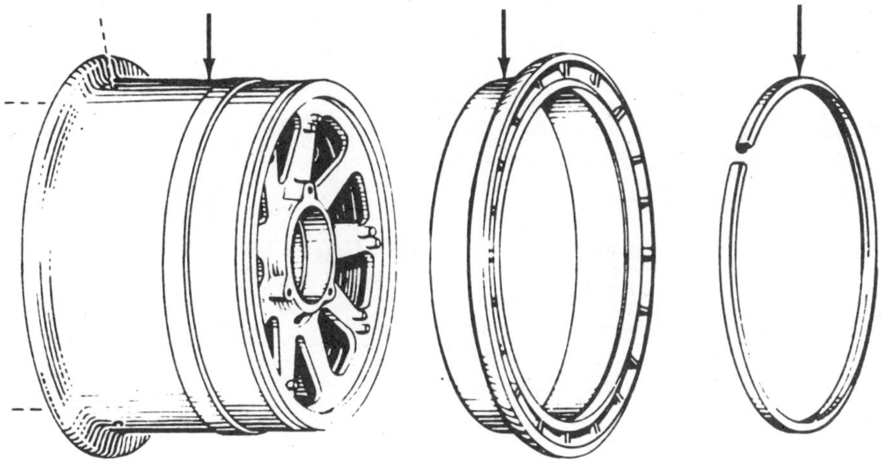

Fig. 8.69 The break up of the main wheel hub assembly

Fig. 8.70 Corrosion pits in the groove (× 2/5)

Specifications

Material used for the main wheel hub component was an American alloy to MIL-M-46062B specification.

Selection of Specimens

The rim cracked portion was cut open and subjected to fractographic examination under a low powered stereoscopic microscope. A section was cut across the fracture for micro-examination under the optical microscope.

Visual Examination of General Physical Features

Examination revealed a 25 cm long crack along the machined groove near the rim that seated the tubular ring (Fig. 8.69). The crack was found branching off on either side. Numerous corrosion spots seen in the groove, on cleaning, were found penetrating deep at places. No other type of serious defect was observed.

Testing Procedure and Results

Surface Examination

Macrofractography The crack along the rim was carefully opened out and the fractured surface covering the entire section thickness was found smooth in a plane square to the zone of cracking in the groove surface and showing distinct 'shell' markings, typical of fatigue failure (Figs 8.70 and 8.71). Origins of fatigue were traced to the several deep corrosion pits on the groove surface, constituting parallel ratchet type fatigue zones (Fig. 8.71) at different places in the initial stages of crack propagation.

Fig. 8.71 Zones of fatigue and their origins in the groove (× 2/5)

Metallography

Microstructural Analysis Micro-examination of a section of the cast wheel material in and around the zone of cracking, revealed no structural abnormalities. The corrosion attack was found intergranular forming deep fissures while fatigue cracking was typically transgranular.

Chemical Analysis

Chemical composition of the wheel hub was found meeting the requirements of MIL-M-46062B specification (Table 8.15).

Table 8.15 Chemical specification (Wt. %) for the American magnesium-aluminium alloy casting (MIL-M-46062B) from which the wheel hub was made and actual composition of the failed part

Element	Min.	Max.	Failed part
Aluminium	8.1	9.3	9.1
Manganese	0.13	-	0.25
Zinc	0.40	1.00	0.65
Silicon	-	0.30	0.27
Copper	-	0.10	0.06
Nickel	-	0.01	0.005

Discussion

Fractographic examination clearly established the fact that the cracking along the rim had occurred due to corrosion fatigue. It is known that magnesium-10% aluminium cast alloy material is resistant to corrosion with chromate coating. In this case occurrence of corrosion indicated that the effectiveness of the chromate protective coating was lost due to prolonged usage. Based on this suspicion an enquiry was made and it was learnt that during maintenance, the tubular rim was often removed and put back and in the process the protective coating on the groove got abraded off at places. This facilitated corrosion attack by the water droplets/moisture from the changing atmospheric conditions to which the wheel assembly is always exposed. It was also ascertained that in many cases, prompt protection of the bare abraded surface was not resorted to by the maintenance crew.

Conclusion and Recommendations

Most Probable Cause The basic cause of the cracking of the rim portion was due to corrosion fatigue (the corrosion attack arising from the atmospheric conditions of the extremely corrosion prone bare magnesium alloy portions and the fatigue due to repeated loadings during landing and take off).

Remedial Action Prompt protection of the abraded regions before insertion of the ring, and periodic check for deep corrosion pits as well as fatigue cracking were recommended.

8.19 FRETTING FATIGUE FAILURE OF INTERNAL BRACKET OF TAILBOOM OF HELICOPTER

Summary and Recommendations

A forged and machined aliminium alloy integral bracket was reported cracked across the LH lug bush hole during inspection.

Macroscopic studies revealed severe fretting marks on the inner surface of the lug around the bush hole. The crack was opened out and the fracture surface showed clear signs of a fatigue crack, originating from the severely fretted zones at the hole edge. Micro-examination further confirmed the fretted zone as the origin of fatigue with numerous strain lines in the vicinity of cracking as an evidence of severe cold working effect. This phenomenon indicated the presence of a relative motion between the mating lugs during flight, operating under complex torsional cum bending stresses. Based on these observations, the failure was attributed to fretting fatigue.

It was recommended to provide adequate shimming between the mating lugs in order to minimise the relative motion.

Background

Applications

The integral bracket was machined out of aluminium copper-nickel-magnesium-iron-manganese alloy forging and was assembled on to sheet metal angles in the rear end of the tailboom of a helicopter, replacing the existing pre-mod bracket. The tail gear box which is mounted onto the bracket lugs imposes complex stresses due to vibration and centrifugal forces developed during flight.

Circumstances Leading to Failure

A crack was noticed on one of the LH lugs of the bracket after it had done 573 hours of service (Fig. 8.72) during periodic inspection.

Specifications

The bracket material was an alloy of aluminium-copper-nickel-magnesium-iron-manganese to a French AU2GN specification.

Performance of Other Parts in Same Service

A number of identical failures were reported almost at the same time.

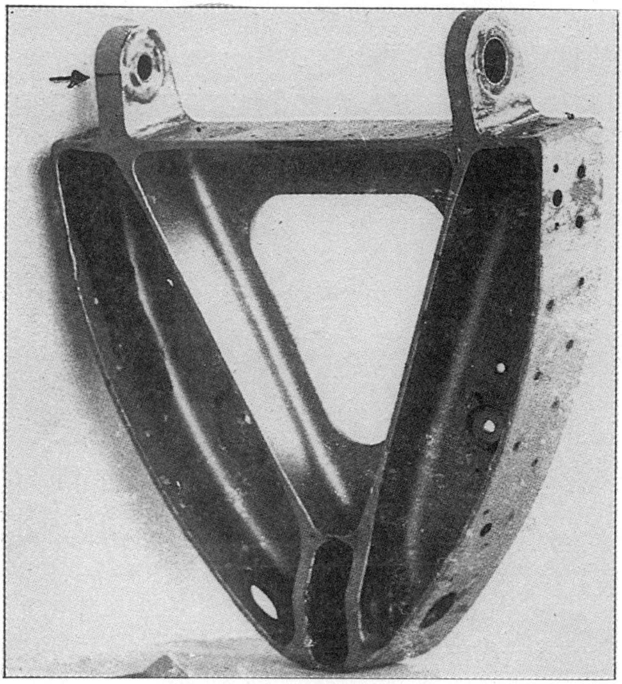

Fig. 8.72 Integral bracket showing crack in the lug (× 1/3)

Selection of Specimens

The crack of the lug was opened out and one portion was examined under a low powered stereomicroscope. A section across the zone of cracking was taken for micro-examination.

Visual Examination of General Physical Features

Examination of the bracket revealed a distinct horizontal crack through the bush hole of the LH lug. At places, severe fretted marks around the hole were noticed on the inner face of the lug (Fig. 8.73). Except for this, no other damages were observed in the bracket.

Testing Procedure and Results

Surface Examination

Macrofractography Examination of the fracture surface revealed a smooth, conspicuous zone of 'shell' markings (typical of fatigue), covering half the cross-section diagonally on either side of the bush hole (Fig. 8.73). The rather concentric shell markings pointed the origins of fatigue in the fretted zones of the inner surface of the lug. The remainder was found rough and fibrous, suggestive of a final fracture when being opened out.

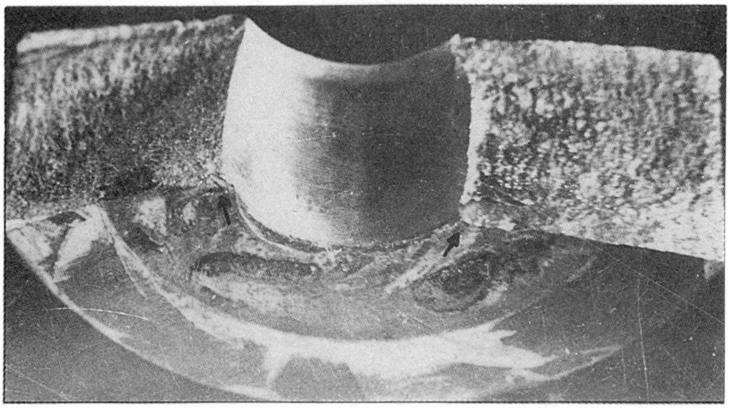

Fig. 8.73 The fatigue cracks originating from the fretted zones (marked by arrow) (× 3)

Metallography

Microstructural Analysis Micro-examination of the aluminium alloy forged lug material across the zone of cracking revealed numerous parallel transgranular cracks, emanating from the fretted surface together with several strain lines in the close vicinity of fatigue origin (Fig. 8.74). Zones away from the fatigue origin showed a satisfactory heat treated structure with no abnormalities.

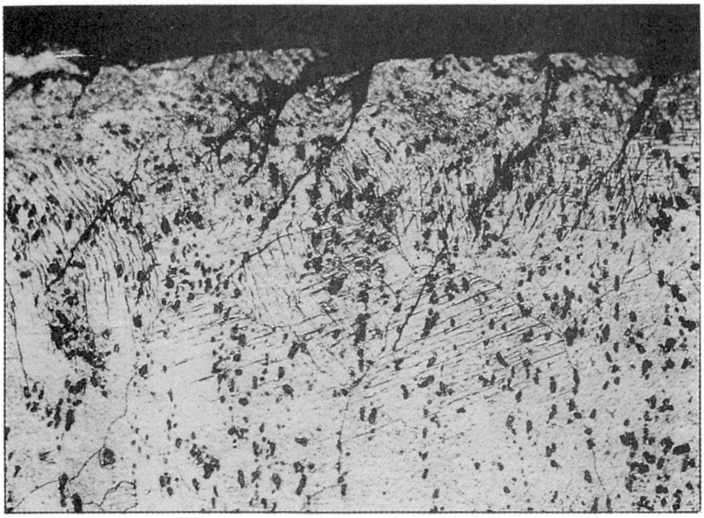

Fig. 8.74 The microstructure with strain lines and transgranular cracks at the zone of failure (× 100)

Chemical Analysis

The bracket forging material was found conforming to AU2GN specification (Table 8.16).

Table. 8.16 Chemical specification (Wt.%) for the French aluminium copper-magnesium-nickel aJlloy bar (Au2GN) from which the internal bracket was made and actual composition of the failed part.

Element	Min.	Max.	Failed part
Copper	2.00	2.60	2.28
Magnesium	1.20	1.80	1.65
Silicon	0.10	0.25	0.15
Iron	0.90	1.40	1.10
Nickel	0.90	1.40	1.25
Titanium	0.05	0.15	0.11

Mechanical Properties

Hardness Hardness check was carried out on the micro-specimen and a satisfactory hardness value of 150 BHN was obtained.

Discussion

Fractographic as well as micro-structural analysis of the broken lug proved that horizontal cracking of the lug had occurred by a combination of fretting corrosion and fatigue mechanism.

Evidence of severe fretting marks on the inner face of the LH lug (Fig. 8.73) around the bush hole clearly indicated that some relative motion had existed between the mating lugs during flight, causing the fretting action to occur. Fatigue had originated at these fretted zones, which acted as stress raisers under complex stresses of centrifugal forces by the rotation of the tail rotor blade assembly, which is in close proximity of the LH lug. Moreover, the strain lines observed in the microstructure near the fatigue origins indicated a severe cold working effect —a deformation process that accelerated the fatigue crack initiation.

Conclusion and Recommendations

Most Probable Cause The cause of failure of the lug by fretting was primarily due to the existence of a relative motion between the lugs, assembled under complex torsional, bending etc., stresses.

Remedial Action In view of the repetitive failure of an identical nature, it was recommended to review design measurements so as to provide adequate shimming between the mating lugs in order to minimise the relative movement.

8.20 DETERIORATION OF CAST MAGNESIUM ALLOY MAIN GEAR BOX HOUSING DURING SERVICE

Summary and Recommendations

An imported main gear box of one of the helicopters was found corroded during routine inspection.

Visual examination revealed deep 'furrow' type corrosion pits filled with heavy powdery deposits. Since X-ray examination revealed no defect in the affected portion of the housing, a texture examination was carried out on the broken surface of the casting wall. This revealed flouride type flux inclusions in the form of black patches at several places after a day of exposure to atmospheric conditions. These black patches were also found increasing in area on longer exposure, confirming the reactive condition of the flux present in the casting. Therefore, it was concluded that the material deterioration was a case of internal chemical reaction of the flouride flux with the moisture of the environment during service.

It was recommended to carry out texture tests on the test coupons for detection of flux inclusions in future melt batches.

Background

Applications

The main gear box, usually cast out of magnesium alloy, is used in helicopters to house a combination of complex gear assemblies for the power transmission system. Since magnesium alloy casting is sensitive to atmospheric corrosion, the housing, immediately after manufacture, is normally protected with chromate and paint coatings. However any leakage in the casting by way of corrosion or heavy porosity may lead to loss of engine oil and in turn, severely affects the flight performance.

Circumstances Leading to Failure

The main gear box housing showed deep corrosion pits with powdery deposits on its outer surface (Fig. 8.75) after a period of service (not known), on routine visual inspection during overhaul.

Specifications

The material of the main gear box housing was a French alloy to RZ5 specification (Table 8.17).

Fig. 8.75 MGB housing outer surface showing furrows on cleaning off powdery deposits in localised zones (× 1/2)

Table 8.17 Chemical specification (Wt.%) for the French magnesium-zinc casting (RZ5) from which the main gear box housing was made and actual composition of the failed part.

Element	Min.	Max.	Failed part
Zinc	4.0	5.0	4.6
Manganese	-	0.15	0.12
Silicon	-	0.01	0.007
Copper	-	0.03	0.01
Iron	-	0.01	0.006
Zirconium	-	0.70	0.62
Rare earths	-	1.50	1.48

Selection of Specimens

Two small strips were cut lengthwise from the intact portion (i.e. without corrosion) of the housing wall for texture examination.

Visual Examination of General Physical Features

Examination, on brushing off the powdery deposits, revealed 2-3 mm deep furrow type pits at places, as if eaten away by worms. These pits were mainly confined to the upper portion of the outer surface of the chromated and painted housing.

Testing Procedure and Results

Non-destructive Evaluation

Radiography Radiographic examination of the entire housing revealed the casting structurally sound all over, even in the vicinity of the deep furrows.

Simulation Tests Suspecting, from the nature of corrosion pits and its powdery deposits, that the corrosion was a product of an internal chemical reaction, a texture examination was carried out on the cut-out strips after breaking them off under bending overload. Although immediately on fracturing no abnormalities were noticed, after a day of exposure to the atmosphere the fresh fracture started developing black patches (Fig. 8.76). These black patches were also found increasing in area with exposure time. These characteristics were typical of corrosion due to flux inclusion.

Fig. 8.76 Black zones developed in fractured suface when exposed to atmosphere, due to the presence of flux inclusions (× 1)

Discussion

From visual examination of the corrosion products alone, it become evident that the magnesium alloy housing had suffered a corrosion attack from within the body. Moreover, formation of heavy powdery deposits was indicative of some internal chemical reaction. It is an established fact that the presence of flouride containing flux (which is generally used for melting of magnesium alloy in the castings) serves as a source for nucleation of corrosion by virtue of its reactive nature with the base metal when in contact with moisture. On this basis, a simulation test was carried out on the unaffected portion of the housing and this confirmed the presence of flux inclusions in the wall thickness, which got entrapped during the melting practice. Over a period of long service, it was these flux inclusions, which were present in close proximity of the surface, that reacted after coming in contact with atmospheric moisture possibly through some porosity.

It is pertinent to note that the presence of such inclusions are not easily detectable by X-ray examination due to the defect being of a lighter density.

Conclusion and Recommendations

Most Probable Cause The basic cause of the material deterioration was attributed to the flouride type flux inclusions present in the casting during melting.

Remedial Action It is important to note that no flouride type of flux inclusion is permitted to be present in the body of the casting, as it deteriorates the chemistry and soundness of the material, specially when they are used for critical applications. Therefore, it was recommended to carry out the texture test on text coupons of the melt batch for detection of such flux inclusions before acceptance.

8.21 FAILURE OF THE EXHAUST PIPE ASSEMBLY OF A PISTON ENGINE DUE TO THE WELD DECAY PHENOMENON

Summary and Recommendations

One exhaust pipe assembly belonging to the piston engine of a trainer aircraft was found cracked at the toe of the weld after a period of service.

Fractographic examination of the cracked surface indicated a coarse crystallined (brittle) failure. Micro-examination of a zone in the failed weld portion of the pipe and the collar revealed distinct intergranular corrosion cracks in the weld as well as HAZ. Chemical analysis of the pipe and the collar (which was arc-welded together to form the assembly) confirmed that 18/8 stainless steel pipe material was not stabilised, whereas that of the collar was stabilised. From microstructural analysis it was evident that the intergranular corrosion at the weld zone had occurred due to depletion of chromium at the grain boundaries in the form of chromium-carbide precipitates. Since the pipe material was a simple 18/8 type austenitic stainless steel without any addition of stabilising element (Ti/Nb), it became susceptible to corrosion attack.

Based on the above facts it was concluded that the cracking was due to a weld decay phenomenon. It was recommended to heat treat the batch, if welded, to dissolve the carbides and retain them in solution by rapid quenching.

Background

Applications

The exhaust pipe is an arc-welded assembly of tubular pipe and collar, made of austenitic stainless steel (Fig. 8.77) and is used as an outlet for hot gases from the engine. The collar end of the pipe assembly is attached to the engine bay area by bolts and is left hanging outside the aircraft.

Circumstances Leading to Failure

During routine inspection of the aircraft, the exhaust pipe assembly was found to have cracked along the toe of the welded joint (Fig. 8.78). The total period of service that the part had undergone prior to failure was not available.

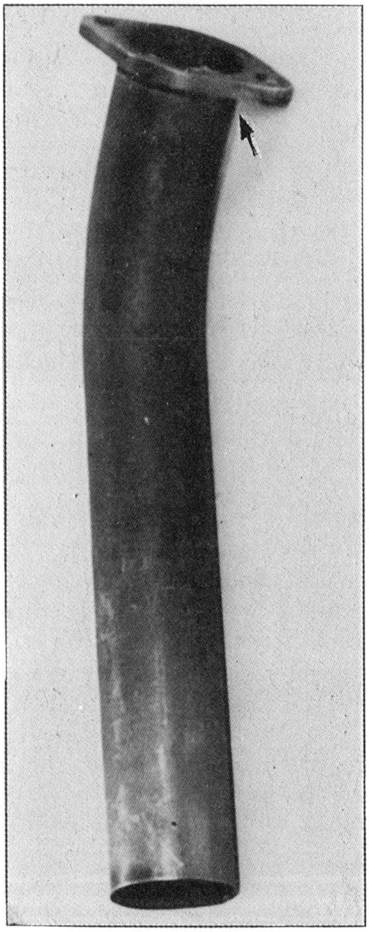

Fig. 8.77 The failed exhaust pipe assembly showing the location of the crack (× 1/3)

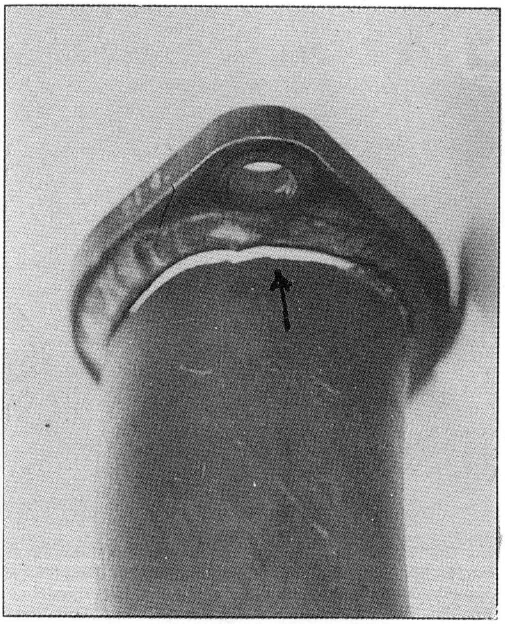

Fig. 8.78 Close up view of the cracked area at the weld toe (× 2/3)

Specifications

The material used for both pipe and collar was an austenitic stainless (stablised) steel and their specifications were reported to be MIL-T-6737-Comp G 321 and MILS-6721, respectively.

Selection of Specimens

The crack was opened out and a section was taken across the fractured portion for micro-examination and the remaining portion was subjected to fractographic examination under the stereoscopic microscope and chemical analysis.

Visual Examination of General Physical Features

The pipe was found cracked/broken circumferentially at the toe of the welded joint (A–B) except for a small portion in the weld 'X' (Fig. 8.79). The OD surface of the tube which was inserted into the bore region of the collar showed heavy oxide scales. However, it did not exhibit any perceptible deformation at the fracture region. Physical damages like nicks, dents, etc., were not observed in the vicinity of the failed region.

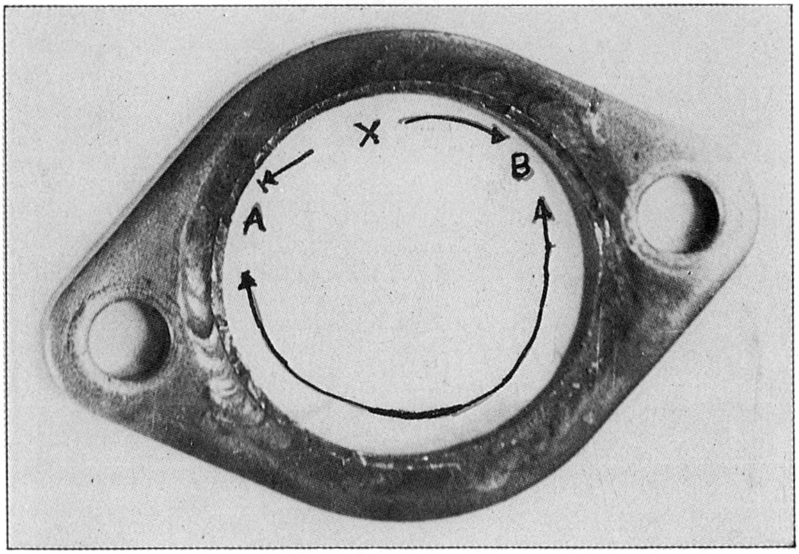

Fig. 8.79 The fractured surface of the cracked portion (A-B). Zone 'X' is fractured at the laboratory (× 1)

Testing Procedure and Results

Surface Examination

Macrofractography Examination of the cracked surface under the stereoscopic microscope at × 20 magnification revealed irregular and crystallined features of a brittle failure, with brownish discolourations (due to hot gases). No indication of any progressive type of failure such as fatigue was discernible on the fractured surface.

Metallography Micro-examination of the section across the failed region of the pipe at × 400 magnification, revealed numerous intergranular corrosion cracks with few of them propagating extensively into the heat affected zone (Fig. 8.80). However, on examination of the same area at × 1600 magnification, chromium carbide particles were observed precipitating along the grain boundaries (Fig. 8.81). The welding characteristics, such as fusion and penetration were found satisfactory.

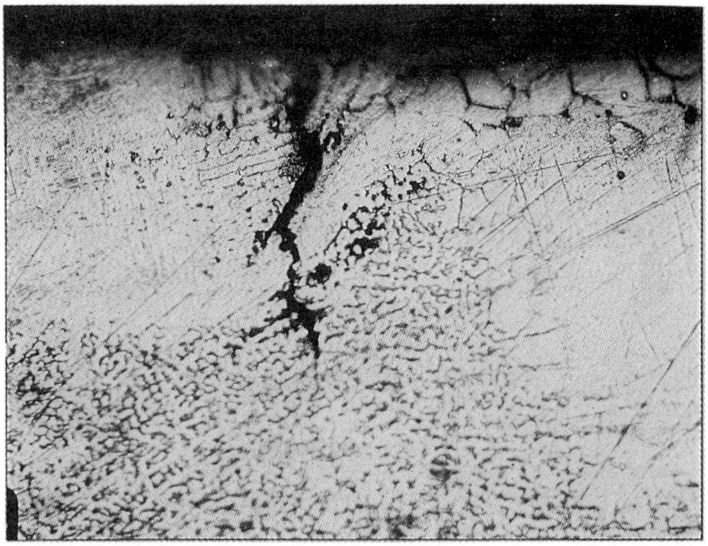

Fig. 8.80 Several intergranular corrosion cracks with some of them penetrating deep (× 400)

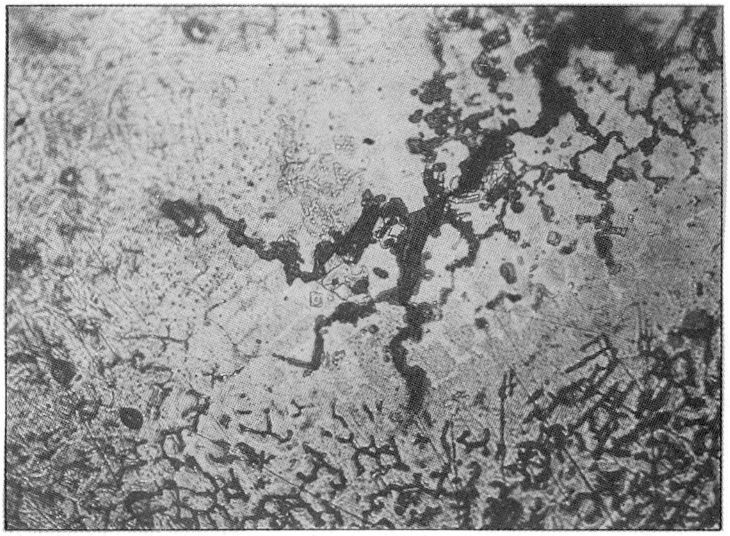

Fig. 8.81 Massive chromium carbide precipitation along the grain boundaries at HAZ (× 1600)

Chemical Analysis

On spectrographic analysis the exhaust pipe material was found to be of 18/8 stainless steel but without any stabilising element (i.e. not conforming to MIL-T-6737 Comp. G 321 specification), whereas, the collar material was identified as titanium stabilised 18/8 stainless steel, conforming to MIL-S 6721 specification.

Discussion

Microstructural studies clearly established the fact that the circumferential cracking along the toe of the weld had occurred due to a weld decay phenomenon (intergranular corrosion cracking). From chemical analysis it was observed that the collar was made out of stabilised 18/8 stainless steel, whereas, the pipe material failed to meet the required specification owing to absence of stabilizing element like titanium or niobium. It is well-known that to resist weld decay phenomenon (which occurs as a result of severe depletion of chromium content in the grain boundaries) atleast 12% chromium is required to be maintained homogenously within each grain. Massive precipitation of chromium carbides along the grain boundaries, both in the weld as well as in the heat affected zones, due to greater affinity between chromium and carbon, and the absence of any stabilizing element, had rendered the tube material more susceptible to intergranular corrosion with chromium content falling below the critical level. This type of weld decay phenomenon is very common in unstabilised austenitic stainless steel, particularly when exposed to sensitising temperature ranges either during welding or heat treatment.

Conclusion and Recommendations

Most Probable Cause

The basic cause of the circumferential cracking at the toe of the weld was attributed to the intergranular corrosion phenomenon (popularly known as 'Weld Decay').

Remedial Action

In view of the above structural deficiency observed in the welded pipe material, the following basic approaches were suggested to avoid weld decay failure in future batches.

- Heat treatment of the pipe (assembly if welded) to dissolve the precipitated carbides, i.e. solution treatment which involves heating to 1050° ± 10 °C followed by rapid cooling (water quenching) to retain the dissolved carbides in solution.
- Inspection to ensure the specification of the incoming raw materials prior to welding.

8.22 STRESS CORROSION CRACKING OF HIGH STRENGTH ALUMINIUM ALLOY MAIN FITTING ATTACHMENT OF FIGHTER AIRCRAFT

Summary and Recommendations

The main fitting attachment of the nose landing undercarriage unit of a fighter aircraft was found cracked through and through within the wall of the hollow housing body during a major overhaul.

Visual examination of the copper bush insert (assembled inside the body) showed several mechanical abuses on the surface in line with the crack. These damages invariably caused localised residual stresses in the zone where crack was found to have originated. Fractographic analysis of the cracked surfaces revealed brittle layers of fractures emanating from a bright 'half moon' zone under the stress corrosion phenomenon. Micro-examination confirmed that the main crack had an intergranular mode of propagation along with numerous secondary parallel cracks (also of intergranular type). These characteristics clearly established the fact that the cause of the failure was due to a stress corrosion mechanism.

It was suggested to examine the contact surface between the bush insert and the corresponding inner surface of the hollow body to ensure that there was no mechanical deformation, etc., before final assembly.

Background

Applications

The main fitting attachment (Fig. 8.82) is a highly stressed component used in the nose landing undercarriage unit of a fighter aircraft and is subjected to varying shock loads/ stresses while in service. The component is a complex shaped forging made out of aluminium-zinc-magnesium alloy and rough machined prior to heat treatment.

Circumstances Leading to Failure

The main fitting attachment was found cracked during periodic inspection, after it had undergone 500 hours of service since the previous major overhaul.

Specifications

The material of the main fitting attachment was a British alloy to DTD 5044 specification.

Selection of Specimens

The crack of the hollow housing body was opened up and one fracture segment was subjected to fractographic studies under the stereoscopic microscope. On disassembly, the copper alloy bush insert was visually examined. Micro-examination of the cut sections in the origin area was carried out using an optical microscope.

Visual Examination of General Physical Features

Examination of the fitting attachment revealed a linear crack emanating from the top rim of the hollow body, tracing vertically through the mid-zone of the attachment lugs down to the bottom radius of the housing and then turning side ways (Fig. 8.82). The fracture path was found in the longitudinal direction of the forged component covering the entire wall thickness of the body.

Fig. 8.82 Main fitting attachment and location of the straight longitudinal crack (× 3/4)

At this stage, it was felt pertinent to examine the copper alloy bush insert (Fig. 8.83) for any damages it might have suffered at the inner surface of the hollow body (i.e., at the location of the insert where maximum straining is expected due to service load). On close examination, evidence of longitudinal damages such as waviness in the top land, deep ingression (appearing as if caused by any hit) on the border land below and a nick and bulge in the bottom raised rim portion (marked as 3-1 respectively in Fig. 8.83) were observed on the outer surface of the bush in the location corresponding to that of the crack.

Examination of the bore surface of the body in the location of the bulge in the bush

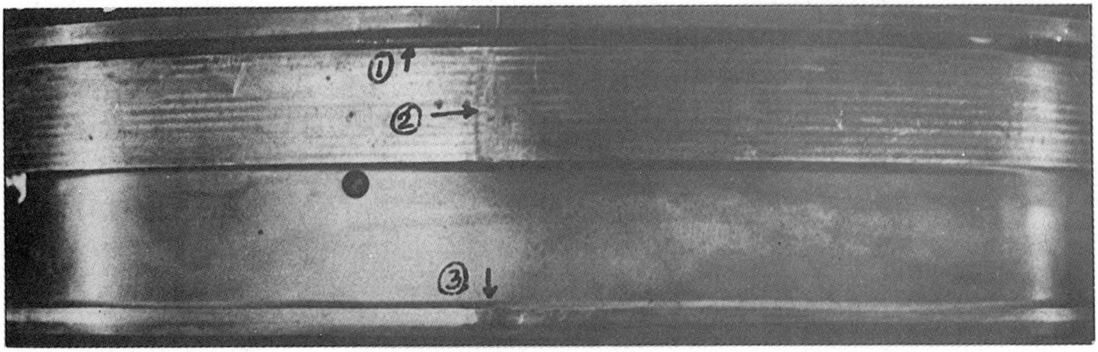

Fig. 8.83 Bush insert showing longitudinal damages on the outer surface (× 1)

showed deep circumferential scoring marks extending to 3-4 cm on either side of the crack (Fig. 8.84).

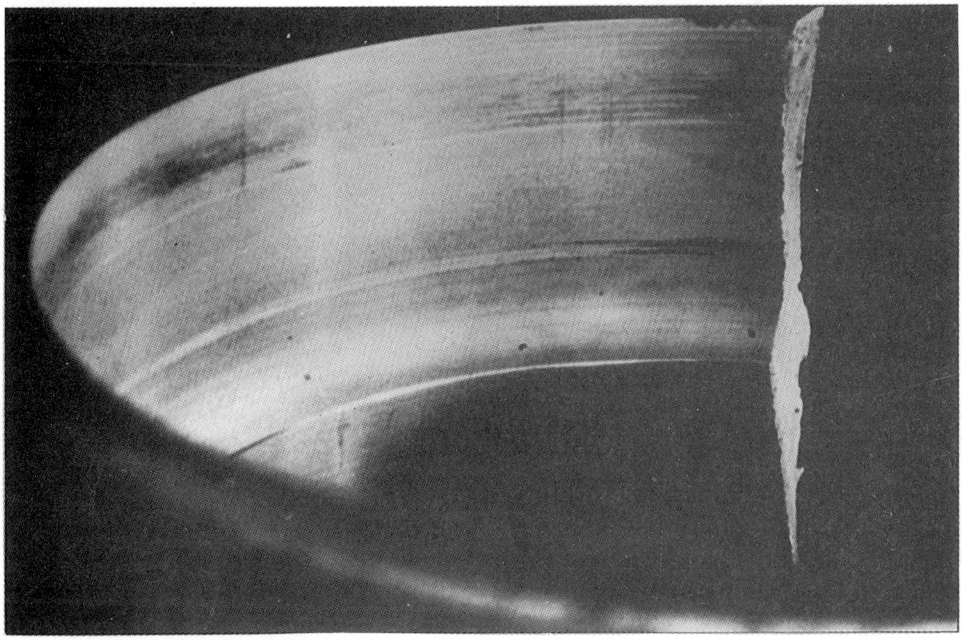

Fig. 8.84 Bush insert showing deep circumferential scoring marks on the inner face (× 1/2)

Testing Procedure and Results

Surface Examination

Macrofractography Fractographic examination of the cracked surface at x 35 magnification, revealed a distinct bright 'half moon' zone at the inner corner of the cylinder wall and brittle layeral surface (Fig. 8.85).

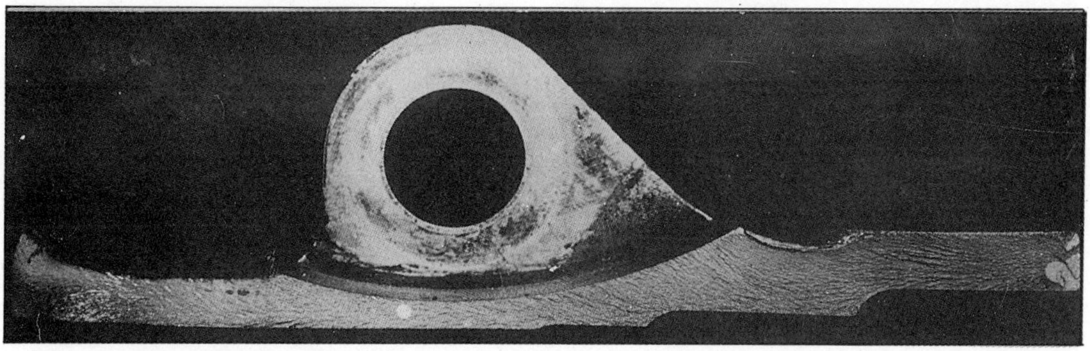

Fig. 8.85 A distinct bright half moon zone and brittle layer surface in the fracture (× 1/4)

Metallography

Microstructural analysis Micro-examination of a section parallel to the zone of fracture initiation revealed that the main fracture was intergranular along with numerous discontinuous intergranular cracks, also running parallel to the main crack (Fig. 8.86). This was a positive indication of cracking under the stress corrosion phenomenon. Other than this the microstructure did not reveal any abnormalities.

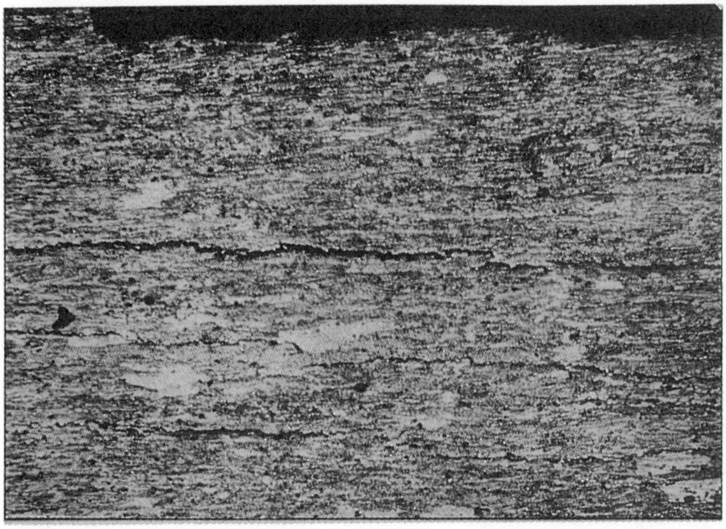

Fig. 8.86 Microstructure showing parallel intergranular cracks (× 100)

Chemical Analysis

Chemical composition of the main attachment fitting material was found conforming to DTD–5044 (Table 8.18).

Table 8.18 Chemical specification (Wt. %) for the British aluminium-zinc-magnesium alloy (DTD 5044) from which main fitting attachment component was made and actual composition of the failed part.

Element	Min.	Max.	Failed part
Copper	0.30	0.70	0.60
Magnesium	2.20	3.20	2.95
Manganese	0.30	0.70	0.45
Silicon	-	0.50	0.42
Iron	-	0.50	0.38
Nickel	-	0.10	0.08
Chromium	-	0.05	0.03
Zinc	5.20	6.20	5.80
Titanium	-	0.30	0.22
Lead	-	0.05	0.03
Tin	-	0.05	0.03

Mechanical Properties

Hardness Hardness check was carried out on the microspecimen and an average hardness value of 155 BHN was obtained.

Discussion

Fractographic studies clearly established the fact that cracking had occurred under stress corrosion phenomenon, despite the fact that the cylinder's hollow body material was satisfactory with respect to strength properties, grainfibres and microstructure.

It is important to note that stress corrosion phenomenon invariably involves either residual or applied stresses or both, while in service. In this case, presence of a bulge of the copper insert as well as circumferential scoring marks on either side of the crack indicated that these mechanical damages had existed prior to the final assembly of the fitting. Formation of these damages had set up some localised (residual) stresses by virtue of deformation in the form of a definite bulge on the outer surface of the bush insert at a location in line with the top inner corner of the cracked cylinder. Moreover, the material used was a high tensile strength aluminium 5% zinc, 3% magnesium alloy which is known to be susceptible to failure by stress corrosion mechanism.

Conclusions and Recommendations

Most Probable Cause The basic cause of the premature failure of the cylinder housing body was due to stress corrosion phenomenon.

Remedial Action Although it appeared to be stray case, it was suggested to examine the contact surface between the bush insert and the corresponding inner surface of the hollow body for mechanical defaults before final assembly.

8.23 FAILURE OF ALUMINIUM-MAGNESIUM ALLOY RIVET HEADS OF MAIN ROTOR BLADE BY STRESS CORROSION CRACKING

Summary and Recommendations

A number of main rotor blades were reported to have undergone severe corrosion attack with almost all the rivet heads severed along the shoulder radius while in service.

Examination of the fracture surface of a broken rivet with the help of stereoscopic as well as scanning electron microscopes, established that the failure was due to stress corrosion cracking. Micro-examination also confirmed the stress corrosion phenomenon, with positive indications of secondary intergranular cracks in the microstructure. Further, it revealed that the main crack was emanating from the region of severe cold working (due to rivetting) in the rivet-shank shoulder radius zone.

The material used for the rivets was aluminium-5% magnesium alloy which is well-known for its susceptibility to stress corrosion cracking, particularly in the cold worked condition (as rivetting in this case) and marine atmosphere.

In view of the severeity of the corrosion environment as well as cold working, it was recommended to change the aluminium 5% magnesium (AG5) to aluminium 4%, copper 1% magnesium (AU 4G) alloy or allow AG5 rivets to undergo a stabilizing treatment before assembly in order to minimise stress corrosion susceptibility.

Background

Applications

A number of main rotor blades belonging to different helicopters used in service for an unspecified period of time, were received for repair work. These helicopters were reported to have flown in the coastal areas under marine atmospheric conditions.

Circumstances Leading to Failure

On opening out the wooden box in which these blades were enclosed, all of them were found extensively corroded. While inspecting the extent of damage, practically all the rivets in the trailing edge portion (i.e. joining the 'U' shaped aluminium 4% copper-magnesium sheets) were found not only corroded, but also brittle. On slightly tapping the rivet heads, they fell apart. This was an unusual situation in which the whole body of the aluminium alloy blade (Fig. 8.87) was found mortgaged to an indiscriminate corrosion attack. This reflected total negligence of preservation of the blades against corrosion.

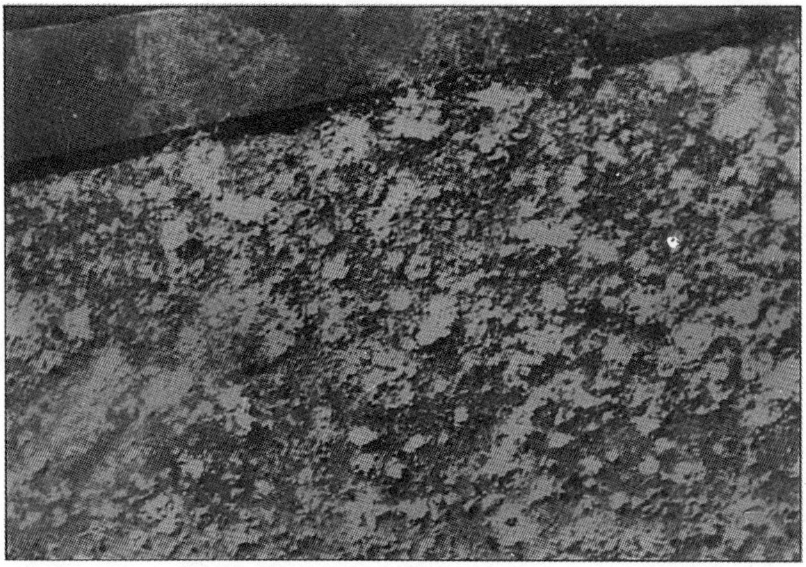

Fig. 8.87 Severe corrosion attack on the surface of the propeller blade (× 1)

Specifications

The rivet material used was a French alloy to AG5 specification.

Performance of other Parts in Same or Similar Service

It is interesting to note that a similar rivet head failure of a tail rotor blade made of the same material, belonging to another helicopter, was also reported prior to this incident. On investigation, the fracture surface of the rivet head broken piece (the only portion available), which was so perfectly cleaned prior to its receipt, pointed towards one of the shearing. As such, results of the investigation were inconclusive as the sample piece was too small to carry out the microstructural and chemical analysis. Also, the background information was inadequate to throw any light on the failure mechanism.

Selection of Specimen

The two rivets which broke into two along the shank radius on slight tapping, were subjected to fractographic studies under the stereoscopic microscope and the scanning electron microscope. One of the broken rivet heads was sectioned and examined under the optical microscope.

Visual Examination of General Physical Features

Examination of the main rotor blade revealed extensive surface corrosion. The corrosion patches (Fig. 8.87) appeared to be atmospheric in nature. Out of all the rivets in the failed blade, only two were found intact, i.e. not separated like others but partly affected by cracking. No corrosion products like those on the blade surfaces were noticed in these two rivet surfaces or in the cracked zones.

Testing Procedure and Results

Surface Examination

Macrofractography Examination of the broken rivet head fractures under the stereoscopic microscope at × 20 magnification, revealed a brittle surface with layeral features (Fig. 8.88). This was suggestive of failure initiated by stress corrosion cracking. Also, the broken portions did not show any perceptible deformation in the region of cracking (on matching both the rivet head and the shank together).

Fig. 8.88 Brittle layeral fracture (× 20)

Scanning Electron Microscopy Fractography Examination of the fracture surface under the scanning electron microscope, at × 1500 magnification, revealed the crack progagating predominantly in an intercrystalline mode (Fig. 8.89), with linear markings on the fracture. These characteristics indicated that the failure was due to the stress corrosion phenomenon.

Matallography

Microstructural Analysis Micro-examination of the longitudinal section taken through the failed zone of one of the two rivet heads revealed that the cracking was by

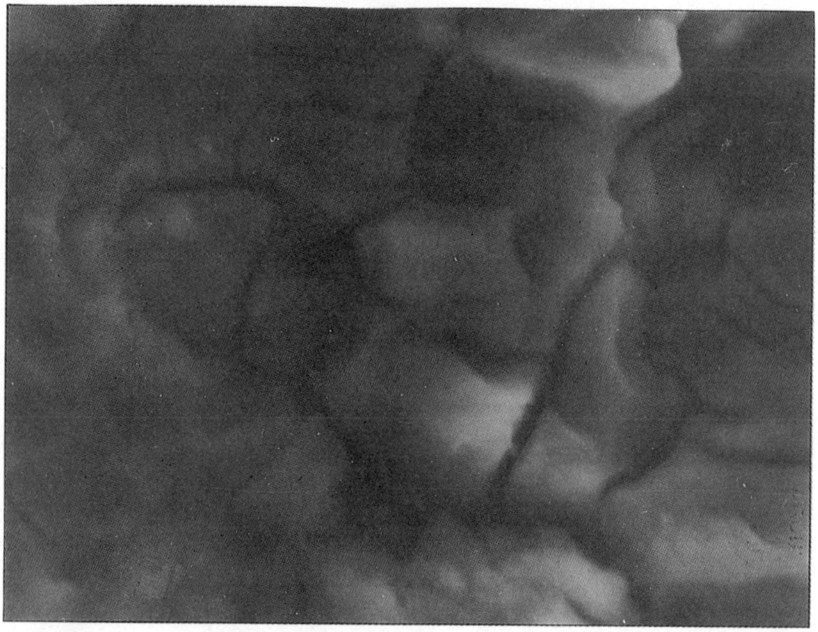

Fig. 8.89 Intercrystalline mode of fracture under the scanning electron microscope ($\times$ 1500)

intergranular mode with multiple secondary cracks running parallel to the main crack (Figs. 8.90 and 8.91). These structural features gave a positive indication of stress corrosion. As seen from the microstructure of the adjoining area of the rivet head, cracks were found emanating from the severely cold worked regions (Fig. 8.91) in the shoulder radius zone between the head and the shank.

Chemical Analysis

The rivet was chemically analysed and the composition of the material was found conforming to AG5 specification. (Table 8.19)

Corrosion Deposits The residual deposit of corrosion products on analysis was found answering for aluminium oxide as reactant from general atmospheric corrosion attack.

Discussion

Fractographic as well as scanning electron microscopic examination of the rivet head fracture clearly established the failure mechanism by stress corrosion phenomenon. Further, micro-examination of the rivet head also confirmed the failure by stress-

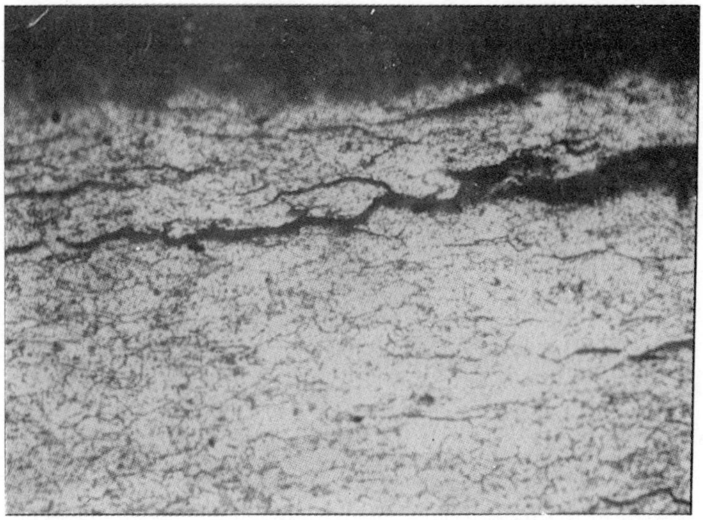

Fig. 8.90 Intergranular mode of fracture with multiple secondary cracks running parallel to the main crack (× 200)

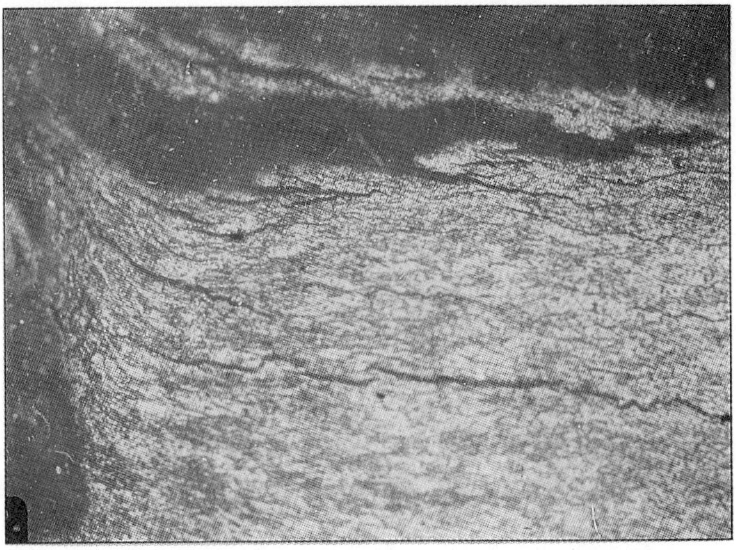

Fig. 8.91 Cracks emanating from the severely cold worked region of the rivet head shoulder radius (× 200)

corrosion with positive evidences of intergranular (parallel) cracks in the zone of the failure.

Table 8.19 Chemical specification (Wt. %) for the French aluminium-magnesium alloy (AG5) from which rivet was made and actual composition of the failed part

Element	Min.	Max.	Failed part
Magnesium	4.5	5.8	4.90
Manganese	0.05	0.07	0.06
Silicon	-	0.30	0.25
Iron	-	0.40	0.38
Chromium	0.05	0.30	0.20
Titanium	-	0.10	0.10

The fact that all the rivets of the main rotor blades had failed and the blade surfaces had badly corroded, indicated severe corrosion in a marine environment. The rivet material which was made out of aluminium 5% magnesium is well-known for its susceptibility to stress corrosion phenomenon. This is particularly so in the cold worked condition such as rivetting as in the present case and more so, if the parts are exposed to a corrosion environment or marine atmosphere.

Conclusion and Recommendations

Most Probable Cause The basic cause of the rivet failure was attributed to the occurrence of stress corrosion cracking.

Remedial Action In view of the severity of the corrosion environment and occurrence of stress corrosion in aluminium 5% magnesium (AG5) rivet alloy material, it was recommended that either a material change from AG5 to AU AG/2024 type alloy or a stabilizing (stress relief) treatment to AG5 rivet material would be necessary to minimise stress corrosion susceptibility.

8.24 BURSTING OF OLEO LEG OF BOMBER AIRCRAFT DUE TO STRESS CORROSION CRACKING

Summary and Recommendations

The oleo leg of a bomber aircraft collapsed while taxing after landing due to bursting of the bottom end portion of the cylinder. The fracture surfaces were examined under the stereoscopic microscope. This showed several bright 'half moon' zones emanating from the corroded pits on the bore surface. Macro and micro-examination of the origin of the fracture revealed several secondary intergranular cracks which were found running parallel to the main crack (also intergranular). This feature was indicative of a characteristic stress corrosion phenomenon. Further, on chemical analysis, the material was found to be a high strength aluminium-zinc magnesium alloy which is known to be highly susceptible to stress corrosion cracking.

The primary cause of the failure was attributed to a stress corrosion attack, with corrosion pits as the source of stress concentration. The development of pits in the enclosed bore surface was suspected to be due to contamination of the hydraulic oil used inside the cylinder for piston operation. It was recommended to check the oil prior to usage for the presence of water and opt for a material which is resistant to stress corrosion.

Background

Applications

The massive oleo leg structure (Fig. 8.92) is machined out of a high strength aluminium-zinc-magnesium alloy forging and heat treated to T6 (solutionised and aged) temper.

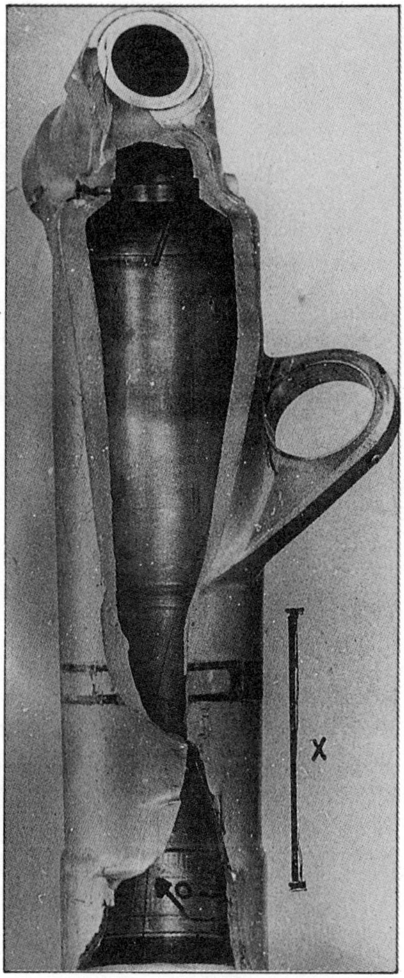

Fig. 8.92 The oleo head after bursting (× 1/6)

The component is used in the main landing gear assembly as a shock absorbing unit during landing of the aircraft. Any failure of the unit would have catastrophic consequences.

Circumstances Leading to Failure

Bursting of the oleo leg took place during taxing after landing, with the consequent collapsing of the landing gear assembly. However, the total service period of the component was not known.

Specifications

Material used for fabrication of the part was a British alloy to DTD 683 specification.

Performance of Other Parts in Same Service

A number of oleo head parts were reported to have failed in service due to stress corrosion cracking. The material used in all the cases was a high strength aliminium 5% zinc, 3% magnesium alloy to DTD 683 specification.

Selection of Specimens

The broken lower half portion of the cylindrical oleo head was selected for fractographic studies. One section adjoining the origin was cut and metallographically polished and etched for microstructural analysis.

Visual Examination of General Physical Features

Examination showed that the bottom end of the oleo cylinder had broken off with three longitudinal cracks emanating from the open end (Fig. 8.92). It also revealed numerous corrosion pits all over the lower half of the bore surface (Fig. 8.93). Further, presence of very conspicuous bright 'half moon' zones at five places (Fig. 8.94) covering almost the entire wall thickness were observed on the fracture, in the region marked 'X' (Fig. 8.92).

Testing Procedure and Results

Surface Examination

Macrofractography The long middle crack (marked 'O' in Fig. 8.92) was carefully opened and the entire fracture surface studied under a stereoscopic microscope at × 10 magnification. Fracture lines and chevron markings indicated that the tear had initiated from the bright moon zones, which were found brittle with layeral cracks (Fig. 8.95). Further, the radial fracture lines were found to have their nuclei at the corrosion pits on the bore surface (Fig. 8.96).

Matallography

Microstructural Analysis Micro-examination of the bore surface adjacent to the nuclei of the bright layerally cracked zones revealed several discontinuous and parallel macro

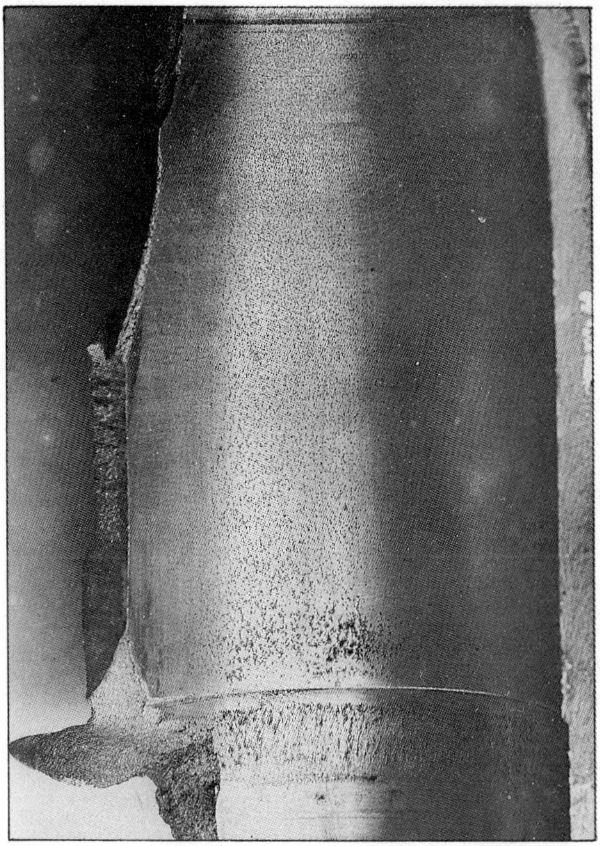

Fig. 8.93 The bore surface of the oleo head in the portion marked 'X' in Fig. 8.91 with numerous corrosion pits (× 1/2)

cracks (Fig. 8.97). Micro-examination of the same section, on polishing and etching, showed that the fracture was intergranular in nature, with numerous secondary intergranular cracks running parallel to the main crack (Fig. 8.98). These microstructural characteristics positively indicated the occurrence of a stress corrosion phenomenon.

Chemical Analysis

Chemical composition of the oleo head material was found conforming to DTD 683 alloy specification (Table 8.20).

Mechanical Properties

Hardness A hardness survey was carried out on the microspecimen, using Brinell hardness testing machine at 62.5 kg load, and an average hardness value of 164 BHN was obtained against the minimum specified requirement of 150 BHN.

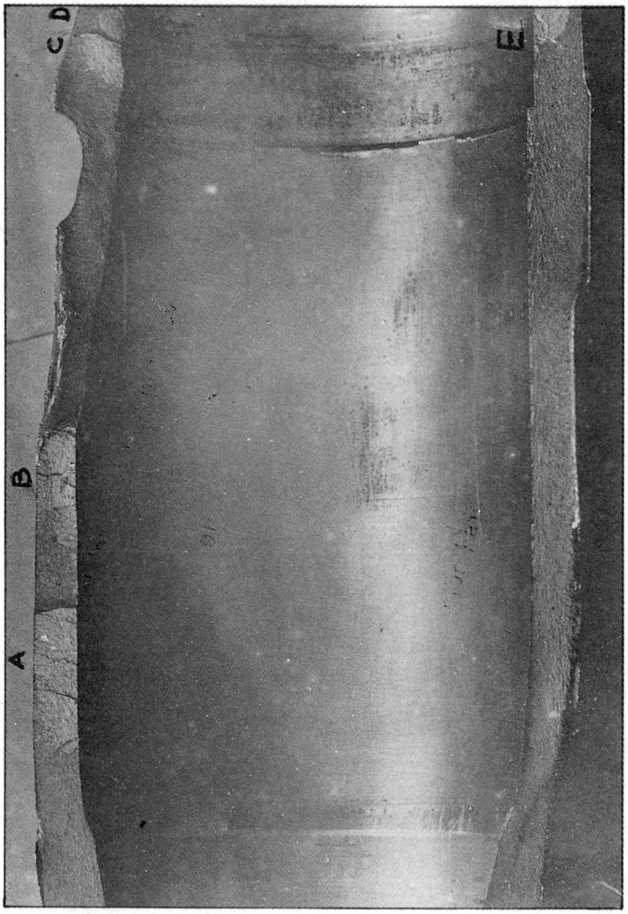

Fig. 8.94 Location of five bright 'half moon' zones marked by A, B, C, D and E (× 1/2)

Discussion

Fractographic as well macro and micro-examination clearly established the fact that the premature failure of the oleo leg had occurred due to stress corrosion attack. Further, examination revealed that the multiple cracks had originated from the bore surface instead of the normal outer surface zone of the cylinder, where applied tensile stresses tend to concentrate maximum for any progressive type of failure to occur.

However, the reason for this was obvious from the presence of numerous corrosion pits on the bore surface, and which in turn, acted as stress raisers. It is pertinent to note that, since the bore surface is an enclosed area and not exposed to atmosphere, the corrosion pits were suspected to have developed due to any contamination of the hydraulic fluid with moisture or water in the cylindrical housing. Another important aspect of the failure was the type of material used for construction of the part. In this

Fig. 8.95 Brittle layeral cracks at location 'A' in Fig. 8.94 (× 2½)

case, a high strength aluminum 5% zinc 3% magnesium type of alloy was chosen, which is known for its greater susceptibility to the stress corrosion phenomenon.

Conclusion and Recommendations

Most Prabable Cause The basic cause of the catastrophic bursting of the oleo leg and subsequent collapsing of the landing gear was attributed to stress corrosion cracking.

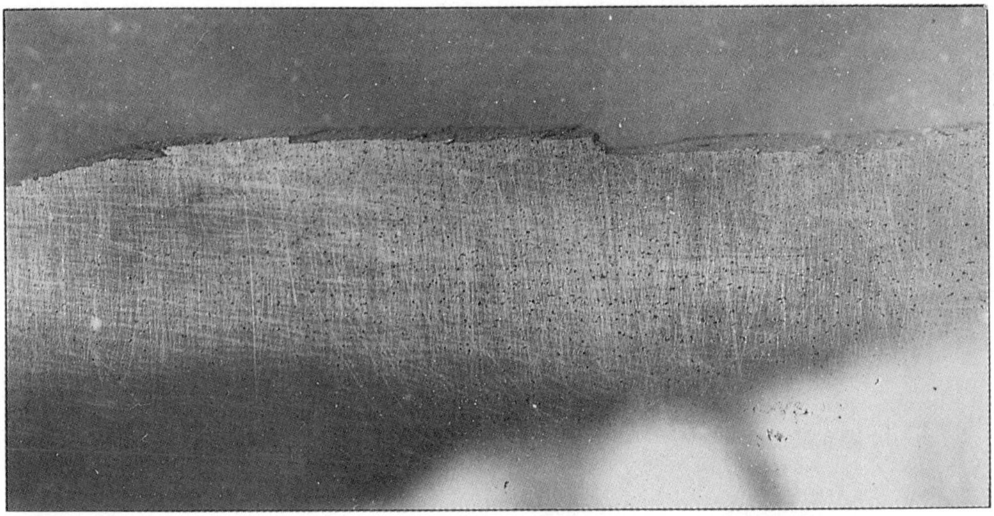

Fig. 8.96 Origins of fracture from corrosion pits

Fig. 8.97 Parallel cracks on the bore surface adjoining the fracture, on etching (× 10)

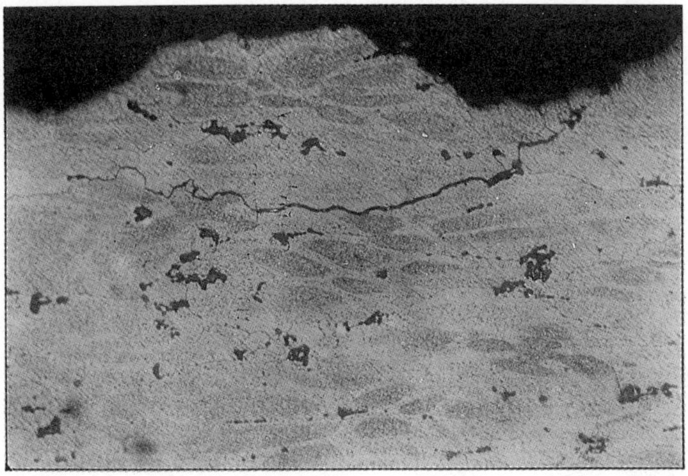

Fig. 8.98 Intergranular fracture and secondary cracks (× 100)

Table 8.20 Chemical specification (Wt. %) for the British aluminium-zinc-magnesium alloy (DTD-683) from which the oleo leg was made and actual composition of the failed part

Element	Min.	Max.	Failed part
Copper	0.30	0.70	0.42
Magnesium	2.20	3.20	2.85
Manganese	0.30	0.70	0.45
Silicon	-	0.50	0.44
Iron	-	0.50	0.42
Zinc	5.20	6.20	5.90
Chromium	-	0.05	0.04
Nickel	-	0.10	0.08
Titanium	-	0.30	0.22

Remedial Action In view of development of corrosion pits on the bore surface, it was suggested to check the mineral oil used for the hydraulic operation of the piston for any contamination prior to use. Also to avoid repetitive failure, a material change from DTD 683 to DTD 5012 (which is resistant to stress corrosion) was recommended.

8.25 FAILURE OF DRIVING GEAR OF A PASSENGER AIRCRAFT ENGINE DUE TO HYDROGEN EMBRITTLEMENT

Summary and Recommendations

Inspection of the driving gear of an engine belonging to a passenger aircraft showed breakage of a tooth and several chipped off metal pieces from the tooth surface in the oil filter. The gear was a case-carburised steel component which had undergone an anodic treatment after grinding.

Macroscopic examination of the tooth revealed numerous criss-cross hairline cracks both on the driving as well as non-driving faces. The broken tooth flank was found severely flaked off. The fracture showed a coarse crystallined surface with 'Neumann bands' indicating that the failure was due to the hydrogen embrittlement phenomenon.

Based on the above facts, it was suspected that a high residual stress was present in the high strength case carburised steel part. This was the result of a lapse either in the stress relieving or de-embrittlement baking treatment and which might have subsequently contributed to the excessive pick-up of hydrogen. In view of the above facts, it was recommended to ensure adequate stress relieving or baking treatment.

Background

Applications

The driving gear (Fig. 8.99) is a carburised component made out of nickel-chromium-molybdenum steel forged blank and is used in the engine of a passenger aircraft. The total service period of the part was not known. However, manufacturing history revealed that the part had been anodically etched after the final grinding and stress relieving operation.

Circumstances Leading to Failure

During overhaul of the engine, metal particles were found adhering to the oil filter. The engine was taken apart piece by piece and on inspection, one single tooth of the driving gear was found broken off.

Specifications

Material used for the gear part was a French alloy to 16 NCD 13 specification.

Selection of Specimens

The broken tooth was subjected to fractographic examination under a stereoscopic microscope.

Visual Examination of General Physical Features

Examination showed that a gear tooth had broken off along the root radius. Metal pieces observed on the filter were found to be the same that had flaked off the broken tooth

Fig. 8.99 An unused driving gear (× 1)

surface. Also, a number of irregular hair line cracks were noticed in a criss-cross manner on the loaded flank as well as on the outer surface of all the gear teeth (Fig. 8.100). Even a few of the non-driving tooth faces showed similar cracks. Some of the flanks showed severe flaked off regions with a transition from unbranched cracks to branched cracking (Fig. 8.101).

Testing Procedure and Results

Surface Examination

Macrofractography Examinations of the fractured surface under the stereoscopic microscope at × 10 magnification, revealed that the cracks were restricted to the depth of the hardened case and had originated from the internal surface. Further, nature of the fracture was found to be coarse crystallined with Neumann bands, typical of a brittle fracture under hydrogen embrittlement (Fig. 8.102).

Chemical Analysis

Chemical composition of the gear material was found conforming to 16NCD13 specification (Table 8.21).

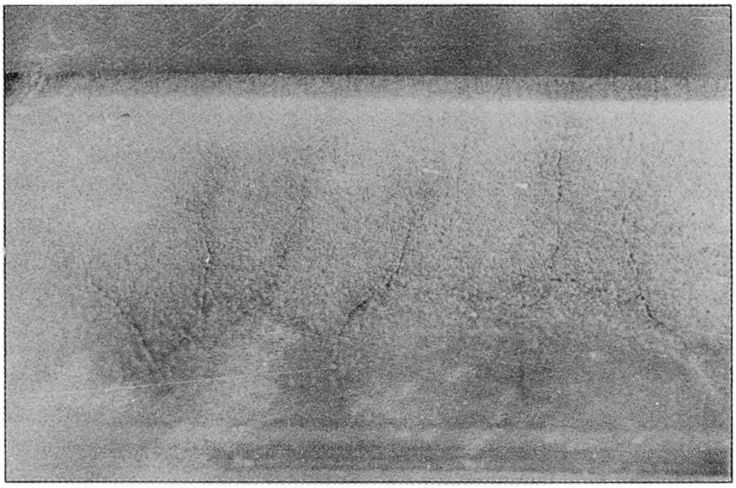

Fig. 8.100 Criss-cross hair line cracks on the outer surface of gear tooth (× 9)

Fig. 8.101 Severe flaking on the surface of a gear tooth (× 9)

Mechanical Properties

Hardness Hardness measurements were made on the case-carburised surface of the broken tooth using mini load hardness tester and an average value of 820 VPN was obtained.

Fig. 8.102 Coarse crystallined fracture surface with 'neumann bands' (×9)

Table 8.21 Chemical specification (Wt. %) for the French nickel-chromium-molybdenum steel (16 NCD 13) from which the driving gear was made and actual composition of the failed part

Element	Min.	Max.	Failed part
Carbon	0.12	0.17	0.15
Manganese	0.30	0.60	0.42
Silicon	0.15	0.40	0.34
Sulphur	-	0.020	0.016
Phosphorus	-	0.025	0.018
Nickel	3.00	3.50	3.10
Chromium	0.80	1.10	1.00
Molybdenum	0.25	0.30	0.27

Discussion

Fractographic analysis clearly established the cause of the fracture of the gear tooth as hydrogen embrittlement. The fact that both the driving as well as non-driving faces of the gear teeth were showing criss-cross hair line cracks and flaking off of metal chunks, the development of cracks had originated from an internal crack nucleation with an intercrystalline mode of propagation. These characteristic features together with the presence of a typical 'Neumann band' on the brittle surface definitely indicated that the failure was due to the hydrogen embrittlement phenomenon. Further, as the manufacturing history of the part indicated that the case-hardened gear had undergone an anodic etching treatment after final grinding and stress relieving operation, it was suspected that mandatory stress relieving after grinding or de-embrittlement baking after anodic etching were not adequately carried out. Hence this high residual stress was a possible source of hydrogen pick up to which high strength steel is known to be extremely susceptible.

Conclusion and Recommendations

Most Probable Cause The basic cause of the gear tooth failure was attributed to embrittlement phenomenon by hydrogen, that had been probably introduced during the anodic etching treatment.

Remedial Action Since the failure indicated a serious lapse in the process operation, it was suggested to ensure adequate backing treatment after grinding and anodic etching without delay between etching and baking.

8.26 EXFOLIATION CORROSION OF AN ALUMINIUM COPPER ALLOY EXTRUDED SPAR CAP OF A TRANSPORT AIRCRAFT

Summary and Recommendations

One aluminium —4% copper alloy extruded spar cap was found corroded during routine inspection after a service period of 10 long years in marine environments.

Visual examination of the corroded zones established the fact that the localised corrosion was accompanied by a powdery deposit which indicated an exfoliation type of corrosion failure. Micro-examination further confirmed that the corrosion attack was due to exfoliation, as evidenced by the layeral flaking off of the grain fibres intergranularly along the extrusion direction. Material hardness property indicated that the spar cap was in a fully heat-treated (T6) condition. It is well-known that aluminium 4% copper alloy of 2024 type is prone to exfoliation corrosion attack specially in the T6 condition and more so in a severe marine environment.

In view of the above, it was recommended to use the alloy material in T851 temper (instead of T6) condition which has better resistance to exfoliation.

Background

Applications

The spar cap is made out of aluminium copper extruded bar stock by machining and is used in the outer wing of the aircraft after heat treatment. It was reported that the aircraft was on the ground for nearly 10 years and had undergone a total 832 flying hours prior to detection of corrosion patches. The transport aircraft had flown for most of its service life in a marine atmosphere.

Circumstances Leading to Failure

Corrosion patches were detected at two places of the spar cap during routine inspection (Fig. 8.103).

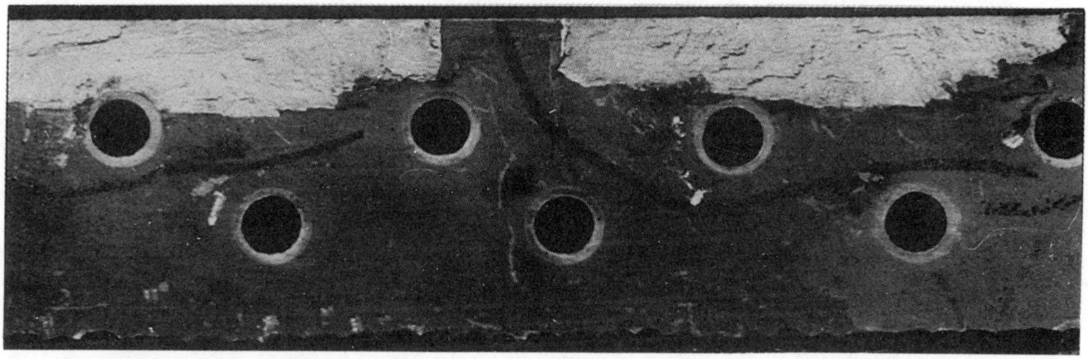

Fig. 8.103 Corroded layeral surface at two zones of the spar cap (× 1)

Specifications

Spar cap material was made out of an American alloy to 2024 specification.

Selection of Specimens

A piece was sectioned out of the spar cap with corrosion patches intact for observation under a low powered stereoscopic microscope. Two micro-specimens were made out of the corroded zones for microstructural analysis, both in the longitudinal and transverse directions.

Visual Examination of General Physical Features

Examination of the cut piece showed severe corrosion patches in two adjacent zones, extending to an identical area of 6 cm × 1 cm with voluminous corrosion products in the powdery form. On removal of the powder, corroded zones were found intact with no evidence of defects such as corrosion pits, etc. However, the corroded surface showed characteristics of a brittle feature (Fig. 8.103).

Testing Procedure and Results

Surface Examination

Macrofractography Examination of the corroded surface at × 25 magnification under the stereoscopic microscope, showed layeral features with numerous secondary cracks, producing a leaf-like appearance.

Metallography

Microstructural Analysis Micro-examination of the sections both in the longitudinal and transverse directions of the corroded zone revealed numerous thin, long and parallel surface layers, flaking off along the extrusion grain fibres. These directional attacks resulted in a pronounced leafing action (Figs 8.104 and 8.105). The leafing action was

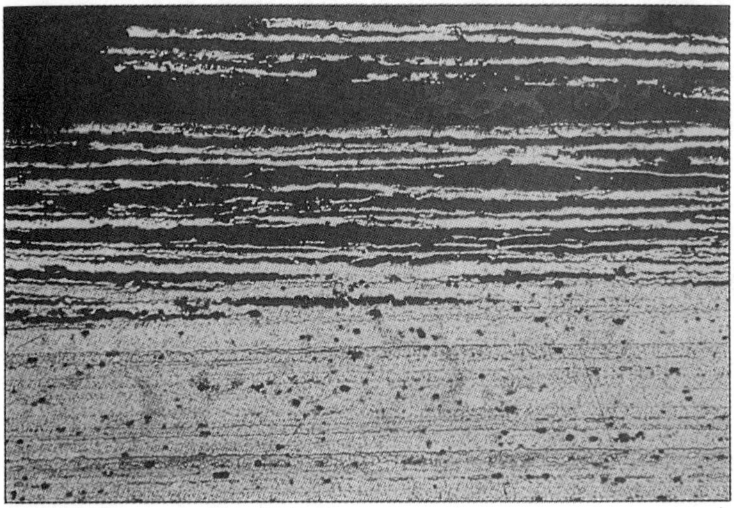

Fig. 8.104 Long parallel surface layers flaked off along the extrusion grain fibres in the microstructure of the corroded zone (× 100)

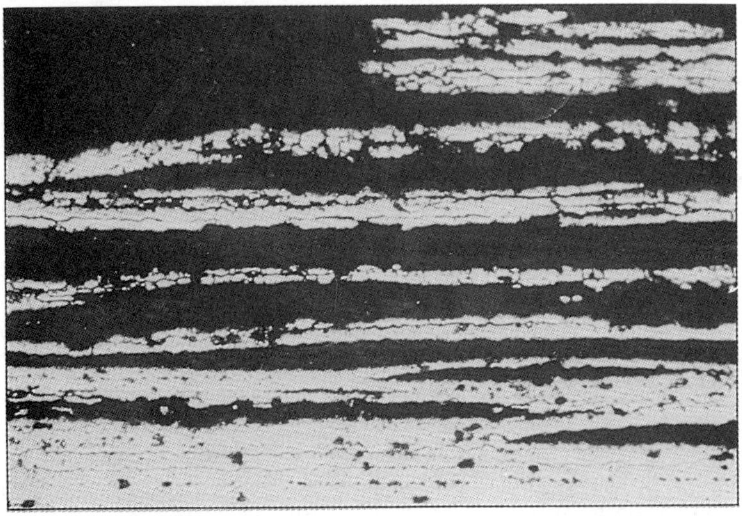

Fig. 8.105 Enlarged view of the layeral cracking (× 200)

found extending to a maximum depth of 1.1 mm in the sections examined. All these features were well indicative of an exfoliation corrosion phenomenon.

Chemical Analysis

The spar cap on chemical analysis confirmed the material to specification 2024 (Table 8.22).

Table 8.22 Chemical specification (Wt. %) for the American aluminium copper alloy extruded bar (2024) from which the spar cap was made and actual composition of the failed part

Element	Min.	Max.	Failed part
Copper	3.80	4.90	4.22
Magnesium	1.20	1.90	1.55
Manganese	0.30	0.90	0.60
Silicon	-	0.50	0.42
Iron	-	0.50	0.41
Chromium	-	0.10	0.08
Zinc	-	0.25	0.20

Mechanical Properties

Hardness Hardness value of the micro specimen was found to be 142 BHN and indicated the spar cap material in the heat-treated (T6) condition.

Discussion

Stereomicroscopic as well as micro-examination of the corroded zones clearly established the failure due to exfoliation corrosion phenomenon to which 2024 aluminium-copper alloy material is prone, particularly in the extruded/rolled condition. It was also noticed that the microstructure was having relatively thin grain compared to their width. Such a microstructural condition usually favoured exfoliation type of attack in aluminium –4% copper alloy wrought products.

Additionally, the formation of voluminous powder as corrosion products was also an attendant feature of exfoliation attack.

It is well-known that 2024 aluminium 4% copper alloy material with T6 (solutionised and aged) condition develops weakness for exfoliation type of corrosion attack. Besides, the transport aircraft had to serve in the marine atmosphere and its long exposure to such a severe corrosion environment had rendered the already susceptible alloy more vulnerable to the exfoliation attack.

Conclusion and Recommendations

Most Probable Cause The basic cause of corrosion failure was attributed to exfoliation attack.

Remedial Action In view of excessive exposure to outdoor environment during service and an inherent proneness of the material to exfoliation attack, it was recommended to use the alloy in T851 temper condition, which has a better resistance to exfoliation.

8.27 FAILURE OF 'D' HEADED BOLT OF JET ENGINE DUE TO CREEP

Summary and Recommendations

Six 'D' headed nimonic bolts used in the assembly of a jet engines were found broken after 36.25 hours of flying. The breakages were found along the thread roots just below the securing nut. Fractographic examination showed the fibrous type fracture of a ductile material. Macro and micro-examination revealed a multiplicity of intergranular cracks at the thread root and across the section thickness of the bolt, besides voids and intergranular fissures in the microstructure. All these features indicated that the failure was due to a creep phenomenon.

The basic cause of creep was attributed to excessive torque applied during assembly, which in turn, increased stresses at the thread root, thereby accelerating the creep phenomenon at the normal operating temperature.

In view of the above, it was recommended not to exceed the design stipulated torque while tightening the assembly.

Background

Applications

The 'D' headed bolt (Fig. 8.106) is a nimonic alloy component of a jet engine. The bolt is used to secure the exhaust cone of the engine to the joining ring.

Fig. 8.106 An unused 'D' headed bolt (× 3)

Circumstances Leading to Failure

Six bolts had failure and were found lying in the front shroud of the rear fuselage after 36.25 hours of flying.

Specifications

The bolt material was a British alloy to nimonic 80A specification.

Selection of Specimens

All the broken pieces of the failed bolts were subjected to fractographic examination under a stereoscopic microscope and one of them was longitudinally sectioned across the fracture for macro and micro-examination.

Visual Examination of General Physical Features

Examination showed that all the bolts had failed at the threads just below the securing nut. No indication of the operating temperature exceeding the limit nor any deformation in the zone of fracture was evidenced.

Testing Procedure and Results

Surface Examination

Macrofractography Examination of the fractured surfaces of all the failed bolts at × 10 magnification showed identical fibrous type of fracture of a ductile material (Fig. 8.107).

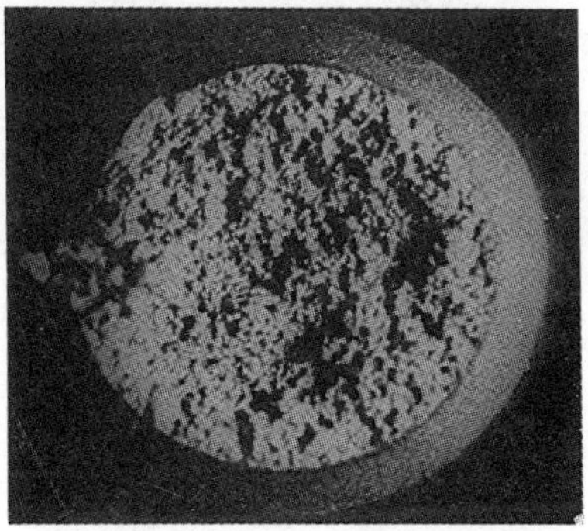

Fig. 8.107 Fibrous fracture (× 10)

Metallography

Microstructural Analysis Micro-examination of the failed bolt in the longitudinal section revealed a multiplicity of cracks originating at the thread roots and propagating throughout the section thickness in an intergranular manner (Fig. 8.108). Micro-examination of the same section showed voids and formation of intercrystalline fissures with

Fig. 8.108 Intergranular cracks across the thread root (× 12)

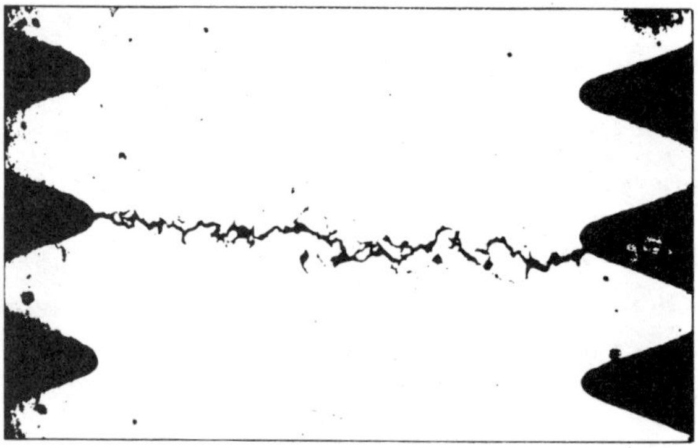

loosening of the grain boundaries (Fig. 8.109), all typical of a creep failure. The structure was otherwise found satisfactory.

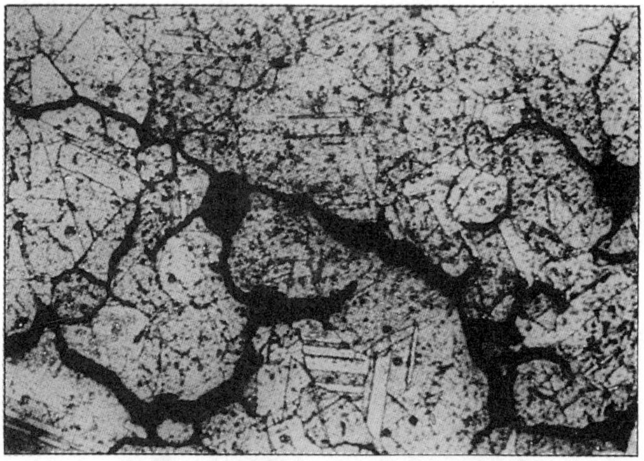

Fig. 8.109 Voids and intercrystalline fissures (× 100)

Chemical Analysis

Chemical composition of the bolt material was found conforming to nimomic 80A specification (Table 8.23).

Mechanical Properties

Hardness A hardness survey using the Vickers pyramid hardness tester was carried out on the micro specimen and an average value of 272 VPN was obtained, against the minimum value of 238 VPN as per specification.

Table 8.23 Chemical specification (Wt. %) for the British nickel-chromium-iron-cobalt-titanium alloy (nimonic 80A) from which the 'D' headed bolt was made and actual composition of the failed part

Element	Min.	Max.	Failed part
Chromium	18.00	21.00	19.85
Iron	-	5.00	4.50
Cobalt	-	2.00	1.70
Silicon	-	1.00	0.82
Manganese	-	1.00	0.75
Titanium	1.80	2.70	2.35
Aluminium	0.50	1.80	1.28

Discussion

Fracture analysis as well as macro and micro-examination clearly established the nature of the premature fracture of the bolt as creep.

Presence of a multiplicity of intergranular cracks across the section thickness and along the thread roots indicated a creep phenomenon. Besides, the resulting microstructural features such as voids, intergranular fissures and triple point cracks confirmed the creep phenomenon without any doubt. Since the operating temperature was found normal and no creep failure with the recommended torque was anticipated, it could be deduced that the failure might have occurred due to accelerated creep as a result of possible excessive tightening of the flanges. This resulted in an increased stress on the bolt thread radius root and thus yielded to creep at lower operating temperatures.

Conclusion and Recommendations

Most Probable Cause The basic cause of the bolt failure was due to creep.

Remedial Action It was suggested to avoid over-tightening during the bolt assembly.

8.28 FAILURE OF THE FRICTION PLATE OF THE UNDERCARRIAGE ASSEMBLY OF A TRAINER AIRCRAFT DUE TO THE CREEP PHENOMENON

Summary and Recommendations

Six tenons of the copper-chromium alloy friction plate were reported broken along the radii, after a fairly long period of service. This part was subjected to high temperature friction during each landing when the brakes were applied.

Fractographic analysis of the broken tenons revealed coarse and irregular fracture features in places where the fracture was intact. Some mechanical damages/pressed marks observed around the base portion of the tenons indicated a forced fitting of the tenon sleeve, introducing additional stresses at the radius zone. Micro-examination revealed a creep phenomenon.

Based on the above, the breakage of the tenons was attributed to failure due to creep.

Background

Applications

The copper friction plate (Fig. 8.110) is used in the brake system of the landing gear assembly. Since the undercarriage assembly was imported, no drawing/specification was available in connection with the investigation. It was reported that the temperature due to friction during landing, although momentary, can rise to about 500-600 °C.

Fig. 8.110 Failed friction plate showing the location of cracks of six tenons (× 1/3)

Circumstances Leading to Failure

Six out of the ten tenons of a friction plate were found broken at the radius zone, after 12 hours of service since the previous overhaul. The total period of service that the part had undergone could not be ascertained.

Selection of Specimens

All failed tenons were subjected to fractographic examination under the stereoscopic microscope. One of them was sectioned for micro-examination and the remainder was subjected to chemical analysis to determine the material composition.

Visual Examination of General Physical Features

On visual and macro-examination, the surfaces of all the broken tenons were found totally covered with heavy black soot, including discolouration on the top surface between the tenons. This was indicative of a high frictional temperature effect. All the fractures were found almost identical in nature, and were in an angular plane with the initiation zone at the radius marked in Fig. 8.111. However, the fracture surface were

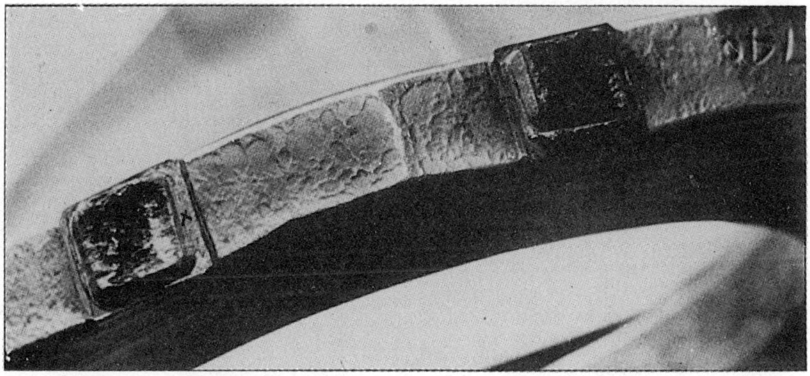

Fig. 8.111 Close-up view of fracture of the tenons showing the origin sites (× 1/2)

found almost rubbed. Further, a close examination under the stereoscopic microscope at × 10 magnification revealed stepwise damages/pressed marks around the base portion of the tenons, due to improper (forced) insertion of the tenon sleeve.

Testing Procedure and Results

Surface Examination

Macrofractography Examination of the broken surfaces at x 20 magnification, (on cleaning with petrol and 1% hydrochloride acidic solution) revealed a coarse and irregular fracture, rubbed at most places. To confirm the presence of any crack in the remaining unbroken tenons, one of them was forcibly broken in the laboratory and the fracture showed the presence of black zones in the radius portion, indicating that some progressive crack had existed in the unbroken tenon.

Metallography

Microstructural Analysis Micro-examination of a section across one of the broken tenons showed a fracture predominantly of the intergranular mode, with a few zones in and around the fracture revealing a wide opened intergranular crack, i.e. by joining the voids along the grain boundary path and having originated at a triple point (Figs 8.112 and 8.113). All these features indicated a failure by creep.

Fig. 8.112 Triple point cracks and creep voids (× 400)

Fig. 8.113 Creep voids and cracks under SEM examination (× 1500)

Chemical Analysis

On spectographic analysis, the friction plate material was found to contain 1.3% chromium and the remainder was pure copper.

Discussion

Microstructural features clearly established the fact that the failure of the tenons was due to creep. The copper-chromium alloy friction plate is subjected to high friction temperature (around 600 °C) during each landing, and over a period of time and a number of landings, creep phenomenon is usually expected to occur at the critical radius zones of the tenons under service stresses. It is pertinent to mention here that presence of stepwise mechanical damages/pressed marks around the base portion of the broken tenons indicated an additional imposed stress besides the service stress. The combined effect of temperature and stress had accelerated the creep phenomenon.

Conclusion and Recommendations

Most Probable Cause The basic cause of the breakage of the tenons of the friction plate was attributed to creep phenomenon.

Remedial Action It was recommended to avoid undue stresses caused by the improper insertion of the tenon sleeves.

Chapter 9

Chemistry and Metallurgical Behaviour

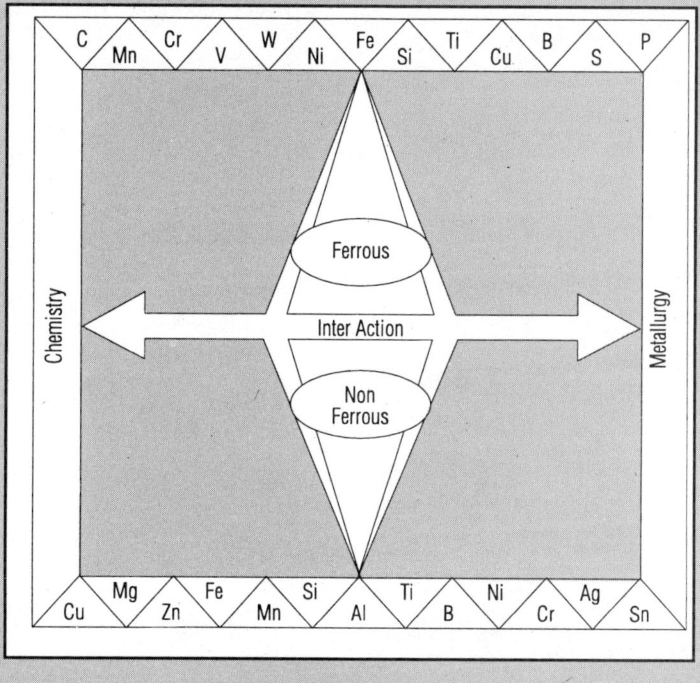

9 Chemistry and Metallurgical Behaviour

Metallurgical properties such as strength, ductility, toughness, fatigue, corrosion, etc., are achieved in metallic products through different processes viz. alloying, manufacturing (mechanical deformation processes), heat treatment, etc. It is needless to mention that the chemistry of a product is important in determining its basic properties. Therefore, any slight deficiency, in the analysis of an element, may alter the desired mechanical as well as allied properties. Inaccurate analysis may arise sometimes due to inadequate knowledge on the part of the analyst about the exact role of alloying elements and their significant effects on properties, even in trace additions. Improper melting practices do contribute to the deficiency of elements, even within the specified limits of chemical composition. Unless the chemical analyst is aware of the true significance of the effects of alloying elements, there is always a possibility of misjudgement in reporting the results and recommending the (specification) requirements for acceptance.

This chapter deals with, apart from description of the influences that the alloying elements exert on the metallurgical properties, the consequences that the product could undergo leading to failure during the manufacturing or service period in the event of an erratic chemical analysis. Quite often it is experienced, during defect investigation, that metal failures were associated with the anomalies between the chemistry and metallurgy of the accepted parts and this has been illustrated through case studies.

The above subject is included in this chapter with a view to make the failure investigator ensure the importance of integrity in chemical characterisation of elements so that authenticity of the actual composition is made foolproof. Hence, an exhaustive approach is made available in the concluding chapter for close interaction between the chemist and the metallurgist.

Effect of Alloying Elements in Steel

Alloying elements are added to steel to effect changes in the properties of the steel. Therefore, it is important to know the effect of their presence in steel, either as an alloying element or a residual element (not intentionally added but generated from the raw materials/steel making practices).

Carbon It is the most important single alloying element in steel. It is essential for the formation of cementite (and other carbides), pearlite, bainite and martensite. Microstructures comprising of one or more phases of the components can provide a wide range of mechanical properties and fabrication characteristics. The relative amounts and distributions of these phases (elements) can be manipulated by heat treatment to alter the microstructure, and therefore render the desired properties to a particular piece of steel. Changes in microstructure, strength, hardness and ductile to brittle transition temperature occur as the carbon content is increased. Toughness and ductility of pearlitic steel is reduced by an increase in carbon content. The hardness of iron-carbon martensite is increased by raising the carbon content of steel, reaching a maximum at about 0.6% carbon. Increasing the carbon content also increases the hardenability.

Manganese It is normally present in all commercial steels. It facilitates hot working of the steel by reducing its susceptibility to hot shortness. Manganese also combines with sulphur to form Mn-sulphide and improves the machinability of steel. It increases the strength slightly, but has a greater effect on the hardenability of the steel. It increases the likelihood of retained austenite in quenched steel. It acts as a deoxidiser.

Silicon It is one of the principal deoxidisers in the manufacture of steel. Fully killed steel contains 0.15 to 30% silicon. It increases the strength of ferrite slightly, without causing any serious loss of ductility. In large amounts, it increases the resistance of the steel to scaling in air (up to about 260 °C).

Chromium It is used in low alloy steel to increase its (a) resistance to corrosion and oxidation, (b) high temperature strength, (c) hardenability and (d) abrasion resistance in high carbon composition. Chromium carbides require high austenitizing temperature for dissolution. Straight chromium steels can be quite brittle and they are more susceptible to temper embrittlement.

Nickel It is used in low alloy steel to improve low temperature toughness and to increase hardenability. It reduces the sensitivity of steel to variations in heat treatment, and to distortion and cracking during quenching. It strengthens ferrite, thus strengthening the steel. Nickel is particularly effective when used in combination with chromium and molybdenum in forming an alloy steel that has high strength toughness and hardenability.

Molybdenum It increases the hardenability. This element when present in amounts between 0.15 to 0.30%, minimises the susceptibility of steel to temper embrittlement. Molybdenum is unique in the extent to which it increases high temperature tensile and yield strengths of steel. It retards the transformation of austenite to pearlite far more than it does the transformation of austentite to bainite, and thus bianite can be produced by continuous cooling of molybdenum containing steel.

Copper It is added to steel primarily to improve its resistance to atmospheric corrosion. Amounts added to steels for this purpose typically range from 0.2 to 0.5%. Copper is deterimental to surface quality and hot working behaviour because it migrates to the grain boundaries of the steel during hot working.

Vanadium It is generally added to steel to inhibit grain growth during heat treatment. It improves strength and toughness. Additions of vanadium of up to about 0.5% increases the hardenability of steel but large additions appear to reduce the hardenability probably because vanadium forms carbides that hardly dissolves in austenite.

Niobium It imparts a fine grain size, retards tempering and increases the elevated temperature strength of steel. Because it forms very stable carbides, it can decrease the hardenability of steel by reducing the amount of carbon dissolved in the austenite during heat treatment.

Titanium It may be added for grain size refinement and toughness. It increases hardenability.

Zirconium and Cerium They are most frequently used in HSLA steel can be used to control the shape of inclusions (primarily sulphate) thereby increasing steel toughness.

Boron It is usually added in amount of 0.0005 to 0.0031%, it significantly increases the hardenability of steel. It is particularly effective at lower carbon levels. Because boron does not affect the strength of ferrite, it can be used to increase the hardenability without sacrificing ductility, formability or machinability of steel in the annealed condition.

Lead It is added to improve its machinability but does not dissolve in the steel. However, it is retained in the form of microscopic gloubules. It can cause liquid metal embrittlement.

Aluminium It is used to control the grain size of steel during hot working and heat treatment. It is used as a deoxidiser.

Cobalt It decreases hardenability when alloyed with other elements. But when alloyed with vanadium, chromium or tungsten it greatly increases red hardness.

Tungsten It imparts red hardness, wear resistance and hot strength to steel.

Residual Elements

Phosphorus It increases the strength and hardenability of steel but severely decreases ductility and toughness. It increases the susceptibility of medium carbon alloy steel, particularly straight chromium steels to temper embrittlement. It may be added to improve its machinability or corrosion resistance. It induces cold shortness.

Sulphur It is detrimental to the transverse strength and impact resistance of steel. It impairs surface quality and weldability. Sulphur is deliberately added to some steels solely for improvement in machinability.

Nitrogen It increases the strength, hardness and machinability of steel, but it decreases the ductility and toughness. In aluminium killed steels, nitrogen forms aluminium nitrite particles that control the grain size of the steel, thereby improving both toughness and strength.

Oxygen It is most likely to be found in rimmed steels, and can slightly increase the strength of steel but seriously reduces toughness.

Hydrogen When it is dissolved in steel during manufacture it can seriously embrittle it by causing flaking during cooling from hot rolling temperature. This effect is not the same as embrittlement that results from electroplating pickling.

Tin It can render steel susceptible to temper embrittlement and hot shortness.

Arsenic and Antimony These also increase susceptibility of steel to temper embrittlement.

Effect of Alloying Elements in Aluminium Alloys

Copper It is one of the most important alloying ingredients for aluminium because of its appreciable solubility and strengthening effect. Copper is added to aluminium alloys in concentrations between 1 and 10 per cent. It imparts substantial precipitation-hardening characteristics, i.e. strength properties. With magnesium, copper addition accelerates and increases the age hardening effects at room temperature. It decreases the corrosion resistance as well as weldability properties. Density of aluminium is increased with its addition.

Magnesium Magnesium addition in the aluminium alloys as aluminium-magnesium alloys forms the basis for an important class of non-heat treatable alloys as well as a few whose characteristics can be altered by heat treatment. Although magnesium has substantial solubility in solid aluminium, the alloys do not show appreciable precipitation-hardening characteristics with concentrations below 7% mg. Magnesium, however, does provide substantial strengthening and high work hardening characteristics. It increases resistance to corrosion, promotes high strength without heat treatment and imparts good weldability. However, above 4% magnesium, the alloy is susceptible to stress-corrosion. Magnesium in combination with a proper ratio of silicon increases the strength properties. Density decreases with the addition of magnesium.

Manganese It is the most significant of all minor additions to aluminium alloys. It raises the recrystallisation temperature, improves uniformity of dispersion hardening and strengthens both dispersion and age-hardening aluminium alloys. A1-Mn alloys have poor casting characteristics but excellent resistance to corrosion. Manganese addition increases the tensile and yield strengths.

Silicon The outstanding effect of silicon additions to aluminium and its alloys is the improvement in casting characteristics. It imparts high resistance to corrosion, good weldability and low specific gravity.

Zinc As an alloying element zinc is mostly used along with magnesium in wrought alloy. The principal use of A1-Zn binary alloys is for electrolytic protection against corrosion. Addition of zinc moderately improves the strength. On the other hand, A1-Zn alloys containing elements such as magnesium, copper offer outstanding advantages in strength and the highest combination of tensile properties in wrought aluminium alloys.

Nickel It increases the strength to a moderate extent but reduces ductility. Nickel, in aluminium alloy, as major constituent imparts resistance to softening at high temperature but gives poor casting and machining characteristics. Addition of nickel to aluminium-copper and aluminium-silicon type alloy system improves hardness and strength at elevated temperature. It improves resistance to corrosion in P/M products.

Iron It forms coarse constituent with A1 and other alloying elements such as Cu, Ni, Mn and Si, reducing ductility and inducing cracking of wrought alloy during fabrication. Iron also reduces the strength and impairs corrosion resistance and fatigue resistance properties.

Chromium In concentration of 0.15 to 0.25% in A1-Zn-Mg-Cu alloy, it improves the resistance to general corrosion and reduces the stress corrosion cracking tendency.

Titanium The outstanding effect of titanium addition to aluminium alloys is to refine the grain size.

Zirconium In small quatities, up to 0.25% zirconium, it produces grain refining effects.

Lithium Addition of lithium to an aluminium copper system improves strength at elevated temperatures, reduces density and increases modulus of elasticity.

Boron In concentration of 0.005 to 0.01% refines the grain size. In combination with small concentration of titanium (0.01%) it intensifies the grain refining effect.

Beryllium In concentration of 20–25% imparts certain desirable characteristics to aluminium, including strength at elevated temperatures, low density, high modulus of elasticity and low confficient of expansion.

Bismuth, Cadmium and Lead These soft, low melting metals improve the machining characteristics, and are more effective in aluminium-copper alloys at concentrations of 0.1% and above. Bismuth alone and in combination with lead in a total concentration of 2% improves strength, elongation and machinability of aluminium-silicon-manganese casting alloys. Cadmium addition of 1-2% or more to aluminium-copper alloy system markedly influences artificial ageing characteristics resulting in increase in hardness and resistance to corrosion and stress corrosion.

Sodium, Calcium, Strontium Modify the A1-Si eutectic structure resulting in improved strength properties and pressure tightness effect.

Effects of Alloying Elements in Magnesium Alloys

Aluminium Forms the basis of the heat treatable alloys. It confers strength and refines cast structure of magnesium, particularly in conjunction with superheating or carbon innoculation. Magnesium-Aluminium cast alloys show some tendency towards microporosity.

Manganese Forms the basis of medium strength Mg-Mn wrought alloy and confers corrosion resistance but little strength in cast magnesium.

Zinc Imparts grain refinement and increase in strength. Mg-Zn alloys are heat-treatable but somewhat brittle and hot-short unless zirconium is added to further refine the grains. Alloys show some tendency towards microporosity.

Zirconium Its presence in concentrations of 0.5 to 0.7% imparts extreme grain refinement with consequent ductility but little increase in inherent strength. This in consequence forms the basis of various alloy types made possible by stiffening up the Mg-Zr matrix in various ways. Wrought zirconium alloys are resistant to grain growth.

Annealing or hot working reduces the room temperature properties.

Rare Earth Metals

They imparts long-term creep resistance up to 250 °C but the cast alloy gives low tensile properties unless grains are refined with zirconium. They are useful in suppressing microporosity and offsetting the brittleness due to zinc in some alloys. Neodymium and Parseodymium show considerably greater solid solubility than Cerium and Lanthanum.

Thorium Confers long-term creep resistance up to nearly 350 °C. It is also effective in suppressing microporosity and embrittlement effect due to zinc and in fact improves ductility.

Silver It forms a heat treatable eutectic system with considerable solid solubility and greatly increases the response to full heat treatment in the presence of rare earth metals.

Cerium Increases strength at high temperature and ductility. It reduces microporosity.

Lithium Confers extreme ductility on magnesium when added in amounts of at least 10%.

Silicon Imparts some grain refinement but little strength.

Calcium It is sparingly soluble in magnesium but provides useful effects on physical and mechanical properties. Addition of 0.35% calcium increases tensile properties of pure magnesium by 50%. In heat treating Mg-Zn alloys, a small addition is advantageous in somewhat raising the low eutectic temperature. It also improves, to some extent, the corrosion properties. Calcium confers some grain refinement though at the expense of ductility and weldability. It increases ductility in wrought Mg-A1-Zn alloys when present in concentrations of 0.1 to 0.3%.

Beryllium It is only soluble to the extent of a few thousandth per cent but it has a strong effect in suppressing oxidation of molten alloys. It coarsens the grain size in Mg-A1 or Mg-Zr type alloys.

Tin It confers some strengthening effect at the expense of ductility.

Iron It has an adverse effect on corrosion resistance of Mg-A1 alloys.

Nickel and Cobalt They have an adverse effect on corrosion resistance of magnesium alloys. 1.5% addition of cobalt is useful in maintaining hardness of Mg-Ce alloys. Increasing the proportions of nickel yields greater hardness but at the expense of ductility and impact strength.

Copper It exerts a slightly adverse effect on corrosion resistance when present in amounts of several tenths per cent, but it however, increases toughness and ductility.

Cadmium It is highly soluble in magnesium and forms several intermetallic compounds. It increases impact toughness and imparts good strength and ductility properties.

Discussions on Case Studies of Problems in Chemical Analysis

Having discussed the effects of alloying elements and their importance in the development of strength and other properties, the interaction between chemistry and metallurgy is best understood through some typical cases of failures in ferrous and non-ferrous products.

Due to Segregation

One rectangular shaped thin walled part belonging to a batch quantity of 10, cast out of medium plain carbon steel (specified carbon between 0.36 -0.40%), was found cracked in the mid-surface of the part immediately on quenching in water after hardening at 860 °C. The most striking feature was that the remaining parts of the batch as well as other previous batches were found unaffected under the same heat treatment conditions. Since it was a stray case, it was initially suspected that the part in question might have been quenched from a relatively higher hardening temperature compared to the specified one.

Microstructural analysis of the part in the cracked zone revealed no effect as is normally indicated by the presence of grain coarsening or brittle widmanstratten structure. Chemical check by wet analysis method on the part, confirmed carbon content as 0.38%, which is well within the specification range. This led to rethinking on the part of the investigator as to why the part should have cracked even though everything seemed to be satisfactory as regards chemistry and heat treated structure. Chemical analysis for carbon was repeated but this time sample drillings were taken from the area close to the cracked region. Surprisingly, carbon content was found to be 0.42%, a little more than the maximum limit of the specified range.

It is a well established fact that any plain carbon steel exceeding the limit of 0.40% C is prone to cracking on quenching in water for high temperature hardening treatment. Although the chemistry of the part with respect to carbon content in general was found well within the specification range, stray cracking had developed due to local segregation of carbon. It is known that among all other elements, carbon is most prone to segregation at the micro-level during solidification of cast products. This case history relates to the importance of proper sampling for analysis, based on the understanding of the role that carbon plays in the heat treatment of steel.

Due to Decarburisation

This is another case of wrong sampling leading to rejection of a large batch of medium carbon (0.4%) low alloy steel rolled bars. These bars meant to be used as forging stock are chemically checked for conformance to specification. Analysis indicated a much lower carbon content (i.e. 0.25%) and this was the basis for outright rejection. Non availability of another batch and emergency for critical items to be made from the same consignment, a review of the chemical check was done and it was discovered that sample drillings were taken superficially confined to the surface, i.e. the surface of each rolled bar. Micro-examination of a section across the surface zone revealed a 0.5 mm deep decarburised layer. The presence of decarburisation implies a depletion of carbon from the original carbon content. Hence, the low percentage of carbon was reported, as the samples taken for analysis were mostly from the area of decarburisation. On re-checking the carbon content in the area, after machining off the decarburised layer, the carbon percentage was found to be meeting the specification requirements (i.e. 0.4%).

The wrong sampling was carried out due to sheer ignorance of the occurrence of decarburisation on the part of the analyst. Hence it is important that the analyst should be aware of the basics of heat treatment and manufacturing problems which may interfere with his accurate/proper chemical analysis.

Due to Carbon Equivalent

Analysis of carbon equivalent in the production of grey cast iron is very important. Chemical analysis of a cast iron melt during pouring indicated total carbon, combined carbon, silicon, phosphorus and other alloying elements well within the specification range. But a test bar, cast separately from the same melt, yielded a lower hardness than the stipulated minimum. Microstructure was found satisfactory with respect to pearlite, ferrite and cementite.

As the reason for low hardness could not be attributed either to chemistry or to structure, on the basis of results reported, the focus was shifted to correctness of carbon equivalent which determines the desired properties. However, on calculation from the results obtained, carbon equivalent was found to be on the lower side of the actual requirement intended for minimum hardness or strength. If the carbon equivalent was determined during the pouring stage, the alteration of chemistry with respect to CE = Total C/ Si+P could have been made to suit the property requirements. This emphasises that the role of the chemical analyst is more on the basic understanding of chemical-metallurgical relationships of cast iron (in terms of carbon equivalent to the microstructural changes) and their effect on the strength property, even though his total analysis tallied with the specification requirements in general.

Due to Absence of Trace Additions

Quite often the repeated failures of welded 18/8 stainless steel by intergranular corrosion (popularly known as weld-decay) were reported either in pre-service (storage) or service condition. On chemical check, the chemistry of the material was found to be 18% Cr-8% Ni stainless steel without any stabilizer such as titanium or niobium as minor addition which prevents intergranular corrosion. Had it been known to the chemical analyst that such a material is susceptible to weld decay phenomenon after welding, they could have suggested the stable one and prevented the recurrence of such failures. One glaring example is that of 18/8 made exhaust pipes which were only chemically checked prior to use and failed prematurely. With prior knowledge of metallurgical behaviour due to absence of minor additions, a chemist could have acted as a metallurgist particularly when tests involved only chemical analysis, to warn the users against the undesirable consequences.

Due to Deficiency of Elements

Wrought aluminium 4% copper alloys are used for high strength structural applications and in these alloys of aluminium-copper system which are strengthened by age-hardening process, copper acts as the main strengthening element. Several cases were reported having low tensile strength property after solutionising and ageing of A1 alloy 4% Cu forged parts. During investigations it was noted that the batches were accepted on the basis of spectrovac chemical analysis on separately cast samples.

A trial heat treatment was carried out as per specification on the tensometer piece made from the part, suspecting that the low tensile strength values might be associated with some lapse in the heat treatment process such as quench delay, inadequate ageing etc. The values obtained were found low. This necessitated the rechecking of chemical analysis on the part which gave copper content slightly less than or the lower limit of the specification range (i.e. 3.8% against 3.9% in the range 3.9–4.5%). Since repeated heat treatment did not improve the values, it was the copper content which was short of the minimum copper limit that caused lowering of strength values, as copper is considered to be the most strengthening element in the age-hardening process of the alloy system. At the first instance, the analysis made on the cast sample seemed to be misleading, in having reported (3.9% Cu) fulfilling the specification requirement either due to incorrect interpretation or some instrumental error. It thus became evident that the percentage error however small mattered, particularly when it dealt with strengthening of elements.

Due to Excess of Elements

We have seen cases where slight deficiency in certain types of chemical constituents can cause problems in achieving optimum properties. Let us now examine the cases where elements exceeding the maximum limit of the specification range could cause a problem of total rejection in terms of brittleness/crack or machining difficulties. A batch of aluminium bronze parts were cast and during the knocking-off process from the mould, the castings cracked in the thin walled section area. Initially it was thought that this was due to some inadvertant high impact during the knocking-off operation. Chemically the material was reported meeting the specification requirements. However, on microstructural analysis the structure was found brittle due to the presence of numerous 'γ' (gamma) constituents which are normally not expected in the specified chemical range. This led to rechecking chemical analysis, as in this alloy, zinc equivalent factor depends on the content of aluminium, iron and manganese, and becomes an important criterian in determining the presence or extent of the gamma phase. As per the binary diagram of Cu-Zn system, with increase in Zn exceeding 40% the gamma phase starts precipitating when zinc equivalence reaches 5%. Even when Zn content is well within the range, any increase in the addition of aluminium (more particularly) than the specified, increases the zinc equivalence approaching the range of gamma formation.

On rechecking, applying this important condition, aluminium percentage was found to be higher (as expected from the microstructural analysis) than the original value (i.e. 3.02% against 2%). This explains why the utmost care has to be taken in analysing such elements, which have got an adverse effect on material properties, and why analysts should be well aware of the chemical-metallurgical consequences in advance, in order to prevent problems including high rejections.

Due to Lack of Analysis of Trace Elements

One interesting case is that of a severely cracked zinc base alloy die casting. It was reported that the die castings numbering about 600 were accepted based on the chemical check of the major alloying elements and stored for six months . During usage cracks were noticed. The microstructure confirmed that these surface cracks had progressed intergranularly (i.e. intergranular corrosion cracking) from corrosion pits. The nature of cracking and the fact that they had occurred prior to service raised some doubts as to the correctness of the chemistry of the material, including trace addition, such as arsenic, antimony, which are known to promote intergranular corrosion in zinc base alloys. On analysis, the material was found to contain trace additions of arsenic.

This emphasises the fact that although the presence of trace elements is due to improper melting practices (may occur without the knowledge of the melter), the personnel concerned were not aware of the effects of such elements and in the process full analysis was not contemplated, resulting in wastage of a large batch quantity of castings.

References

1 Raghuram A.C. and R.V. Krishnan, 'Failure Analysis': A Real Time Test; *Bangalore ICASTAD Proceedings*, July 29–August 3, 1990.
2. Louthan M.R. 'Fault Tree Techniques and Failure Analysis'; *Metallography* II, 1978, 33–42.
3. Shamala A.R. and A.C. Raghuram, 'Fault Tree Techniques for Failure Analysis', *Trans. of Indian Institute of Metals*, Vol. **35**, 1982, 1–16.
4. Raghuram A.C. and A.R. Shamala, *Failure Analysis: Present Concepts and Future Perspectives,* (unpublished paper), National Aeronautical Laboratory, Bangalore, 1985.
5. Collings J.A., B.T. Hagam and H.M. Bratt, The Failure Experience Matrix –A Useful Design Tool, *Trans. of ASME Journal of Engineering for Industry*, **98**, 1976, 1074–1079.
6. Srinath L.S., *Concepts in Reliability*, 2nd ed., Affiliated East West Press, Pvt. Ltd, New Delhi, 1974.
7. Das A.K., K.G.B. Nair, L. Mallaiah, 'Sources of defects and their effects in Metallic Materials,'*HTS Digest*, Vol. **24**, July–Dec. 1983.
8. Das A.K. and Prem Baveja, *In-Process Failure Investigation in Aeronautics*, Bangalore, Hindustan Aeronautics Limited, 1988.
9. *Aluminium,* Vol. **1**, Properties, Physical Metallurgy and Phase Diagrams, Metals Park, Ohio, American Society for Metals, 1967.
10. *Metals Handbook,* Vol. **5**, 8th ed., 'Forging and Casting,' Metals Park, Ohio, American Society for Metals, 1970.
11. Das A.K., 'Blistering Phenomena and their Case Studies in Aluminium Alloy Wrought Products', *HTS Digest*, Vol **26–27** April–Sept, 1984.
12. *Metals Handbook,* Vol. 3, 8th ed, 'Machining', Metals Park, Ohio, American Society for Metals, 1967.
13. Nair K.G.B., A.K. Das *et al*, *Trouble Shooting for Non Attainment of Strength Properties in Heat Treatment Shop, Tool and Alloy Steels*, May 1981.
14. Chatterjee B., K.G.B. Nair, A.K. Das and G.S.B. Chari,' *Failure Investigations of Aircraft Components* (Part I), Bangalore, Central Laboratory, HAL, 1980.
15. *Metals Handbook,* Vol. **10**, 8th ed., 'Failure Analysis and Prevention', Metals Park, Ohio, American Society for Metals, 1975.
16. *Metals Handbook,* Vol. **9**, 8th ed.,'Fractography and Atlas of Fractographs', Metals Park, Ohio, American Society for Metals, 1974.
17. Prasad J. and H.S. Ganesha,' *Treatise on Non–Destructive Testing and Evaluation: Theory and Practice of Surface Inspection,* Vol. **3**, Bangalore NDT Centre, Central Laboratory, HAL, 1983.
18. Prasad J. amd H.S. Ganesha, *Treatise on Non–Destructive Testing and Evaluation: Theory and Practice of Industrial Radiography,* Vol. **1**, NDT Centre, Central Laboratory, HAL, 1983.
19. Prasad J., T. Rangachary and B.N.S. Murthy, *Treatise on Non Destructive Testing and Evaluation: Theory and Practice of Ultrasonic Testing*, Vol. **2**, Bangalore NDT Centre, Central Laboratory, HAL, 1983.

20. Radhakrishnan S., 'Techniques of Failure Analysis: Inspection and Testing,' *Proceedings of Workshop on Fracture, Fatigue and Failure Analysis*, Bangalore, National Aeronautical Laboratory, 1979.
21 Krishnan R.V., 'Techniques of Failure Analysis: Microscopic Examination, *Proceedings of Workshop on Fracture, Fatigue and Failure Analysis*, Bangalore National Aeronautical Laboratory, 1979.
22. Das A.K. , K.G.B. Nair and C.G.K. Nair, ' Identification, Rating and Acceptance Limits of Inclusion in Aircraft Steel,' *Tool and Alloy Steels*, Annual Issue, Dec. 1976.
23. Asundi M.K., 'Failure Mechanism, '*Proceedings of Workshop on Fracture, Fatigue and Failure Analysis*, Bangalore, National Aeronautical Laboratory, 1979.
24. Engel Lother and Klingele Hemann, 'An Atlas of Metal Damage', Vienna, Wolfe Publishing Ltd., 1981.
25. Nambodhiri T.K.G., ' Hydrogen Embrittlement of Metallic Materials,' *Trans. of the Indian Institute of Metals*, Vol. **37**, No. 6, Dec., 1984.
26. Kolachev B.A., 'Hydrogen Embrittlement of Non–ferrous Metals, *Israel Programme for Scientific Translations*. Jerusalem, 1968.
27. Zapffe C.A. and Sims C.E., *'Trans,. AMIE*, 145, 1941, 225.
28. Kazinczy, F. De. J., *Iron and Steel Inst*. (London), 177, 1954.
29. Tetelman, A.S., *'Fundamental Aspects of Stress Corrosion Cracking*, National Association of Corrosion Engineers, Houston, 1967, 446.
30. Petch N.J. and P. Stables, *Nature*, 169, 1952, 842.
31. Frohmberg R.P., W.J. Bernett and A.R. Troiano, *'Trans. ASM*, 47, 1955, 892.
32. Oriani R.A. and P.H. Josephic , *Acta Metall*, 25, 1977, 979, also 27 (1979), 997.
33. Beachem C.D., *Metall Trans.*, 3, 1972, 437.
34. Lynch S.P., *Metals Forum*, 2, 1979, 184.
35. Hirth J.P., *Metall Trans.*, 11A, 1980, 861.
36. Hack J.I. and G.R. Leverent, *Metall Trans.*, 13A, 1982, 1729.
37. Westlake D.G., *Trans. ASM*, 62, 1969, 1000.
38. Koike S. and T. Suzuki, *Acta Metall*, 29, 1981, 553.
39. Cavannaugh M.A.,T.P. Fehlner, J.A. Kargol and N.F. Fiorie, *Scr. Metall.*, 16, 1982, 709.
40. Sandoz G, *Metall Trans.*, 3, 1972, 1169.

Glossary

Abrasion The process of rubbing, grinding or wearing away by friction.

Abrasive A hard particle or substance (such as rock, sand, emery, corundum, aluminium oxide, silicon carbide, etc.) used for grinding, lapping, polishing or blasting under pressure to wear out the surface.

Abrasive blasting A process of cleaning or finishing which involves impingement of abrasives at high velocity against the workpiece.

Abrasive wear The removal of material from the surface by abrasive particles under pressure through sliding or rolling action.

Accuracy The closeness of approach of a measurement to the true value of the quantity measured.

Adhesion The force of attraction between the molecules or atoms of two phases.

Adsorption The formation of a thin molecular film of a substance (fluid or solid) on the surface of another.

Age-hardening Hardening by ageing at ambient or moderately elevated temperature, usually after rapid cooling or cold working.

Ageing A change in the properties of a metal which occurs slowly at room temperature and rapidly at slightly elevated temperatures, usually after drastic thermal or mechanical treatment.

Alclad A composite sheet produced by bonding either corrosion resistant aluminium alloy or aluminium of high purity to base metal of structurally stronger aluminium alloy. It is anodic to the core material and provides protection against corrosion. The thickness of coating is normally 5% of the total sheet thickness.

Allotropy The reversible phenomenon by which certain metals may exist in two or more different forms with different physical properties (structure, density, etc.), but retaining the same chemical composition.

Alloy A combination of two or more chemical elements of which at least one is a metal. Most pure metals can be strengthened by a small addition of other metals or non-metals. The most striking example is steel which is essentially an alloy of iron and carbon.

In preparing glossary the author acknowledges the assistance taken from Metals Handbook, Desk Edition, edited by Howard E. Boyer, Timothy L. Gall, American Society for Metals, Ohio, May 1985 and Dictionary of Metallurgy, D. Birchon, Royal Naval Scientific Service, London, 1965.

Alloying element An element which is intentionally added to improve or modify the properties of another element.

Alpha iron The body centered cubic form of pure iron stable below 910 °C.

Aluminising A process for improving the oxidation resistance of mild and low alloy steels at temperature up to 950°C, by forming coating of aluminium or aluminium alloy on the metal surface by hot dipping, hot spraying or diffusion.

Aluminium A soft white metal of density 2.70 g/ml and melting point 660°C. Pure aluminium is characterised by its light weight, ductility, corrosion resistance and good electrical and thermal conductivity. The corrosion resistance of aluminium owes to the rapid formation of a thin adherent layer of aluminium oxide ($A1_2O_3$) on the surface.

Annealing A high temperature heat treatment process consisting of heating to and holding above austenitizing temperature followed by slow cooling in the furnace to achieve primarily the material in soft condition for machining purpose.

Angstrom A unit of linear measurement equal to 10^{-10} m or 0.1 m.

Anodic The term applied to the component of an electro-chemical cell which has lower electrode potential. Under corroding conditions the attack occurs at the anode area.

Anodic coating A film on the metal surface resulting from an electrolytic treatment at the anode.

Anodic protection Reduction of the corrosion rate of a metal by making its potential sufficiently electropositive so as to render it passive. This is controlled usually by e.m.f.

Anodising A conversion coating process which artificially thickens the layer of aluminium oxide on the surface of aluminium and its alloys. The process involves passing of an electric current through the chromic or sulphuric acid bath in which the article acts as anode, where nascent oxygen evolves and results in thickening of oxide film. The electrical resistance.of the film sets a limit to thickness of the order of 0.001 inch. It provides protection against corrosion and specially abrasion resistance when anodised by sulphuric acid process.

Arc welding A group of welding processes that fuse metals together by the heat generated by an electric current between an electrode and the work piece, in the form of an arc. Welding can be done with or without the pressure and with or without the filler rod.

Asperites A name applied to the high spots of nominally flat surface.

Austenite A solid solution of one or more elements in face centered cubic iron (F.C.C).

Austenitic steel A alloy steel whose structure is normally austenitic at room temperature, such as, Hadfield manganese steel, stainless steel.

Austenitising Heating an alloy steel to a high temperature zone where previous structure transforms to austenite.

Baking A low temperature heating operation for removal of gases present in the finished products, particularly after completion of plating process. It is also applied for curing or hardening the surface coating such as, paints.

Banded structure A segregated inhomogeneous structure consisting of alternating parallel bands of different composition (phases), mostly aligned in the direction of hot working

Base metal The metal present in maximum proportion in an alloy.

Beach marks Progressive shell or conch type marks on a fatigue fracture surface indicating the successive locations of the advancing crack front.

Bend angle The angle through which a bending operation is carried out.

Bend radius The inside radius of a bent section.

Biaxial stress A state of stress in which only one of the principal stresses is zero, the other two being in tension.

Billet A solid semifinished or square product produced through hot working by forging, extrusion or rolling. Also it is referred to wrought starting stock for forging or extrusion.

Bimetallic corrosion The corrosion of two metals in electrical contact and in exposure to an electrolyte in which one metal initiates attack on the other which is less noble metal.

Binary alloy An alloy having only two component elements.

Blasting Cleaning the metal surface by impingement with abrasive particles at high velocity.

Blister Often a dome shaped (raised) area resulting either from the loss of adhesion between a coating and the basis metal or expansion of entrapped gases under pressure during heat treatment of a forged or extruded section.

Blow hole A hole in a casting or a weld caused by gas entrapment during solidification.

Bore Producing a hole or cylindrical cavity with single point or multipoint tools, other than a drill.

Brinell hardness testing Determination of the hardness of material by forcing a hard steel ball of specified diameter into it under a standard load. The result is expressed as the Brinell Hardness Number (BHN), which is obtained by dividing the applied load in kilograms by the surface area of the resulting impression in square millimetres.

Brinelling Occurrence of evenly spaced dents in a race way of a roller bearing. When bearing assembly is subjected to heavy force or impact, it causes the rolling elements to indent the race way surface.

Brittle crack propagation A very sudden propagation of crack without the absorption of energy.

Brittle fracture Separation of two metallurgically bonded structures of a metallic body, practically without plastic deformation. Brittle fracture occurs by rapid crack propagation.

Brittleness It is the property of material by which crack propagation occurs with little evidence of ductility.

Burning A permanent damage caused to a metal by exposure to a high temperature. It causes either incipient melting or intergranular oxidation.

Carbide Carbon chemically combined with one or more other alloying elements, such as, iron carbide (Fe_3C).

Caburising Absorption and diffusion of carbon into the solid surface of ferrous alloys by heating to a temperature usually above AC_3 (the critical temperature for transformation of alpha iron into gamma iron, i.e. austenite) in contact with carbonaceous material.

Case hardening A generic term for several processes applicable to steel that brings about a change in chemical composition of the surface layer by absorption of carbon, nitrogen or a mixture of two and simultaneously by diffusion which creates a concentration gradient.

Cast structure The metallographic structure of a casting evidenced by shape and orientation of grains.

Catastrophic failure Sudden failure of a component or assembly leading sometimes to mishaps (including destruction of property and life).

Cementite A compound of iron and carbon formed in steel as iron carbide. It is hard and brittle in nature.

Chevron A characteristic pattern of radial marks appearing as nested 'V', generally observed in a brittle fracture surface. Also called sometimes a heringbone pattern. The points of the chevrons can be traced out to the fracture origin.

Chromate treatment The production of a conversion coating of tri and hexavalent chromium compounds on the metal surface. It provides good adhesion for organic finishes such as, paints and resins and affords a good measure of corrosion resistance.

Cleavage Splitting of a crystal on a crystallographic plane of low index.

Cleavage fracture A fracture in which most of the grains have failed by cleavage resulting in bright crystalline facets, commonly observed in low-energy brittle failure.

Coherent precipitate A crystalline precipitate that forms from solid solution with an orientation that maintains coherency between the crystal lattice of the precipitate and the lattice of the matrix, usually accompanied by some strain in both lattices.

Cold shortness Lack of ductility at ambient temperature.

Cold shut A casting defect in the form of a discontinuity on the surface of a cast metal as a result of failure of two streams of liquid metal to unite. The defect is often caused by low casting temperature and entrapped oxide films due to turbulence in pouring. This is also observed as a lap on the surface of a forging that was closed without fusion during deformation.

Cold work It is a plastic deformation of a metal at a temperature low enough to result in permanent strain hardening.

Contact fatigue Pitting and subsequent fatigue cracking of a surface subjected to alternating stresses such as those under rolling contact or combined rolling and sliding action.

Corrosion Chemical or electro-chemical reaction between a metal and its environment usually resulting in deterioration in the properties and/or appearance of the metal. Corrosion occurs in contact with an electrolyte (rain, sea water, etc.), in which driving force is electro-chemical action between local anodes and cathodes or chemical dissolution.

Creep A continuous or semi-continuous deformation with time under constant stress at constant temperature. The creep strain occurring at diminishing rate is shown as primary creep and at constant rate it is a secondary creep, which are more important while designing.

Crystal A solid made of atoms, ions or molecules arranged in repetitive geometric pattern.

Crystalline fracture A fracture showing bright facets.

Cup and cone fracture Fracture often resulting from tension—test specimen exhibits a central depressed flat face region surrounded by a shear Olip in one of the mating fracture surfaces, while the other surface looks like a miniature truncated cone. It involves a mixed fracture mode where the central portion undergoes plane strain fracture and the surrounding region undergoes plane stress mode.

Current density The average electric current distribution over the surface of an electrode. The unit is usually amps/ft^2 or amps/decimetre^2.

Deburring Removal of burrs, sharp edges or fins from the semi-finished or finished parts by filing, grinding or rolling.

Decarburisation Loss or carbon from the surface layer of steel due to reaction with chemical substances. It occurs during high temperature heating in air and results in a reduction of fatigue strength.

Deep drawing Drawing a cup shaped form of large depth/diameter ratio from sheet material.

Defect A departure from the specified quality level where the product will not fulfill its desired function.

Degassing Removal of dissolved gases from liquids, usually practised on molten metal by adding deoxidising agents or bubbling chlorine gas.

Degreasing A surface treatment which cleans oil or grease from a surface.

Delayed fracture Fracture occurring after a time during which metal has undergone a static stress, which itself is of low magnitude to cause instant failure.

Dendrite A crystal that has a tree like branching pattern usually observed in casting.

Deoxidiser A material which when added to the molten metals removes dissolved and combined oxygen, such as aluminium, silicon, etc.

Depth of fusion In welding the distance covered by fusion into the base metal.

Dezincification A form of corrosion attack to which some brasses are susceptible, particularly when exposed to warm sea water. The attack has little effect on the surface and results in deep copper red colour at the surface.

Diamond pyramid hardness tester Measuring the hardness of a material by determining its resistance to indentation. The vicker's diamond is square based pyramid of 136° included angle, usually applied under 30 kg load for steel. The hardness number is determined as the load/area of impression in kg/mm^2. Hardness is described as VHN.

Die A tool usually containing a cavity that imparts shape to material by forging, die casting or as a die into which powder metals are compressed.

Die casting Casting molten metal into a metal die to give accurate finish and dimension under gravity or pressure.

Die forging A forging whose shape is determined by impression in specially prepared dies.

Diffraction The formation of interference pattern in a beam of radiation. It is used in X-ray diffraction of metal lattices to measure lattice spacing and determine space group.

Diffusion The spontaneous movement of atoms or molecules to new sites within a material.

Directional solidification Solidification of molten metal which occurs in such a manner that it provides continuously the molten metal to solidifying front as long as necessary to feed the casting. When perfectly controlled it results in elimination of shrinkage cavities.

Discontinuity Any interruption as a form of crack, lap, seams, inclusion or porosity in the normal physical structure of a part.

Dislocation A linear imperfection (line defect) in a crystalline array of atoms. This may be present as a closed ring bounding a slipped area of the lattice or anchored at its ends to other dislocation, grain boundaries or a surface. Dislocations move in their slip plane under the influence of an applied stress, permitting consecutive translation of atoms on either side of the slip plane.

Double ageing Two different ageing treatment are employed to control the type of precipitate formed from a super saturated matrix in order to obtain desired properties.

Dross The oxide deposit (often contaminated by other impurities) on the surface of molten metal.

Ductile fracture Fracture characterised by tearing of metals and associated with gross plastic deformation.

Ductile crack propagation Slow crack propagation accompanied with perceptible plastic deformation.

Duralumin A group of aluminium alloys based essentially upon 2-4.5% copper and 0.3-1.2% each of manganese, silicon, iron and magnesium and is age hardenable.

Dye penetrant Thin fluid used in non-destructive testing for detection of surface defects.

ECM An abbreviation for electro-chemical machining.

Eddy current Electric current set up in a conductor by a changing magnetic flux.

Eddy current testing An electromagnetic non-destructive testing method in which eddy current flow is induced in the test piece. Changes in flow caused by variation in the object are reflected into the coil for detection and measurement by suitable instrument.

EDM An abbreviation for electro-discharge machining.

Electro-chemical machining Techniques of material removal in which electrolytic dissolution is used alone or in combination with some other techniques.

Electro-discharge machining Removal of stock from an electrically conductive material by rapid, repetitive spark discharge through a dielectric fluid flowing between the work piece and a shaped electrode.

Electrolyte A substance containing ions which can be used to migrate under the application of an electric potential. Aqueous solutions of salts are common electrolytes.

Electro-motive force Electrical potential.

Electro-beam welding A welding process that uses a stream of electrons accelerated from a heated filament by a high voltage and focussed into a narrow beam, to provide a heat source for welding can be used foı welding dissimilar material as well as reactive and refractory materials. Unlike other welding processes, this process provides a narrow heat affected zone.

Elongation The extension of the gauge length of a tensile specimen, divided by the original gauge length, measured after the specimen breaks.

Embrittlement Loss of ductility in a metal. The embrittlement may be caused by one of several factors. Examples include brittle fracture, caustic embrittlement, hydrogen embrittlement, temper brittlement, over heating.

Endurance limit The ratio of fatigue limit to ultimate tensile strength which is usually of the order of 0.5 for steels. In other words, it is the maximum stress below which a material can sustain for infinite number of stress cycles.

Equiaxed grain structure A structure in which grains exhibit similar dimensions in all directions.

Equilibrium A state of balance, i.e a stable condition.

Erosion Physical removal of a material from a surface by mechanical means such as flowing water, steam or gas. When corrosion occurs simultaneously. It is called "Erosion-Corrosion".

Etching It is a preferential chemical or electrolytic attack on the surface of a metal to reveal structural details for metallographic examination.

Eutectic melting Melting of localised microscopic areas whose composition corresponds to that of the eutectic in the system.

Exfoliation A type of corrosion that attacks the grain boundary intergranulary and progresses approximately parallel to the outer surface of the metal with layeral separation in the form thin parallel leaves.

Extrusion Conversion of an ingot or billet into lengths of uniform cross-section by forcing metal to flow plastically through a die orifice.

Face-centered cubic A cubic metal lattice structure in which the atoms composing the corners of one lattice of cubes lie at the centres of the faces of another lattice of cubes and vice versa. The structure of austenite, aluminium, copper, nickel belongs to FCC category and is characterised by ductility.

Facet A flat, usually highly reflective surface on a crystal or fractured surface.

Failure A general term used to imply that a part is not capable of carrying out the intended function.

Fatigue Failure of a material by repeated application of stresses, each of which is individually insufficient to cause failure. Fatigue fractures are progressive, beginning as minute cracks that grows under the action of cyclic stresses.

Fatigue life The number of cycles of stress that can be sustained prior to failure under a specified load.

Fatigue limit The maximum stress that leads to fatigue fracture in a specified number. stress cycles.

Fatigue strength The maximum stress that can be sustained for a specified number of cycles without failure.

Fatigue striation Small parallel lines in bands observed in electron microscope fractograph of fatigue fracture surfaces.

Ferrite The body-centered cubic (alpha) form of iron, which is stable below 760 °C in pure form.

Fibre The characteristic directionality of wrought products.

Fibrous fracture A fracture whose surface is characterised by a dull gray or silky appearance.

Filler In welding, the rod used to deposit metal in the joint.

Filliform Applied to fibre or thread like materials or phenomenon such as standard wires or the thread like corrosion sometimes observed on the surface of painted metals, caused by superficial corrosion under the surface of coating.

Fillet A radius imparted to inside meeting surfaces.

Fillet weld A weld joining two surfaces essentially at right angles to each other.

Finish grinding The final grinding operation with primary objective of achieving surface finish and dimensional accuracy.

Flakes Discontinuous and short internal cracks in steel due to release of hydrogen. Also called snowflakes, fisheyes and shatter cracks.

Flaring Forming an outward accurate-angle flange on a tubular part.

Flash Excess metal that is forced out in forging through the opening between the upper and lower dies.

Flash line The line of location of flash formed around a forging.

Flaw A term often used to describe a crack like discontinuity.

Flow lines Texture revealed by deep etching a longitudinal section of a mechanically hot worked component, showing the direction of metal flow. The lines are caused by differential etching of parallel layers having slightly different composition.

Flow stress The uniaxial true tensile stress required to cause plastic flow at any particular temperature.

Fluidity The ability of liquid metal to flow into and fill a mold cavity.

Flux In metal refining, a material used to remove undesirable substances (sand, ash, dirt, etc.) from the melt, as a molten mixture. It is also used as protective coating for certain molten bath.

Forging Plastic deformation of a metal by squeezing or hammering, to produce a required shape. It may be open or die forging.

Forging stock A rod/bar used to make forging.

Formability The relative ease with which a metal can be shaped through deformation.

Fractography Fractography is the microscopic inspection of failure surfaces to gain information about the events involved in fracture.

Fracture mechanics Essentially it is a technique that involves a study of the stresses and strains at the tips of sharp crack or defects and relates them to the applied loading. Under elastic condition this stress field is described in terms of a single parameter known as the stress intensity factor K.

Fracture toughness The critical value of K, necessary for crack growth under a static load is called K_c— the fracture toughness.

Fretting A type of wear occurring between the surfaces in close contact when subjected to cyclic relative motion. Fretting is usually accompanied by corrosion, product of which are fine wear debris. This is known as fretting corrosion.

Fretting fatigue Fatigue fracture initiating from the fretted zone in the surface.

Galvanic corrosion It is electro-chemical corrosion in which the current is generated by the corrosion processes. When two dissimilar metals are electrically coupled together in the presence of an electrolyte, the current produced results in dissolution of the anode.

Gamma iron The face-centered cubic form of pure iron. Stable form from 910 °C to 1400 °C. Example is austenite.

Gas holes Casting defects in the form of small, evenly distributed cavities, usually circular, with bright surfaces. Caused by the release of dissolved gases during solidification.

Gas porosity Fine holes or pores within a metal that are caused by entrapped gas or by evolution of dissolved gases, primarily hydrogen, during solidification.

Grain flow Fibre like lines appearing on polished and etched surface or section of forgings in the direction of working. Proper grain flows improve the mechanical properties of the forging.

Grain growth An increase in the average size of the grains, usually when exposed to an elevated temperature.

Grain size The size attained by the grains of steel when heated to the austenite formation range.

Grinding Removal of material by rubbing a surface with abrasive belt or grinding wheel.

Grinding cracks Shallow cracks formed on the ground hard surface as result of generation of excessive heat.

Grinding stress Residual stresses either in the form of tensile or compressive, in the surface layer of the worked piece.

Hardenability The relative ability of a steel to form martensite when quenched from the austenitic condition. The property is related to the size of section considered and is not concerned with the ultimate hardness value obtainable. It is a measure that includes the depth of hardening.

Hairline cracks Cracks are sometimes developed within steels which have undergone cooling fairly rapidly through the critical temperatures. Also, presence of hydrogen within the metal can lead to such cracks.

Heat affected zone The zone within the parent metal of a welded joint in which heating has caused structural changes without melting.

Heat checks A pattern of parallel surface cracks formed by alternate rapid heating and cooling. This phenomenon is generally observed in the surface of thermal fatigue related failure. Also in extruded and forged surfaces.

Heat treatment Heating and cooling a metal or alloy in such a manner so as to develop the desired mechanical properties.

Herringbone pattern Same as Chevron pattern.

Heterogeneity Variable composition or structure within a material.

Homogenizing Holding at high temperature to eliminate or minimise chemical segregation by diffusion and break down as cast structure for better workability.

Hot shortness Also termed as red shortness property of certain materials which separate along the grain boundaries when subjected to deformation in the near melting zone. The presence of the low melting constituent, even in minute quantity can cause hot shortness in steel.

Hydrogen damage A general term used for the embrittlement, cracking, blistering and hydride formation that can occur in the presence of hydrogen.

Hydrogen embrittlement Loss of ductility of toughness in a metal as a result of the presence of hydrogen within the material.

Impact energy The amount of energy required to fracture a material, usually measured by izod or charpy test.

Impact test A test to determine toughness behaviour of materials under high rates of loading by a single blow as in charpy and izod tests.

Impurities Elements or compounds which deteriorates the material property.

Inclusion They may be either non-metallic or foreign particles. Non-metallic inclusions include oxides, sulphides, silicates and alumina. They are harmful if present more than the specified requirement.

Ingot A solid metal casting prepared for subsequent hotworking, normally rectangular in section and tapering slightly from top to bottom.

Ingotism The presence of excessively large dendrites in a casting, due to too high a casting temperature.

Intercrystalline Relating to effects occurring between grains or crystals, i.e. in the grain boundaries. This is also called as intergranular.

Interference fit A fit in which the size of the male part is slightly greater than that of the female, requiring force to fit them together.

Intergranular corrosion Corrosion that occurs preferentially at the grain boundaries without affecting the adjacent grains much.

Intermediate phase In an alloy system a distingusihable homogeneous phase whose composition does not extend to include any of the pure components of the system.

Intermetallic compound A compound of two or more metals which has distinctive crystallographic structure and definite composition. Normally hard and brittle. They form the principal hardening influences in the alloy systems, i.e. Cu A12 and Mg2 Si in aluminium alloys.

Intracrystalline Within or across the crystals or grains of a metal.

Joint efficiency The ratio of the strength of a joint (riveted or welded) as compared with the strength of the parent metal, usually measured in tension.

Joint penetration The minimum depth to which a flange weld extends from its face into joint, including root penetration.

Lap A surface imperfection, similar to seam in rolled products, caused by the presence of fins or sharp corner, which on being rolled or forged remain unwelded to the body of the part.

Lattice A geometrical pattern in space. A metal lattice is the regular arrangement of the atoms.

lattice defects Any aberrations in an otherwise perfect geometrical array of atoms in a regular three-dimensional network.

Lattice distortion Distortion of the regular geometrical array of lattice from any cause such as substitutional impurities, intersitials, precipitates, etc. since local distortion impedes dislocation movements, lattice distortion is always accompanied by hardening and usually increases the loss of ductility.

Low-alloy steel Typically containing maximum alloy content up to 5%.

Lubricants Materials used to reduce frictional forces and prevent seizure between metal faces subject to contact pressures and relative motion.

Linear elastic fracture mechanics A method of fracture analysis that can determine the stress requiring to induce fracture instability in a structure having a flaw of known size.

Machinability The relative ease with which a metal may be machined.

Machining Removing material from a metal part using cutting tools.

Macro-etching Etching a polished metal surface to observe the gross structural details such as grain fibre, porosity or crack with unaided eye or under very low magnification.

Macrograph A graphic reproduction of the surface of a prepared specimen at magnification lower than ten diameters.

Macrostructure The structure of metals as revealed by macroscopic examination of the etched surface.

Magnetic particle inspection It is one of non-destructive methods of surface inspection for detection of cracks and other imperfections in ferro magnetic materials. Finely divided magnetic particle applied to the magnetized part are attracted to and outline the pattern of any magnetic leakage fields created by discontinuities.

Magnification In microscopy, the ratio of the length of the image to the length of the object (often referred to as 'so many' diameters). For most practical purposes it is the product of the magnification of the eye piece and objective used with the draw tube set to the appropriate tube length.

Martensite The name was originally coined for the hardest phase formed in steels during the breakdown of austenite or rapid cooling or quenching in water. It is characterised by an acicular (needle like) structure.

Master alloy A mixture or alloy of elements which are added to a charge of molten metal in order to provide a desired composition or a fine grain size, etc.

Metallography The science of studying the constitution and structure of metals and alloys under optical and electron microscopes.

Metallurgy The science and technology of metals and alloys.

Micrograph A graphic reproduction of the surface of a prepared specimen after etching at a magnification higher than ten diameters.

Microhardness Determination of hardness of a material under very light load enabling the small indentation to be measured with a microscope. Its advantage is that it can determine the hardness of different phases within a structure.

Microprobe An instrument for selective analysis of a microscopic feature in which an electron beam bombards the point of interest.

Microstructure The structure of a metal as revealed by microscopic examination of the etched surface of a polished specimen.

Molecule The smallest particle of a material that retains the physical properties of the original material.

Natural ageing Spontaneous ageing of a super saturated solid solution at room temperature (see Ageing).

Necking Reduction of the cross-sectional area of a metal in a localised area by strectching.

Neumann band Mechanical twin in ferrite.

Nitriding Introducing nitrogen into the surface layer of suitable ferrous alloys at suitable temperature (usually 500 °C) in contact with nitrogenous material (ammonia molten cyanide). It develops a surface hardness of 750-1100 VPN depending upon composition of steel.

Normalising Heating a ferrous alloy to a suitable temperature above the transformation range and then cooling in air. The process refines the structure.

Notch brittleness Susceptibility of a material to brittle fracture at points of stress concentrations.

Orange peel A surface roughening in the form of a pebble-grained pattern normally occurring in a metal of abnormal coarse grain size when stressed beyond its elastic limit.

Overageing Ageing for longer times and/or at higher temperatures than required to achieve the optimum properties in an age-hardening system.

Oxidation Combination of the material with oxygen or loss of electrons by a substance during a chemical reaction. Heating a metal in air, oxidation occurs in the form of oxide film.

Passivation The changing of a chemically active surface of a metal to a much less reactive state.

Pearlite The lamellar carbide containing constituent present in annealed and normalised steels.

Peening Introducing plastic deformation into a surface by hammering or bombarding with hard round shot. It hardens the surface and introduces compressive residual stress improving fatigue life.

Penetration The depth by which the fused zone of a weld penetrates into the parent material or previous weld deposits.

Permanent set Plastic deformation that remains on release of the stress that produces deformation.

Phase A chemically homogeneous and thermodynamically stable constituent of an alloy system.

Pickling Immersion of a metal part in a solution which will preferentially remove oxide and mill scale. Typical solution for plain carbon and low alloy steels are 15% hydrochloric acid at 15-50 °C, 10% sulphuric acid at 70 °C and 50% phosphoric acid at room temperature.

Pitting Forming small sharp cavities in a metal surface by corrosion.

Plane-strain The stress condition in linear elastic fracture mechanics in which there is a zero strain in a direction normal to both the axis of applied tensile stress and the direction of crack growth (i.e. parallel to the crack front,) most nearly achieved in loading thick plates along a direction parallel to the plate surface.

Plane-stress The stress condition in linear elastic fracture mechanics in which the stress in the thickness direction is zero most nearly achieved in loading very thin sheet along a direction parallel to the surface of the sheet. Under plane-stress condition, the plane of fracture instability is inclined 45° to the axis of the principal tensile stress.

Precious metal One of the relatively scarce and valuable metals, gold, silver and platinum group metals.

Precipitation heat treatment Artificial ageing in which a constituent precipitates from a supper saturated solid solution.

Progressive crack A slow moving crack.

Quality Quality can be expressed in three fashions. One by degree of excellence by comparison against a standard, the other by an quantitative evaluation of characteristics of a product and third by the assurance of its ability to satisfy a specified requirement.

Quench ageing Changes in mechanical properties due to ageing at room or slightly elevated temperatures after high temperature quenching.

Quench cracks Cracks resulting from the inability of the hard structure (i.e martensite on quenching a steel) to absorb strains produced by the rapid transformation and thermal effects.

Quenching Rapid cooling.

Radial marking Radial lines emanating from the fracture origin and easily obserable on the fracture surface at low magnification.

Radiography A method of non-destructive inspection to detect primarily the cast defects. In this method the object is exposed to a beam of X-rays for gamma rays which records the photographic image of the resulting shadow in the film placed under the object.

Ratchet marks Lines that protrude out in the surface of a fatigue fracture, result at the intersection and connection of fatigue propagation planes from the multiple origins.

Recrystallisation Involving formation of a new strain free set of grain structure, altogether different from the existing one in the cold worked state. Usually accomplished by heating.

Refining Purifying impure materials.

Residual elements Elements that exist in a metal other than intentional addition during melting.

Resolution The ability to recognise the closely related physical features of phases under given test method.

River pattern A characteristic pattern of cleavage steps usually observed in the brittle fracture surface.

Riveting A mechanical jointing process for two or more members of a structure by means of metal rivets.

Rosette A rounded constituent in a microstructure.

Rotary swaging Hitting in succession on the bar or tube material by the inward radial blows. Reduces diameter and/or wall thickness.

Rough machining Implying a major machining without finish.

Scaling A thick layer of oxidation products on the metal surfaces, formed at high temperature.

Scuffing When parts in close contact with each other are subjected to a relative motion scuffing occurs as a form of adhesive wear by superficial scratching or a high polish on the rubbed off surfaces.

Seam A crack like discontinuity resulting from the unwelded lap, usually observed in rolled products.

Season cracking Intercrystalline cracking of metals resulting from the combined effects of corrosion and internal stress. This type of cracking in brasses is known as stress corrosion.

Secondary hardening Hardening by precipitation of carbides during tempering/ageing at moderate temperature in certain steels, i.e high speed steel, maraging steel, and precipitation hardening stainless steel.

Segregation Concentration of alloying elements or non-metals (c,s,p, etc.) in the localised areas of a casting.

Seizure Local welding or galling between mating/sliding surfaces.

Sensitisation Susceptibility to preferential corrosion attack in austenitic stainless steel (unstabilised), when heated to a temperature range of 550-850 °C (generally in welding). It results in chromium depletion due to precipitation of chromium carbides in grain boundaries and hence decreases its corrosion resistance property.

Shear lip An area of shear fracture at the edges of a fracture surface, usually recognised by a narrow slanting ridge.

Shim A thin metal sheet piece used as a packing in the gap between the two contacting surfaces or assemblies.

Shrinkage The difference between the size of a metal casting after it had cooled to room temperature and the size of the mould into which the liquid metal was poured, due to any contraction of the liquid during freezing.

Shrinkage cavity A cavity resulting from the shrinkage during solidification of casting.

Shrinkage crack An irregular (zig-zag) and interdendritic crack in a casting, resulting from non-uniform contraction. This is usually referred to as hot tear.

Shock resistance The ability of a material to resist fracture under sudden application of a load.

Shot blasting A process used for cleaning and descaling the metal surfaces.

Shot peening Strengthening the metal surface by cold working under metal shot impingement, resulting in increase in surface hardness and surface compressive stresses for improvement in fatigue resistance property.

Slip Due to plastic deformation shearing or displacement of a part of a crystal occurs over the remainder of the crystal in definite crystallographic direction by the movement of dislocation within the lattice.

Slip band A group of closely spaced parallel slip layers, appearing as a single line under optical observation.

Slip plane The crystallographic plane in which slip occurs in a crystal.

S-N diagram A curve representing the relationship between stress (S) and the number of cycles (N) prior to fracture during fatigue testing.

Stabilizing treatment A low-temperature heat treatment followed by slow-cooling to relieve internal stresses and hasten minor precipitation process to promote dimensional stability.

Stainless steel A term used to describe a wide range of steels containing 12 to 30% chromium as the major alloying element and supplemented by the addition of nickel as the second major alloying element, to enhance corrosion resistance.

Stiffness The ability of a metal to resist elastic deflection.

Strain A measure of the relative change in the size or shape of a metallic body as a result of some external force.

Strain hardening A plastic deformation which induces an increase in hardness and strength at temperature below the recrystallisation range.

Stress Stress is force per unit area. There are three kinds of stress-tensile, compressive and shear. Bending introduces both tensile and compressive and torsion introduces shear stress.

Stress concentration Any notch, crack, hole, corner, groove or change in geometry acting as interruption to the smooth flow of stress and strain in a part, introduces a concentration of stress.

Stress amplitude One-half the algebraic difference between the maximum and the minimum stresses in one cycle of a repetitively varying stress.

Stress concentration factor A multiplying factor (K_1) for applied stress that allows for the presence of a discontinuity (a notch or hole in a structure).

Stress corrosion cracking A cracking phenomenon which occurs under combined action of corrosion and stress, stress may be either applied or residual.

Stress-intensity factor Usually denoted by the symbol K and is used in linear elastic fracture mechanics to relate to the intensification of applied stress at the tip of a crack of known size and shape.

Stress raisers Notches, holes or any changes in the geometry of a component that cause stress concentration, i.e. local increases in stress.

Stress ratio The ratio of maximum to minimum stress in fatigue of one cycle, considering tensile stresses as positive and compressive stresses as negative.

Stress-relieving Low-temperature heat treatment to reduce residual stresses by providing enough thermal activation for tightly bound dislocation to overcome obstacles in the slip planes.

Sub-zero treatment Refrigeration of steel to promote transformation of retained austenite

Swaging See rotary swaging.

Temper The term is notably referred to the condition or hardness of a product (i.e soft temper = annealed, Hard temper = heavily cold worked, etc.). The term is also used to describe softening of hardened steel (tempering).

Temper brittleness Centain low-alloy steels notably containing nickel and chromium become brittle if slowly cooled in the range 600-300 °C or if reheated in this range after quenching from 600 °C. The fracture is largely intercrystalline and therefore associated with grain boundary brittleness.

Tempering A low-temperature reheating treatment to fully hardened steels to reduce high internal stresses developed by the martensite formation. High internal stresses can develop craking or small dimensional changes with time.

Tensile strength Calculated as the ratio of maximum load to the original cross-sectional area. Also known as ultimate tensile strength.

Test piece A piece of material usually carefully prepared and of a closely controlled size and shape, upon which tests are conducted to determine physical properties, typical of the bulk material.

Texture The state of distribution of crystal orientations.

Thermal fatigue Fatigue fracture occurs from the cyclic stresses due to the presence of temperature gradients that vary with time.

Thermal shock The development of combined high stresses and steep temperature gradient in a structure.

Thermal stresses Stresses resulting from non-uniform temperature distribution.

TIG welding Tungsten inert gas welding.

Tolerance The specified permissible deviation from a specified nominal dimension.

Titanium A silvery-white corrosion resistant metal of density 4.5 g/m1 and melting point 1660 °C, which exists in close-packed hexagonal and body centred cubic lattices.

Torque The product of a force and the perpendicular distance from its line of application of the axis about which it acts or a twisting action resulting in shear stress and strains.

Toughness Ability to absorb energy by plastic deformation before fracture.

Transcrystalline Across the crystals of a metal.

Transition temperature A temperature below which brittle fracture is increasingly likely in a steel structure.

Tungsten A steel grey metal of density 19.3 g/m1 and melting point 3380°C. It has a body centred cubic lattice.

Twining A process of deformation within a crystal, in which parallel planes of atoms slip consecutively over each other by some friction of inter atomic spacing.

Ultimate tensile strength It is the ratio of the maximum load sustained during the tensile test, to the original cross-sectional area of the test specimen. Often abbreviated to UTS.

Ultrasonic cleaning Cleaning in a solvent in which ultrasonic vibration are excited.

Under bead crack A crack in the heat-affected zone of a weld.

Upsetting A hot forging operation in which the diameter of a bar is locally increased.

Wear Deterioration of a surface due to relative motion between it and another.

Weld decay An intercrystalline corrosion to which austenitic stainless steels may be susceptible.

Welding A process of joining metals together by fusion so that there is continuous crystallisation across the joint.

Widmanstatten structure It is a characteristic structural formation in which the transformation of solid solution into two phases occurs in such a way that one phase forms plates and the boundaries of the phases have coherency. This results in sets of parallel plates, mutually inclined to each other at equal angles. Originally observed by Widmanstatten in 1808 and hence the nomenclature.

Index